THERAPEUTIC NANOMATERIALS

THERAPEUTIC NANOMATERIALS

Edited by

MUSTAFA O. GULER
Institute of Materials Science and Nanotechnology
National Nanotechnology Research Center (UNAM)
Bilkent University
Ankara, Turkey

AYSE B. TEKINAY
Institute of Materials Science and Nanotechnology
National Nanotechnology Research Center (UNAM)
Bilkent University
Ankara, Turkey

WILEY

Published by John Wiley & Sons, Inc., Hoboken, New Jersey
Published simultaneously in Canada

Library of Congress Cataloging-in-Publication Data

Therapeutic nanomaterials / edited by Mustafa O. Guler, Ayse B. Tekinay.
pages cm
Includes bibliographical references and index.
ISBN 978-1-118-98745-2 (cloth)
1. Nanostructured materials–Therapeutic use. 2. Medical technology. I. Guler, Mustafa O., 1978– editor. II. Tekinay, Ayse B., editor.
R857.N34T44 2016
610.28–dc23

2015034452

Set in 11/13pt Times by SPi Global, Pondicherry, India

Printed in the United States of America

10 9 8 7 6 5 4 3 2 1

1 2016

CONTENTS

LIST OF CONTRIBUTORS

Elif Arslan

Hakan Ceylan

Goksu Cinar

Aykutlu Dana

Mevhibe Gecer

Melis Goktas

Mustafa O. Guler

Gulcihan Gulseren

Nuray Gunduz

Busra Mammadov

Rashad Mammadov

Didem Mumcuoglu

Alper Devrim Ozkan

Melis Sardan

Berna Senturk

Melike Sever

Gulistan Tansık

Ayse B. Tekinay

Ahmet E. Topal

Gozde Uzunalli

I. Ceren Yasa

Seher Ustun Yaylacı

Aygul Zengin

Affiliation (all contributors): Institute of Materials Science and Nanotechnology, National Nanotechnology Research Center (UNAM), Bilkent University, Ankara, Turkey

PREFACE

Interdisciplinary approaches through contributions from chemistry, biology, materials science, physics, engineering, and medicine offer a new generation of therapeutic methods, which can be used for the early diagnosis and treatment of many diseases and injuries that ail human population today. This book aims to provide a general perspective about nanomaterials and their use for therapeutic purposes for scientists, clinicians, patients, students, and novices in the field. It also provides detailed information on types of nanomaterials and their biomedical application areas for experts in nanosciences. Here we discussed how nanomaterials can be used for biomedical applications in addition to understanding side effects of these materials to humans and environment, and we also cautioned the lawmakers to make the necessary regulations.

This book discusses new materials for treatments of different types of tissues and organs. In addition to therapy, new methods for diagnosing diseases are briefly described. Main treatment methods were discussed under regeneration of tissues in situ. With the increase in aging population in the world, especially in developed countries, there is also an increased prevalence of degenerative disorders. Both degenerative disorders and accidental injuries can cause detrimental changes in various tissues, which result in not only deterioration of life quality of patients and caregivers but also a considerable amount of financial burden on the health systems of individual countries. Current treatment options for many of these injuries are insufficient. We believe that new generation of therapeutic materials will be utilized extensively in the regenerative medicine field. Although there are many species that can perfectly regenerate the injuries in their tissues, *Homo sapiens* is not one of them. Therefore, when humans get injured, they need external help for repair of their injuries, with the fact in mind that they cannot fully regenerate. The deficiency of tissue repair mechanisms is more evident in some tissues like brain or cartilage, and the ability to repair is known to decrease with age.

Mustafa O. Guler and Ayse B. Tekinay

1

NANOMATERIALS FOR MEDICINE

Mustafa O. Guler and Ayse B. Tekinay

Institute of Materials Science and Nanotechnology, National Nanotechnology Research Center (UNAM), Bilkent University, Ankara, Turkey

1.1 INTRODUCTION

Nanotechnology is an interdisciplinary research area that studies the characteristics of materials at nanometer scale and developing new materials with new functionalities. Advances in nanotechnology enable us to develop new molecules and materials with more controlled chemical, physical, and biological properties. The new techniques and materials produced by using nanotechnology provide a vast array of opportunities for diagnosis and therapy of many diseases that are still considered extremely challenging by medical professionals such as cancer, Alzheimer's disease, Parkinson's disease, diabetes, and aging-related disorders. In addition, nanotechnology provides us tools to study the in-depth mechanisms of the biological machinery enabling us to learn more about the pathophysiology of the diseases. These detailed analyses can be utilized to pinpoint the exact causes behind these diseases and correct the defects in the biological machinery. Since biological machinery works at nanoscale (e.g., the diameter of DNA is 2 nm; a typical ribosome's diameter is 20–30 nm; individual collagen fibers of the extracellular matrix are ~1.5 nm in diameter and 300 nm in length), it can best be manipulated by using nanoscale materials with controlled functionalities.

Therapeutic Nanomaterials, First Edition. Edited by Mustafa O. Guler and Ayse B. Tekinay.

Thus, nanomaterials with controlled physical, chemical, and biological characteristics can be used for the therapy of the specific causes of the diseases.

Overall, nanomaterials serve two important purposes for medical applications: They can be utilized to understand the pathophysiology of the diseases by enhancing detailed knowledge of biological machinery and increasing diagnosis efficiency, and they can provide us novel approaches to interrupt or correct the regular biological activity depending on the disease type and the treatment strategy.

1.2 NANOSCALE MATERIAL PROPERTIES

Nanoscale is generally considered as dimensions between 0.1 and 100 nm, and nanomaterials can display extraordinary characteristics compared to their micro- or macroscale counterparts. New synthesis techniques can control shape and function of materials at the nanometer level. There are several ways to develop new materials in nanometer scale. Mainly, top-down and bottom-up approaches are the two major techniques to produce nanomaterials. In the top-down techniques, bulk materials are tailored into specific shape and size with recent high-tech tools. For example, soft-lithography techniques can craft bulk surfaces into nanostructured textures to create a high surface area and molecular contact points with the biological materials. In bottom-up approaches, small molecular building blocks are used to form more complex and higher-scale nanometer-sized materials. Both techniques have advantages and disadvantages in terms of their fabrication method and product function. In theory, it is desired to utilize both techniques in conjunction so that we can eliminate the weaknesses of each technique. Depending on the application area, either one or both of these approaches can be used to develop materials that can be used in studying pathophysiology of diseases and their diagnosis and therapy. Especially, bioinspired and biomimetic strategies yield products that can replace or accommodate activities of the natural biomolecules. Nevertheless, for effective diagnosis and therapy of diseases, it is almost crucial to first understand the molecular reasons behind disease development.

1.3 NANOMATERIALS FOR UNDERSTANDING DISEASE PATHWAYS

Biological machinery is known for its perfect balance, and runs within a complex network, which enables it to tolerate irregularities up to a certain level. Diseases occur when these irregularities cannot be tolerated, and several reasons might cause this, which are generally classified as hereditary or environmental reasons. In most cases, both of these components are the culprits behind medical problems, and it is always important to understand the changes in molecular level to decide the most appropriate treatment. For example, when an irregular activity of a protein, which can result in a disease, is detected, the necessary precautions can be taken or developed for the appropriate treatment. In some cases, protein production mechanism can be targeted to discontinue the disease-related activity. In other cases, the specific protein could be targeted and blocked; therefore the protein can be inactivated to stop the undesired activity. Since biological machinery works at the molecular level, these mechanisms can be best understood by using techniques that provide the highest sensitivity. Many of the current techniques that are used in biomedical research utilize microtechnology, which not only require higher amounts of biomolecules for analyses but also are only sensitive at microscale. On the other hand, techniques that utilize nanotechnology have recently been introduced in biomedical research and have revolutionized particular research areas. Developing DNA sequencing strategies for personalized medicine, biosensors with higher sensitivity that can be used for detection of low levels, or biomolecules and even nanoparticles that can be used for isolation for biomacromolecules such as DNA, RNA, or proteins are some of the examples of recent use of nanotechnology in understanding disease pathophysiology. On the other hand, there is an enormous amount of research in the recently published literature on developing better technologies for understanding biological events and pathways including nanomaterials for biocompatible labeling of biomolecules and cells for more efficient monitoring of activity, for tailoring nanomaterials for enhanced targeting ability (compared to regularly used antibodies), and for targeted blocking of biomolecular activity to understand their functionality in more detail. Application of these methods to biomedical research will yield in gaining more knowledge in the working mechanisms of biological machinery, and pathophysiology of diseases, and for enhancing diagnostic capabilities, all of which will in turn provide more opportunities for therapy.

1.4 NANOMATERIALS FOR THERAPY

To cure diseases with synthetic materials, the materials should be able to interact with specific biological actors in their natural environment. These biological actors can be cell surface receptors, which are mostly composed of proteins and carbohydrates; extracellular elements, such as growth factors, cytokines, or structural components like collagens; or intracellular elements, such as DNA, ribosomes, RNA, enzymes, etc. The optimal venue of interaction with biomolecules would be similar to the way they interact with their natural binding partners, so that the balance of the biological machinery can be reinstated. Therefore, the materials to be used should carry physical properties to meet the requirements for appropriate interactions. In addition to these, the materials should be functionalized with bioactive molecules. The interaction between the bioactive domain of the material and the target protein determines the stability of the complex and determines the fate of the biological activity.

Nanomaterials are used for therapy of diseases through several ways such as targeted drug/gene delivery approaches and induction of regeneration of damaged tissues by using nanomaterials. For targeted drug delivery, nanomaterials can be used as targeting molecules, as carrier systems, or as the bioactive drug itself. Aptamers, for example, are one example of how tailored nanomaterials can be used for targeting purposes. On the other hand, most of the research on nanomaterials for drug delivery has focused on developing carrier systems such as liposomes, polymeric nanoparticles, or metal-based nanoparticles. Although small-molecule drugs are the most commonly used therapeutics used for drug delivery approaches, there have been serious advances in producing tailored nanomaterial drugs, mostly in the form of small peptides or their conjugates.

The nanomaterials can be also used in regenerative medicine applications. To regenerate the tissue defects caused by diseases, materials can form an artificial three-dimensional environment to fill the gap with the bioactive signals derived from the natural healing process. The soluble factors can diffuse inside this network, and the cells in the proximity can migrate to the defect side. If correct signals and the optimum environment are provided, the tissue defect can be healed and function of the tissue can be recovered. Many polymeric materials have previously been tailored to mimic the natural biomacromolecules both physically and chemically. These materials have also been further functionalized through addition of natural biological molecules such as growth factors. On the other hand, there is a growing area of nanomaterials that are synthesized by using natural biomaterials such as peptide nanofiber systems, which can be produced through bottom-up approaches. These nanomaterials can be specifically designed to mimic natural proteins and carbohydrates to distinctively interact with particular biomacromolecules so that they induce differentiation of stem cells into specific lineages and induce functional tissue regeneration.

1.5 CHALLENGES AND FUTURE PROSPECTS

Although there have been extensive advances in developing nanomaterials for biomedical purposes, only few of them have been translated into clinics. The major limitations behind this delay are about the biocompatibility and biodegradability of nanomaterials.

One of the desired properties of the nanomaterials in the biological environment is their physicochemical stability. When a nanomaterial is injected into the blood vessels, there are several biological macromolecules that can interact with it in the environment. The noncovalent interactions including hydrogen bonds, electrostatic interactions, and van der Waals forces cause the undesired interactions in the blood. These may cause problems in the blood flow, or simply the nanomaterials cannot travel in the blood vessels, and they fail to reach to the target. In some cases, the interaction of these random molecules in the blood changes the surface chemistry or bioactivity of the nanomaterials, and they may cause undesired side reactions.

Undesired accumulation of the nanomaterials in the body and side products produced by degradation of the nanomaterials is another drawback in the use of nanomaterials for therapeutic purposes since these may cause side effects. Major areas where nanomaterials are accumulated in the body are the liver, spleen, and kidneys, which might result in metabolic problems associated with these organs, which eventually can cause organ failure.

Beyond many advances in the field of molecular biology and medicine, most molecular interactions between biomacromolecules are unknown, and our knowledge pathophysiology of diseases and the mechanisms of tissue regeneration are limited. Thus, one of the major challenges in developing and using nanomaterials for therapeutic purposes lies in the lack information on appropriate target molecular mechanisms or pathways. With more advancement in understanding of these interactions and better control on production of nanomaterials, biocompatible and bioactive nanomaterials with tightly regulated characteristics can be developed to interact with biomolecules to correct and regulate the natural biological interactions to cure diseases in the future. Beyond diagnosis, these advances can also be used to design and fabricate nanomaterials that can deliver drugs or trigger natural key reactions for regeneration purposes.

It is important to stay up to date on how nanomaterials can be used for diagnostic and therapeutic purposes by presenting specific examples from the literature. The research on biomedical nanomaterials can be classified according to their medical applications. Since nanotechnology is a fairly new technology with many unknowns, several examples of nanomaterial–biological organism interactions in terms of nanotoxicology research were demonstrated in order to stress that although nanomaterials provide a vast array of opportunities for the diagnosis and treatment of diseases, the consequences of using these new types of materials should be carefully weighed prior to their use in medical practice.

2

NANOSIZED DELIVERY SYSTEMS FOR TISSUE REGENERATION

Goksu Cinar, Didem Mumcuoglu, Ayse B. Tekinay, and Mustafa O. Guler

Institute of Materials Science and Nanotechnology, National Nanotechnology Research Center (UNAM), Bilkent University, Ankara, Turkey

2.1 INTRODUCTION

Repair and regeneration of damaged tissue is an important clinical need since millions of people all over the world are suffering from tissue and organ failure (Rice et al., 2013; Wei and Ma, 2008). Although tissue regeneration processes and components are quite different depending on tissue properties, the fundamentals of regeneration process involves cells, cellular microenvironment, and biological signals. Hence, the strategies in tissue regeneration focus on cell-based therapies, regenerative biomaterials as scaffolds, and delivery of biological signals to regenerating site. The integration of these strategies and the components with natural tissue healing stages is important to achieve successful therapies in clinical applications (Howard et al., 2008).

Controlled delivery of biologics such as growth factors (GFs), cytokines, nucleic acids, or siRNA is the focus of growing interest in tissue regeneration applications since these biomacromolecules serve highly specific and complex functions in cellular processes compared to small-molecule

Therapeutic Nanomaterials, First Edition. Edited by Mustafa O. Guler and Ayse B. Tekinay.

synthetic drugs (McCall et al., 2011; Vermonden et al., 2012). Although therapeutic approaches in tissue regeneration focus on biologics, the clinical applications are highly limited due to the fragile nature and instabilities of the molecules. Delivery systems can increase therapeutic applicability of biologics sustaining a suitable environment for three-dimensional conformations and protection from enzymatic degradation during transportation to the regenerating site. In addition, these systems aim to mimic natural release mechanisms, conditions, and therapeutic dosage of biologics, which are controlled by distinct patterns and enzymatic reactions in cellular microenvironment (Tessmar and Göpferich, 2007). Stimuli–response of the nanosized carriers is also an important property to decrease side effects and effective dosage of the molecules necessary for the regeneration in tissue-specific conditions (Alvarez-Lorenzo and Concheiro, 2014).

Advances in material science and integration with nanotechnology provide us a new generation of biodegradable and biocompatible nanosized delivery systems with control over size, shape, and multifunctionality (Panyam and Labhasetwar, 2003; Zhang et al., 2013). Nanosized delivery systems can be designed as lipid-based, inorganic, polymeric, or multifunctional hybrid systems including liposomes, polymeric micelles, mesoporous silica nanoparticles, nanogels, or nanocomplexes with sizes varying between 1 and 200 nm (Khandare et al., 2012). Nanosized delivery systems provide tissue penetrating ability, reduced toxicity, and enhanced permeation and retention (EPR) in regenerating tissues (Gu et al., 2011). These systems can be injected to the bloodstream and extend the release of biologics protecting them from proteolytic cleavage and chemical degradation. EPR effect also provides passive targeting for the nanosized carriers increasing circulation time in the bloodstream. Improved solubility of hydrophobic compounds can be obtained via nanosized delivery systems (Mishra et al., 2010). In addition, nanosized delivery systems can be integrated into three-dimensional scaffolds sustaining controlled release and biocompatible microenvironment for regenerating cells (Wei and Ma, 2008).

Multifunctional nanosized delivery systems for biologics have been developed using both natural and synthetic polymers. Synthetic polymers enable sustained release of biologics over a period of days to several weeks based on different release mechanisms including concentration, degradation, affinity or stimuli controlled, and also combinations of them. FDA-approved synthetic polymers such as poly(lactide-*co*-glycolide) (PLGA) have been used for developing nanosized delivery systems due to their biocompatibility and tunable physical and chemical properties (Golub et al., 2010). PLGA nanoparticles can be hydrolyzed into biodegradable metabolites: lactic and glycolic acid at acidic conditions (Kumari et al., 2010). In addition, there are examples of polymeric nanoparticles developed as biological delivery systems using poly(lactic acid) (PLA), poly(D,L-glycolide) (PLG), poly-ε-caprolactone (PCL), and poly(alkyl cyanoacrylates) for tissue regeneration (Kim et al., 2014). Although some of these synthetic nanosized polymeric systems reveal biodegradable properties, degradation products can be toxic compounds, and degradation process takes longer time periods. Moreover, formulation conditions and encapsulation steps may require organic solvents and high ionic strength, creating a harsh environment for biologics.

On the other hand, natural polymers can form biodegradable polymeric nanostructures for biomacromolecule delivery at mild conditions, and these nanostructures are functionalized with different delivery strategies for targeting and internalization. For these purposes, both protein-based natural polymers such as gelatin, collagen, albumin, and elastin and hydrophilic polysaccharides including alginate or chitosan have been used to obtain colloidal nanostructures. Compared to synthetic polymeric nanostructures, natural nanocarriers can be easily degraded by digestive enzymes and degradation products not harmful. However, the hydrophilic nature of these nanosized carriers leads to burst release of biologics when biologics are immobilized within the system via noncovalent interactions or short-term release profiles can be obtained due to easier biodegradability compared to synthetic polymeric nanosized delivery systems. In addition, contamination risk is higher since the sources of these compounds are natural organisms.

Molecular assemblies of both synthetic and natural compounds are intriguing nanosized delivery structures for biologics. Colloidal nanostructures such as liposomes, polymeric nanoparticles, block copolymer micelles, and dendrimers can be designed as stimuli responsive, and delivery mechanisms can be enhanced via internal and external factors including pH, temperature, or redox microenvironment (Fleige et al., 2012; Ganta et al., 2008). Intravenous administration of these colloidal nanosized systems also makes them suitable carriers for tissue regeneration applications (Alyautdin et al., 2014). In addition, the stimulated effects of gold nanoparticles between 20 and 50 nm size on osteogenic differentiation of stem cells and osteoblast-like cells have been showed in different studies (Heo et al., 2014; Ko et al., 2015).

In this chapter, we focus on advanced delivery of biologics including GFs, cytokines, genes, or siRNAs using a variety of nanosized systems for different regeneration applications focusing on bone, cartilage, nervous system, and muscle regeneration strategies. The limitations of biologics delivery and alternative strategies for overcoming recent problems are underlined presenting recent examples from the literature. In addition, specific targeting and cellular internalization strategies of biologics delivery for tissue regeneration are discussed for providing future perspectives to the readers in this field.

2.2 DELIVERY OF PROTEIN THERAPEUTICS WITH NANOCARRIERS FOR TISSUE REGENERATION

2.2.1 GFs and Cytokines

As biological regulatory signals in variety of cellular responses, GFs are important components for controlling and directing tissue formation, maintenance, and regeneration. In addition, these biological signals direct crucial tissue regeneration processes such as angiogenesis and bone or granulation tissue formation (Eichmann and Simons, 2012; Laurencin et al., 2014; Ponte et al., 2007). Although the functions and importance of many GFs in tissue regeneration are well known, the clinical applications and therapeutic efficiency of these biologics are limited due to their short lifetime, production costs, and safety concerns for immunogenic responses (Martino et al., 2014; Rice et al., 2013). To increase their lifetime and facilitate controlled release of biologically active GFs over an extended time period for therapeutic applications, nanosized delivery strategies have been developed using different materials.

In living organisms, GFs are stabilized in extracellular matrix (ECM) via interactions of highly sulfated ECM components such as glycosaminoglycan (GAG) side chains. GFs can be immobilized on nanosized carriers by either noncovalent or covalent interactions (Chen et al., 2010). Polymeric nanoparticles with functional groups can be designed by sustaining these specific interactions for immobilization and affinity-controlled GF delivery (Wang and von Recum, 2011). Natural polyanionic polysaccharides such as heparin and chondroitin sulfate can form nanocomplexes with natural or synthetic polycationic polymers, and polyelectrolyte complex nanoparticles are used to deliver heparin-binding GFs including fibroblast growth factor (FGF) family and transforming growth factor-β (TGF-β) superfamily (Place et al., 2014). Natural polymers including proteins and polysaccharides can be modified with several functional groups and form colloidal nanostructures, which are also suitable candidates for delivering GFs via noncovalent interactions or covalent crosslinking. Gelatin is a well-known natural protein-based polymer with excellent biocompatibility and controllable biodegradability. It can be modified as either negatively or positively charged and enable polyion complexation with several GFs (Young et al., 2005).

Angiogenesis is a critical process in tissue restoration and constitutes of establishing a vascular system to supply required oxygen and nutrition to the regenerating site. Vascular endothelial growth factor (VEGF), which is a member of the cystine knot family, is an important biological factor for mediating angiogenesis in wound healing and myocardial ischemia treatments (Crafts et al., 2015). Different delivery strategies are developed for sustained release of bioactive VEGF to the regenerating site. In one study (Golub et al., 2010), VEGF-loaded PGLA nanoparticles enhanced blood vessel growth via sustained delivery for cardiovascular medicine applications. Polyelectrolyte nanosized complexes formed by coacervation of VEGF-bound dextran sulfate with different polycations such as chitosan, polyethylenimine, or poly-L-lysine were shown to have high encapsulation efficiency for the regeneration applications (Huang et al., 2007). In addition, it was shown that GAG-based polyelectrolyte nanocomplexes are quiet stable delivery systems at physiological conditions and show resistance to high ionic strength for *in vivo* applications (Novoa-Carballal et al., 2014). In another example, *in vivo* therapeutic revascularization was obtained via VEGF-loaded heparin-functionalized nanoparticle–fibrin complexes in a rabbit ischemic hind limb model (Chung et al., 2010).

Bone regeneration consists of cascades of complex biochemical processes that are coordinated via cells, ECM, and bioactive molecules such as osteogenic, angiogenic, inflammatory, and systemic GFs (Vo et al., 2012). The enhanced therapeutic effects of different GFs such as osteoinductive bone morphogenetic protein-2 (BMP-2) and mitogenic platelet-derived growth factor-BB (PDGF-BB) have been shown in clinical trials for bone defect treatments (Nevins et al., 2013; Shah et al., 2014). BMP-2 is an important member of TGF-β superfamily that includes major modulators of osteogenesis, which play important roles in the commitment and differentiation of osteoprogenitors. BMP-2 loaded 2-*N*,6-*O*-sulfated chitosan-based nanoparticles (S-NPs) were developed via complex coacervation of oppositely charged polyelectrolyte solutions as a delivery system for bone regeneration applications (Cao et al., 2014). In another study (Gan et al., 2015), pH-responsive chitosan-functionalized mesoporous silica nanoparticles were used for dual delivery of BMP-2 and dexamethasone (Dex) for osteoblast differentiation and bone regeneration at *in vitro* and *in vivo* conditions.

Neurotrophic factors (NF) including nerve growth factor (NGF), glial-derived neurotrophic factor (GDNF), brain-derived neurotrophic factor (BDNF), insulin-like growth factor-1 (IGF-1), or basic FGF-2 are regulatory biological signals for promoting the development, survival, and regeneration of neurons (Harvey et al., 2014; Ziv-Polat et al., 2014). The alterations in cellular structure and metabolism of damaged neurons lead to impaired regeneration (Sivak et al., 2014). Biological factors can induce endogenous repair, enhance neural regeneration, and hinder inhibitor signals for the regeneration process (Donaghue et al., 2014). Delivery of these factors to the central nervous system (CNS) is a developing approach for treatments of traumatic brain, spinal cord, and peripheral nerve injuries and neural degenerative disorders (Mohtaram et al., 2013).

Different strategies have been developed for therapeutic delivery of biologics into the CNS such as liposomes, nanospheres, nanocapsules, dendrimers, and polymeric micelles (Orive et al., 2009). The design of nanosized delivery systems for the CNS focus on two important requirements: long circulation time of the nanocarrier systems consisting regenerating signals and ability to penetrate blood–brain barrier (BBB), which is formed by tight junctions preventing crossing of large neurotrophic protein molecules (Thorne and Frey II, 2001; Zhong and Bellamkonda, 2008). BDNF-bound magnetically guided nanoparticles overcame impermeability of BBB and decreased morphine-induced apoptosis in the CNS restoring the spine density and promoting regeneration of synaptic connections (Pilakka-Kanthikeel et al., 2013). In another study, bFGF-loaded PEG–PLGA nanoparticles were functionalized with *Solanum tuberosum* lectin (STL) for targeted delivery to brain tissue and enhanced spatial learning and memory of rats with Alzheimer's disease (AD) promoting the survival and neurite growth of neurons (Zhang et al., 2014).

The delivery systems loaded with multiple GFs can mimic complex *in vivo* conditions releasing different therapeutic biologics for effective tissue regeneration (Chen et al., 2010). Time-dependent controlled releases of multiple GFs at different stages of tissue regeneration via nanosized delivery systems enhance the biochemical processes sequentially. In one study (Perez et al., 2013), bone regeneration was induced by initial release of the angiogenic factors like VEGF or bFGF and then supported via BMP-2 release as an osteogenic factor.

Platelet-rich plasma consist of biologically active proteins including several GFs, such as the isoforms of platelet-derived growth factor (PDGF), transforming growth factor (TGF), and FGF (Intini, 2009). The lysates of platelet-rich plasma are natural sources of multiple GFs that are important in bone regeneration (Visser et al., 2009). Hence, controlled delivery of plasma lysates via chitosan–chondroitin sulfate nanoparticles prepared by polyelectrolyte complexation was intriguing for bone tissue engineering applications promoting osteogenic differentiation via multiple factors (Santo et al., 2012).

Cytokines are small signaling molecules with sizes ranging between 8 and 40 kDa and are responsible for regulating the immune response, inflammation, neoangiogenic processes, and cellular differentiation (Ioannidou, 2006). Cytokines are also important biochemical factors in tissue regeneration controlling cellular activity (Gelain et al., 2010). However, short half-lives and serum-mediated degradation of cytokines prevent therapeutic applications and lead to developing different strategies for delivery of these small biologics. Cytokine delivery to injury site by controlled delivery systems can direct tissue regeneration in the absence of the transplanted cells (Roche et al., 2013).

Inflammation process is one of the critical steps of regeneration, and this complex step is controlled by secretion of GFs and pro- or anti-inflammatory cytokines. Cytokine-mediated signaling of inflammation involving activation and proliferation of satellite cells is an essential component of muscle repair, regeneration, and growth (Tidball, 2005). Delivery of cytokines such as tumor necrosis factor-α (TNF-α) (Chen et al., 2007), interferon-γ (Cheng et al., 2008), or interleukin-6 (IL-6) (Serrano et al., 2008) mediating inflammation in muscle regeneration processes can be an alternative approach for the treatments of acute or chronic muscle damages.

2.3 GENE AND siRNA DELIVERY WITH NANOCARRIERS FOR TISSUE REGENERATION

2.3.1 Gene Delivery

Gene therapy provides sustained expression of bioactive molecules including GFs and cytokines required for tissue regeneration (Bonadio et al., 1999) and can assist stem cells to differentiate into a variety of different lineages and cells types for regenerative medicine (Chen et al., 2011b). Low uptake across the cell membrane, limited stability of DNA molecules, and lack of nuclear targeting are the main difficulties in DNA delivery (Luo and Saltzman, 2000). Controlled delivery of DNA using nanosized carriers promotes gene delivery and extended transgene expression. In addition, targeted nanosized carriers for gene delivery can avoid side effects including immune response or distribution to the nontargeted tissue and cells (Pannier and Shea, 2004).

Delivery of DNA encoding for inductive biologics can spatially influence on cellular behavior and enhance the formation of complex architectures for tissue regeneration (De Laporte and Shea, 2007). Although viral and retroviral vectors have been showed to be highly efficient transfection agents for *in vivo*, nonviral systems provide lower immune response and controllable chemical and biological properties for gene delivery (Elangovan et al., 2014;

Leong et al., 1998). Nonviral delivery nanocarriers for DNA are developed using a variety of synthetic and natural polymers such as PLL, PLG, PVA, PEG, poly(ethyleneimine) (PEI), collagen, hyaluronic acid, gelatin, or chitosan (Pannier and Shea, 2004).

Cationic lipids consisting of a positively charged head group, a hydrophobic chain, and a linker that joins the polar and nonpolar regions can interact with negatively charged DNA molecules and form complexes called "lipoplexes" (Bhattacharya and Bajaj, 2009; Gao and Hui, 2001; Tros de Ilarduya et al., 2010). Similar to cationic lipids, cationic polymers can also form complexes with DNA and form "polyplexes," which are capable of gene delivery into targeted cells (Zhang et al., 2004). PEI is a highly cationic polymer due to the presence of amino groups, and it can form nanoparticles by complexation with negatively charged plasmid DNA (pDNA) and prevents pDNA from lysosomal nuclease degradation (Pérez-Martínez et al., 2011). Chondrogenic differentiation of hMSCs was facilitated by the delivery of Sox9 gene, an important transcription factor in the process of chondrogenesis, complexed with PEI on PLGA nanoparticles (Jeon et al., 2012). In another study, pDNA encoding Runt-related transcription factor 2 (RUNX2) was delivered in liposomes, which were immobilized at the surface of polycaprolactone (PCL) nanofiber meshes sustaining physical support for hMSCs during osteogenic differentiation (Monteiro et al., 2014). Bone regeneration in rat cranial defects was also enhanced via localized delivery of PEI/pDNA nanocomplexes inducing expression of bone morphogenetic protein-4 (BMP-4) (Huang et al., 2005). On the other hand, cationic DNA/polymer nanocomplexes have some drawbacks including instabilities in physiological conditions and aggregations due to binding of serum proteins on the surfaces. Also cationic lipids and polymers (PEI) can be toxic. To eliminate these drawbacks, targeted delivery and internalization strategies have been developed for enhanced therapeutic activities.

Dendritic nanostructures are also promising gene delivery systems due to their highly branched structure, shape, and multivalency. These structures can be designed using biodegradable polymers providing non-toxicity and efficiency for gene delivery (Luo et al., 2014). Especially, polycationic dendrimers such as polyamidoamine (PAMAM) were shown to be particularly interesting in nucleic acid delivery providing high multivalent surface moiety, biocompatibility, and low cytotoxicity (Lee et al., 2014). Human bone morphogenetic protein-2 (hBMP-2) gene-containing PAMAM dendrimers were used as gene transfer carriers for inducing osteogenic differentiation of mesenchymal stem cells (MSCs) for bone regeneration applications (Santos et al., 2009).

2.3.2 siRNA Delivery

Short interfering RNAs (siRNA) are a short double-stranded biomacromolecules that silence target gene expression as a consequence of binding to the corresponding messenger RNA. Gene therapy provides permanent genetic alteration, but only transient actions of protein therapeutics can be needed for tissue regeneration applications. Hence, siRNA-based strategies provide a safer alternative compared to the gene therapy. Delivery systems are required for siRNA due to siRNA's inability to diffuse easily to cell membrane related with its polyanionic nature, poor solubility, and readily degradation by enzymes (Yoo et al., 2011). Similar to DNA nanocarriers, different nonviral nanosized carriers have been developed for siRNAs using liposomes, lipoplexes, polymers, and also peptides (Gao and Huang, 2008).

Similar to pDNA/PEI complexation, siRNA can also interact with cationic polymers and form stable nonviral nanosized delivery systems. PEG-PEI/siRNA nanoparticles were used for suppressing the expression of Nogo receptor (NgR), which has an important role in CNS controlling neural inflammation, arresting the neurite growths of injured neurons, and the differentiation of NSCs for CNS regeneration (Liang et al., 2012). Another siRNA delivery-based regeneration strategy was developed to increase the endogenous BMP levels through RNA silencing to downregulate BMP-2 antagonist NOGGIN glycoprotein using different nanosized carriers (Bassit et al., 2015).

2.4 SYSTEMIC TARGETING AND CELLULAR INTERNALIZATION STRATEGIES FOR TISSUE REGENERATION

2.4.1 Targeted Delivery

When small molecules or peptide/protein ligands bind to a receptor, they generally start a cascade of signaling events in cellular microenvironment. Systemic targeting of delivery systems can be sustained by ligand–receptor mediated or locally activated delivery via internal factors (specific enzymes or pH change) and external forces such as light or temperature (Bae and Park, 2011). There are specific ligands that can only bind to one type of receptor on a specific cell or tissue, and these specific ligands render the targeting of that specific tissue or cell type. The specificity also depends on the presence and expression of receptors on certain cells. In the natural process of regeneration, GFs bind to certain receptors on the cell surface and initiate a cellular response such as inflammation, recruitment of stem cells to the tissue, or angiogenesis. In addition, proteins take part in cell-to-cell signaling, which is crucial for a synchronized regeneration process. This receptor–protein interaction is so specific that only a certain protein or a protein family can recognize and bind to its receptor. Therefore, it is important to understand the natural regeneration process and have the knowledge of the expression patterns in each tissue at different regeneration stages to be able to engineer materials presenting specific ligands. These ligands will bind to certain target cells only, making it possible to induce a unique specific response of the body. As another advantage, the localization of the biologics in the target site will decrease the side effects of the treatment. One general concern about administration of biomacromolecules such as GFs is the induction of proliferation of cells and formation of benign tumors. This risk would be decreased via targeted delivery approaches, which can mimic the release of the natural GFs at therapeutic dosage and localization. Chemical ligands, antibodies, specific peptide sequences, and proteins are common strategies to target specific cells or microenvironments for sustained delivery of biomacromolecules via nanosized carriers.

2.4.1.1 Chemical Ligands There has been an intense effort to explore and synthesize chemical ligands that will target a specific tissue, and the knowledge of ECM composition in different tissues is invaluable for targeting. For example, bone tissue contains mineralized matrix in the form of hydroxyapatite, differently from other tissues, whereas cartilage contains high amounts of polysaccharides such as hyaluronic acid. Skin tissue contains elastin and fibrin for the formation of elastic fibers (Uitto et al., 1989). Using the knowledge about ECM composition, chemical molecules are synthesized to target different tissues. In one of the efforts, researchers synthesized a fullerene molecule that will target calcium phosphate mineral in hydroxyapatite of the bone. The synthesized molecule demonstrated high affinity to hydroxyapatite, and it was proposed as a potential chemical targeting ligand in bone regeneration (Gonzalez et al., 2002).

For cartilage targeting, derivatives of quaternary ammonium (QA) have been investigated in different studies. When the biodistribution of QA–glucosamine (antirheumatic drug) conjugates was investigated, the conjugate of glucosamine demonstrated specificity for cartilage in a short time after injection (Giraud et al., 2000). Moreover, QA has been proposed to be effective for cartilage targeting of naproxen and ibuprofen in the treatment of arthritis (Pawar et al., 2012). Based on these studies, cartilage-targeting molecules might have an application in tissue engineering for the specific delivery of molecules.

The proteins upregulated or exposed to cell surface after myocardial infarction (MI) were investigated for a potential in targeting the infarct area (Sy and Davis, 2010) for cardiac regeneration. p38 MAPK inhibitor, the chemical substance designed to bind p38, was delivered with polymeric microspheres to target p38 protein overexpressed after MI. The delivery system was shown to be effective in inhibiting cardiac dysfunction (Sy et al., 2008).

2.4.1.2 Antibodies Antibodies designed to bind a specific cell surface receptor hold great promise in the delivery of biologics for tissue regeneration. Antibody-targeted delivery is one of the focuses in drug delivery due to the specificity of antigen–antibody binding. There are many monoclonal antibody-based pharmaceutics in the market, but most of them were developed for cancer treatment (i.e., trastuzumab, bevacizumab, ibritumomab tiuxetan). On the other hand, their applications in tissue engineering are also expanding due to the several advantages of antibody targeting.

One of these approaches is the immobilization of an antibody to a scaffold for enhanced GF binding. With this strategy, not only cell-specific targeting but also spatiotemporal regulation of endogenous GFs can be mimicked. To mimic the spatial property, the local injection of the growth factor might be a basic strategy. On the other hand, a better strategy is to modulate the presence of the GF in the target tissue for a certain period of time. In a recent study, anti-BMP-2 monoclonal antibody was immobilized on different biomaterials for the purpose of capturing endogenous BMP-2 and inducing bone regeneration at the graft site (Ansari et al., 2014).

2.4.1.3 Peptides and Proteins Peptide ligands are also specific in terms of binding and might be more advantageous than protein ligands due to ease of synthesis and conjugation to the nanosized delivery systems. Peptide phage screening libraries provide the exploration of targeting peptides for a specific molecule and more specific purposes of use. Arginine–glycine–aspartic acid (RGD) is one of the common ligands that have been used in tissue regeneration, which can increase the adhesion of cells by interacting with integrins on the cell surface.

Collagen-binding domain (CBD) is a promising peptide sequence for targeted delivery of nanosized carriers to collagen. GFs that are conjugated to CBD can be used to promote specific binding of GF to collagen. CBD-fused basic fibroblast growth factor (bFGF) was used for the regeneration of abdominal wall defects, and it was demonstrated to promote vascularization for 30 days after surgery (Shi et al., 2011). Similar to CBD, heparin-binding domain is used for targeted delivery purposes. The abundance of heparin in cartilage provides the targetability of cartilage with heparin-binding domain. In one study (Loffredo et al., 2014), IGF-1 was targeted to cartilage with heparin-binding domain and shown to be effective in osteoarthritis treatment *in vivo*.

Delivery of protein therapeutics such as NF to the CNS is highly limited due to BBB preventing diffusion of the macromolecules through the brain tissue. Cell-penetrating peptides (CPPs) can also be used as a brain-targeting strategy. 39-amino acid peptide derived from the rabies virus glycoprotein (RDP) containing specific nerve binding region was shown as an efficient protein carrier for targeting and delivering macromolecular proteins across the BBB (Fu et al., 2012).

Carbohydrate-binding proteins such as lectins are highly specific to monosugars or oligosaccharide structures (Wu et al., 2009). Lectin-conjugated nanosized delivery systems can bind to glycosylated cell membrane components, and lectin–cell membrane interactions improve vesicular transport of the delivery systems enhancing penetration through biological barriers such as BBB, nasal mucosa, lung, buccal cavity, or eye (Bies et al., 2004). The surface modification of PEG–PLA nanoparticles with lectins enhanced the brain uptake after nasal administrations of the delivery system (Gao et al., 2007).

2.4.2 Cellular Internalization Strategies

Controlled delivery of biologics such as GFs, cytokines, nucleic acids, or siRNA is a promising approach in tissue engineering due to several advantages. However, the delivered biomacromolecules with nanosized carriers need to be internalized, and desired internal or external stimuli should be expressed from the target cell only after the internalization. A variety of internalization strategies have been developed since the internalization of biologics into cells is difficult due to electrostatic repulsion and the low uptake efficiencies of the nonviral carriers. The internalization strategies for nonviral carriers can be designed focusing on multiple uptake mechanisms including macropinocytosis, clathrin-mediated endocytosis, or caveolae-mediated endocytosis to promote transduction of the specific genes (Pérez-Martínez et al., 2011) required for the regeneration processes. Cellular internalization is also important for the delivery of the biologics into specific intracellular organelles such as the nucleus.

Peptide domains containing positively charged amino acids such as arginine and lysine have been reported to enhance uptake and cellular internalization of nanosized carriers including PAMAM dendrimers conjugated with L-arginine (Choi et al., 2004), arginine-chitosan/DNA self-assembled nanoparticles (Gao et al., 2008), arginine-grafted bioreducible poly(disulfide amine) polymers (ABP) for siRNA delivery (Beloor et al., 2012), oligoarginine-modified liposomes (Furuhata et al., 2006), and self-assembled oligoarginine–PEG–lipids for gene delivery (Maitani and Hattori, 2009). In addition, integration of CPP to the vectors or peptide nucleic acid conjugates can be used for internalization (Jung et al., 2011). Recently, CPP-decorated self-assembled peptide amphiphile (PA) nanospheres complexed with antisense oligonucleotides increased the internalization of the nanosized delivery system compared to nonbioactive PA nanostructures (Mumcuoglu et al., 2015).

Therapeutic proteins delivered with CPPs can be internalized in cells in the scope of tissue regeneration. Low molecular weight protamine was used as a CPP for the delivery of a transcription factor, transcriptional coactivator with PDZ-binding motif (TAZ). TAZ transcription factor was used for osteogenic differentiation of MSCs, and CPP-conjugated protein could translocate the cell membrane and conserved the functionality (Suh et al., 2014). In addition, production of induced pluripotent stem cells (iPSC) was shown to be possible with induction of CPP-conjugated proteins (Kim et al., 2009).

Cell-penetrating proteins also facilitate delivery of antisense oligonucleotides, which are designed for the modulation of expression levels of a specific gene. Antisense oligonucleotides and small interfering RNA (siRNA) have a great potential in tissue engineering since they can be engineered to target any gene. One antisense oligonucleotide was shown to accelerate the rate of wound healing by modulating expression of connexin 43 (Qiu et al., 2003). GFs can also be targeted by antisense oligonucleotides (Uchio et al., 2004), and the modulation of their expression might provide another way to control the presence of GFs during different stages of regeneration. CPP antisense oligonucleotide conjugates provide higher cellular internalization compared to naked oligonucleotides, and therefore, CPPs would render the antisense oligonucleotide-based therapy more successful. Antisense oligonucleotide designed to alter dystrophin transcript was delivered with arginine-rich CPPs, and the conjugate was systemically administered to restore muscle and cardiac dystrophin expression for promoting muscle and cardiac function (Yin et al., 2008). The peptides containing nuclear localization sequence (NLS) can be used as an alternative internalization strategy for nonviral nucleic acid delivery systems. Incorporation of the self-assembling SV40 peptide having T-antigen NLS with cationic liposomes for nuclear delivery promoted internalization and nuclear accumulation of the DNA (Rea et al., 2009).

2.5 FUTURE PERSPECTIVES

Integration of nanotechnology, material science, and engineering has pioneered developments in advanced targeted and sustained nanosized delivery systems with enhanced therapeutic efficacies for tissue regeneration purposes. Targeted delivery, reduction in therapeutic dosage of biologics, controlled release, and multifunctionality are ultimate goals of developing nanosized delivery systems. The nanosized carriers for tissue regeneration provide significant health benefits including elimination of surgery or requirements of transplantations and low risks of immune responses.

Although different strategies and formulations have been examined for nanosized delivery systems in the preclinical studies, successful regeneration using external factors for different types of tissues in clinical applications is still challenging. There are different aspects that prevent the progress in clinical applicability of the designed nanosized delivery systems for tissue regeneration purposes. Manufacturing costs and problems with reproducible and scale-up production of advanced delivery systems prevent commercialization (Kearney and Mooney, 2013; Zhang et al., 2013) and clinical usage of the carriers for tissue engineering. Moreover, although chemical and physical properties of the nanosized delivery systems are characterized in detail, behavior of nanoscale carriers in *in vivo* conditions cannot be controlled effortlessly due to complex and dynamic nature of the cellular microenvironment (Ruenraroengsak et al., 2010). Successful regeneration can be induced via orchestrated neural, vascular, biochemical, and biophysical components of the tissues (Discher et al., 2009). On the other hand, the inadequate knowledge on repair processes and connections between overall signaling pathways delays the progress of sustained regeneration strategies using delivery systems (Rice et al., 2013).

To eliminate these drawbacks, different strategies have been studied to increase efficacy, stability, and targeting of the nanosized carriers. Since the release of single biologics is not sufficient to sustain continuous regeneration for complex tissues, the combined use of multiple factors synergistically provides a strategic advantage for overcoming the limitations (Borselli et al., 2010; Chen et al., 2010, 2011a; Hwang and Kloner, 2010; Shah et al., 2014). In addition, the delivery of multiple factors and developments of the multicomponent systems including delivering the cells and biologics together using suitable scaffolds promote therapeutic effectiveness and tissue regeneration capacity (Donaghue et al., 2014; Park et al., 2013).

Further improvement can be prolonged via incorporation of ECM-mimetic scaffolds with nanosized delivery systems of biologics sustaining both biochemical and physical factors required for tissue repair and regeneration (Chung et al., 2007). Combination of the biological factor-loaded nanocarriers with three-dimensional networks provides long-term controlled release of the bioactive molecules required for regeneration (Jeon et al., 2008). The incorporation of GF-loaded nanoparticles or nanosized delivery systems with three-dimensional scaffolds also enhances the efficiency of the released molecules for *in vivo* applications, preventing initial high burst release and enhancing bioactivity. For example, VEGF-encapsulated nanoparticles incorporated into three-dimensional delivery systems such as PLGA scaffolds and Matrigel™ hydrogels enhanced angiogenesis *in vivo* (des Rieux et al., 2011). In another application, VEGF-loaded chitosan/heparin polyelectrolyte nanocomplexes immobilized into decellularized scaffolds promoted vascularization in mouse subcutaneous implantation model (Tan et al., 2011). Acidic FGF (aFGF)-loaded mesoporous silica nanospheres shelled by PLA were embedded into collagen scaffold, and the nanocarriers containing scaffold implanted in a rat subcutaneous tissue induced significant proliferation of osteoblast cells (Kim et al., 2014). In addition to the GFs, BMP-2 pDNA/chitosan nanoparticles were encapsulated into the fibrous network for slower release rate and enhanced biological activity (He et al., 2014; Nie and Wang, 2007). Furthermore, engineering of the bioactive molecules such as GFs with superaffinity to the scaffolds or carriers can be an alternative approach to prolonged delivery and increased efficacy and reducing the dosage required for the regenerative medicine applications (Martino et al., 2014).

Developments in fabrication techniques, understanding of tissue regeneration processes, and the roles of biological factors in detail will contribute to progress in this field. Combination of advantages of advanced nanosized delivery systems with the power of biological factors has also great potential to solve the recent problems in tissue regeneration applications. Industrial collaboration with scientific community can synergize the commercialization of the developed systems for therapeutic applications. Overall, it is believed that advanced nanosized delivery systems integrated with multicomponent designs will open new opportunities in delivery technologies and strategies for tissue regeneration.

REFERENCES

Alvarez-Lorenzo, C. and Concheiro, A. (2014). Smart drug delivery systems: from fundamentals to the clinic. Chemical Communications *50*, 7743–7765.

Alyautdin, R., Khalin, I., Nafeeza, M.I., Haron, M.H., and Kuznetsov, D. (2014). Nanoscale drug delivery systems and the blood–brain barrier. International Journal of Nanomedicine *9*, 795.

Ansari, S., Freire, M.O., Pang, E.-K., Abdelhamid, A.I., Almohaimeed, M., and Zadeh, H.H. (2014). Immobilization of murine anti-BMP-2 monoclonal antibody on various biomaterials for bone tissue engineering. BioMed Research International *2014*, 940860.

Bae, Y.H. and Park, K. (2011). Targeted drug delivery to tumors: myths, reality and possibility. Journal of Controlled Release *153*, 198.

Bassit, A., Moffatt, P., Gaumond, M., and Hamdy, R. (2015). The potential use of nanoparticles for noggin siRNA delivery to accelerate bone formation in distraction osteogenesis. Journal of Nanomedicine and Nanotechnology *6*, 2.

Beloor, J., Choi, C.S., Nam, H.Y., Park, M., Kim, S.H., Jackson, A., Lee, K.Y., Kim, S.W., Kumar, P., and Lee, S.-K. (2012). Arginine-engrafted biodegradable polymer for the systemic delivery of therapeutic siRNA. Biomaterials *33*, 1640–1650.

Bhattacharya, S. and Bajaj, A. (2009). Advances in gene delivery through molecular design of cationic lipids. Chemical Communications, 4632–4656.

Bies, C., Lehr, C.-M., and Woodley, J.F. (2004). Lectin-mediated drug targeting: history and applications. Advanced Drug Delivery Reviews *56*, 425–435.

Bonadio, J., Smiley, E., Patil, P., and Goldstein, S. (1999). Localized, direct plasmid gene delivery in vivo: prolonged therapy results in reproducible tissue regeneration. Nature Medicine *5*, 753–759.

Borselli, C., Storrie, H., Benesch-Lee, F., Shvartsman, D., Cezar, C., Lichtman, J.W., Vandenburgh, H.H., and Mooney, D.J. (2010). Functional muscle regeneration with combined delivery of angiogenesis and myogenesis factors. Proceedings of the National Academy of Sciences *107*, 3287–3292.

Cao, L., Wang, J., Hou, J., Xing, W., and Liu, C. (2014). Vascularization and bone regeneration in a critical sized defect using 2-N, 6-O-sulfated chitosan nanoparticles incorporating BMP-2. Biomaterials *35*, 684–698.

Chen, S.-E., Jin, B., and Li, Y.-P. (2007). TNF-α regulates myogenesis and muscle regeneration by activating p38 MAPK. American Journal of Physiology-Cell Physiology *292*, C1660–C1671.

Chen, F.-M., Zhang, M., and Wu, Z.-F. (2010). Toward delivery of multiple growth factors in tissue engineering. Biomaterials *31*, 6279–6308.

Chen, F.-M., Wu, L.-A., Zhang, M., Zhang, R., and Sun, H.-H. (2011a). Homing of endogenous stem/progenitor cells for in situ tissue regeneration: promises, strategies, and translational perspectives. Biomaterials *32*, 3189–3209.

Chen, J., Chen, H., Li, P., Diao, H., Zhu, S., Dong, L., Wang, R., Guo, T., Zhao, J., and Zhang, J. (2011b). Simultaneous regeneration of articular cartilage and subchondral bone in vivo using MSCs induced by a spatially controlled gene delivery system in bilayered integrated scaffolds. Biomaterials *32*, 4793–4805.

Cheng, M., Nguyen, M.-H., Fantuzzi, G., and Koh, T.J. (2008). Endogenous interferon-γ is required for efficient skeletal muscle regeneration. American Journal of Physiology-Cell Physiology *294*, C1183–C1191.

Choi, J.S., Nam, K., Park, J.-y., Kim, J.-B., Lee, J.-K., and Park, J.-s. (2004). Enhanced transfection efficiency of PAMAM dendrimer by surface modification with L-arginine. Journal of Controlled Release *99*, 445–456.

Chung, Y.-I., Ahn, K.-M., Jeon, S.-H., Lee, S.-Y., Lee, J.-H., and Tae, G. (2007). Enhanced bone regeneration with BMP-2 loaded functional nanoparticle–hydrogel complex. Journal of Controlled Release *121*, 91–99.

Chung, Y.-I., Kim, S.-K., Lee, Y.-K., Park, S.-J., Cho, K.-O., Yuk, S.H., Tae, G., and Kim, Y.H. (2010). Efficient revascularization by VEGF administration via heparin-functionalized nanoparticle–fibrin complex. Journal of Controlled Release *143*, 282–289.

Crafts, T.D., Jensen, A.R., Blocher-Smith, E.C., and Markel, T.A. (2015). Vascular endothelial growth factor: therapeutic possibilities and challenges for the treatment of ischemia. Cytokine *71*, 385–393.

De Laporte, L. and Shea, L.D. (2007). Matrices and scaffolds for DNA delivery in tissue engineering. Advanced Drug Delivery Reviews *59*, 292–307.

Discher, D.E., Mooney, D.J., and Zandstra, P.W. (2009). Growth factors, matrices, and forces combine and control stem cells. Science *324*, 1673–1677.

Donaghue, I.E., Tam, R., Sefton, M.V., and Shoichet, M.S. (2014). Cell and biomolecule delivery for tissue repair and regeneration in the central nervous system. Journal of Controlled Release *190*, 219–227.

Eichmann, A. and Simons, M. (2012). VEGF signaling inside vascular endothelial cells and beyond. Current Opinion in Cell Biology *24*, 188–193.

Elangovan, S., D'Mello, S.R., Hong, L., Ross, R.D., Allamargot, C., Dawson, D.V., Stanford, C.M., Johnson, G.K., Sumner, D.R., and Salem, A.K. (2014). The enhancement of bone regeneration by gene activated matrix encoding for platelet derived growth factor. Biomaterials *35*, 737–747.

Fleige, E., Quadir, M.A., and Haag, R. (2012). Stimuli-responsive polymeric nanocarriers for the controlled transport of active compounds: concepts and applications. Advanced Drug Delivery Reviews *64*, 866–884.

Fu, A., Wang, Y., Zhan, L., and Zhou, R. (2012). Targeted delivery of proteins into the central nervous system mediated by rabies virus glycoprotein-derived peptide. Pharmaceutical Research *29*, 1562–1569.

Furuhata, M., Kawakami, H., Toma, K., Hattori, Y., and Maitani, Y. (2006). Intracellular delivery of proteins in complexes with oligoarginine-modified liposomes and the effect of oligoarginine length. Bioconjugate Chemistry *17*, 935–942.

Gan, Q., Zhu, J., Yuan, Y., Liu, H., Qian, J., Li, Y., and Liu, C. (2015). A dual-delivery system of pH-responsive chitosan-functionalized mesoporous silica nanoparticles bearing BMP-2 and dexamethasone for enhanced bone regeneration. Journal of Materials Chemistry B *3*, 2056–2066.

Ganta, S., Devalapally, H., Shahiwala, A., and Amiji, M. (2008). A review of stimuli-responsive nanocarriers for drug and gene delivery. Journal of Controlled Release *126*, 187–204.

Gao, K. and Huang, L. (2008). Nonviral methods for siRNA delivery. Molecular Pharmaceutics *6*, 651–658.

Gao, H. and Hui, K. (2001). Synthesis of a novel series of cationic lipids that can act as efficient gene delivery vehicles through systematic heterocyclic substitution of cholesterol derivatives. Gene Therapy *8*, 855–863.

Gao, X., Chen, J., Tao, W., Zhu, J., Zhang, Q., Chen, H., and Jiang, X. (2007). UEA I-bearing nanoparticles for brain delivery following intranasal administration. International Journal of Pharmaceutics *340*, 207–215.

Gao, Y., Xu, Z., Chen, S., Gu, W., Chen, L., and Li, Y. (2008). Arginine-chitosan/DNA self-assemble nanoparticles for gene delivery: *in vitro* characteristics and transfection efficiency. International Journal of Pharmaceutics *359*, 241–246.

Gelain, F., Unsworth, L.D., and Zhang, S. (2010). Slow and sustained release of active cytokines from self-assembling peptide scaffolds. Journal of Controlled Release *145*, 231–239.

Giraud, I., Rapp, M., Maurizis, J.-C., and Madelmont, J.-C. (2000). Application to a cartilage targeting strategy: synthesis and in vivo biodistribution of 14C-labeled quaternary ammonium-glucosamine conjugates. Bioconjugate Chemistry *11*, 212–218.

Golub, J.S., Kim, Y.-t., Duvall, C.L., Bellamkonda, R.V., Gupta, D., Lin, A.S., Weiss, D., Taylor, W.R., and Guldberg, R.E. (2010). Sustained VEGF delivery via PLGA nanoparticles promotes vascular growth. American Journal of Physiology-Heart and Circulatory Physiology *298*, H1959–H1965.

Gonzalez, K.A., Wilson, L.J., Wu, W., and Nancollas, G.H. (2002). Synthesis and in vitro characterization of a tissue-selective fullerene: vectoring $C_{60}(OH)_{16}$ AMBP to mineralized bone. Bioorganic & Medicinal Chemistry *10*, 1991–1997.

Gu, Z., Biswas, A., Zhao, M., and Tang, Y. (2011). Tailoring nanocarriers for intracellular protein delivery. Chemical Society Reviews *40*, 3638–3655.

Harvey, A.R., Lovett, S.J., Majda, B.T., Yoon, J.H., Wheeler, L.P., and Hodgetts, S.I. (2014). Neurotrophic factors for spinal cord repair: which, where, how and when to apply, and for what period of time? Brain Research *1619*, 36–71.

He, C., Nie, W., and Feng, W. (2014). Engineering of biomimetic nanofibrous matrices for drug delivery and tissue engineering. Journal of Materials Chemistry B *2*, 7828–7848.

Heo, D.N., Ko, W.-K., Bae, M.S., Lee, J.B., Lee, D.-W., Byun, W., Lee, C.H., Kim, E.-C., Jung, B.-Y., and Kwon, I.K. (2014). Enhanced bone regeneration with a gold nanoparticle–hydrogel complex. Journal of Materials Chemistry B *2*, 1584–1593.

Howard, D., Buttery, L.D., Shakesheff, K.M., and Roberts, S.J. (2008). Tissue engineering: strategies, stem cells and scaffolds. Journal of Anatomy *213*, 66–72.

Huang, Y., Simmons, C., Kaigler, D., Rice, K., and Mooney, D. (2005). Bone regeneration in a rat cranial defect with delivery of PEI-condensed plasmid DNA encoding for bone morphogenetic protein-4 (BMP-4). Gene Therapy *12*, 418–426.

Huang, M., Vitharana, S.N., Peek, L.J., Coop, T., and Berkland, C. (2007). Polyelectrolyte complexes stabilize and controllably release vascular endothelial growth factor. Biomacromolecules *8*, 1607–1614.

Hwang, H. and Kloner, R.A. (2010). Improving regenerating potential of the heart after myocardial infarction: factor-based approach. Life Sciences *86*, 461–472.

Intini, G. (2009). The use of platelet-rich plasma in bone reconstruction therapy. Biomaterials *30*, 4956–4966.

Ioannidou, E. (2006). Therapeutic modulation of growth factors and cytokines in regenerative medicine. Current Pharmaceutical Design *12*, 2397–2408.

Jeon, O., Song, S.J., Yang, H.S., Bhang, S.-H., Kang, S.-W., Sung, M.A., Lee, J.H., and Kim, B.-S. (2008). Long-term delivery enhances in vivo osteogenic efficacy of bone morphogenetic protein-2 compared to short-term delivery. Biochemical and Biophysical Research Communications *369*, 774–780.

Jeon, S.Y., Park, J.S., Yang, H.N., Woo, D.G., and Park, K.-H. (2012). Co-delivery of SOX9 genes and anti-Cbfa-1 siRNA coated onto PLGA nanoparticles for chondrogenesis of human MSCs. Biomaterials *33*, 4413–4423.

Jung, M.-R., Shim, I.-K., Kim, E.-S., Park, Y.-J., Yang, Y.-I., Lee, S.-K., and Lee, S.-J. (2011). Controlled release of cell-permeable gene complex from poly (L-lactide) scaffold for enhanced stem cell tissue engineering. Journal of Controlled Release *152*, 294–302.

Kearney, C.J. and Mooney, D.J. (2013). Macroscale delivery systems for molecular and cellular payloads. Nature Materials *12*, 1004–1017.

Khandare, J., Calderón, M., Dagia, N.M., and Haag, R. (2012). Multifunctional dendritic polymers in nanomedicine: opportunities and challenges. Chemical Society Reviews *41*, 2824–2848.

Kim, D., Kim, C.-H., Moon, J.-I., Chung, Y.-G., Chang, M.-Y., Han, B.-S., Ko, S., Yang, E., Cha, K.Y., and Lanza, R. (2009). Generation of human induced pluripotent stem cells by direct delivery of reprogramming proteins. Cell Stem Cell *4*, 472.

Kim, T.-H., Eltohamy, M., Kim, M., Perez, R.A., Kim, J.-H., Yun, Y.-R., Jang, J.-H., Lee, E.-J., Knowles, J.C., and Kim, H.-W. (2014). Therapeutic foam scaffolds incorporating biopolymer-shelled mesoporous nanospheres with growth factors. Acta Biomaterialia *10*, 2612–2621.

Ko, W.-K., Heo, D.N., Moon, H.-J., Lee, S.J., Bae, M.S., Lee, J.B., Sun, I.-C., Jeon, H.B., Park, H.K., and Kwon, I.K. (2015). The effect of gold nanoparticle size on osteogenic differentiation of adipose-derived stem cells. Journal of Colloid and Interface Science *438*, 68–76.

Kumari, A., Yadav, S.K., and Yadav, S.C. (2010). Biodegradable polymeric nanoparticles based drug delivery systems. Colloids and Surfaces B: Biointerfaces *75*, 1–18.

Laurencin, C.T., Ashe, K.M., Henry, N., Kan, H.M., and Lo, K.W.-H. (2014). Delivery of small molecules for bone regenerative engineering: preclinical studies and potential clinical applications. Drug Discovery Today *19*, 794–800.

Lee, J., Jung, J., Kim, Y.-J., Lee, E., and Choi, J.S. (2014). Gene delivery of PAMAM dendrimer conjugated with the nuclear localization signal peptide originated from fibroblast growth factor 3. International Journal of Pharmaceutics *459*, 10–18.

Leong, K., Mao, H.-Q., Truong-Le, V., Roy, K., Walsh, S., and August, J. (1998). DNA-polycation nanospheres as non-viral gene delivery vehicles. Journal of Controlled Release *53*, 183–193.

Liang, Y., Liu, Z., Shuai, X., Wang, W., Liu, J., Bi, W., Wang, C., Jing, X., Liu, Y., and Tao, E. (2012). Delivery of cationic polymer-siRNA nanoparticles for gene therapies in neural regeneration. Biochemical and Biophysical Research Communications *421*, 690–695.

Loffredo, F.S., Pancoast, J.R., Cai, L., Vannelli, T., Dong, J.Z., Lee, R.T., and Patwari, P. (2014). Targeted delivery to cartilage is critical for in vivo efficacy of insulin-like growth factor 1 in a rat model of osteoarthritis. Arthritis & Rheumatology *66*, 1247–1255.

Luo, D. and Saltzman, W.M. (2000). Synthetic DNA delivery systems. Nature Biotechnology *18*, 33–37.

Luo, K., He, B., Wu, Y., Shen, Y., and Gu, Z. (2014). Functional and biodegradable dendritic macromolecules with controlled architectures as nontoxic and efficient nanoscale gene vectors. Biotechnology Advances *32*, 818–830.

Maitani, Y. and Hattori, Y. (2009). Oligoarginine-PEG-lipid particles for gene delivery. Expert Opinion on Drug Delivery *6*, 1065–1077.

Martino, M.M., Briquez, P.S., Güç, E., Tortelli, F., Kilarski, W.W., Metzger, S., Rice, J.J., Kuhn, G.A., Müller, R., and Swartz, M.A. (2014). Growth factors engineered for super-affinity to the extracellular matrix enhance tissue healing. Science *343*, 885–888.

McCall, J.D., Lin, C.-C., and Anseth, K.S. (2011). Affinity peptides protect transforming growth factor beta during encapsulation in poly (ethylene glycol) hydrogels. Biomacromolecules *12*, 1051–1057.

Mishra, B., Patel, B.B., and Tiwari, S. (2010). Colloidal nanocarriers: a review on formulation technology, types and applications toward targeted drug delivery. Nanomedicine: Nanotechnology, Biology and Medicine *6*, 9–24.

Mohtaram, N.K., Montgomery, A., and Willerth, S.M. (2013). Biomaterial-based drug delivery systems for the controlled release of neurotrophic factors. Biomedical Materials *8*, 022001.

Monteiro, N., Ribeiro, D., Martins, A., Faria, S., Fonseca, N.A., Moreira, J.N., Reis, R.L., and Neves, N.M. (2014). Instructive nanofibrous scaffold comprising runt-related transcription factor 2 gene delivery for bone tissue engineering. ACS Nano *8*, 8082–8094.

Mumcuoglu, D., Sardan, M., Tekinay, T., Guler, M.O., and Tekinay, A.B. (2015). Oligonucleotide delivery with cell surface binding and cell penetrating peptide amphiphile nanospheres. Molecular Pharmaceutics *12*, 1584–1591.

Nevins, M., Kao, R.T., McGuire, M.K., McClain, P.K., Hinrichs, J.E., McAllister, B.S., Reddy, M.S., Nevins, M.L., Genco, R.J., and Lynch, S.E. (2013). Platelet-derived growth factor promotes periodontal regeneration in localized osseous defects: 36-month extension results from a randomized, controlled, double-masked clinical trial. Journal of Periodontology *84*, 456–464.

Nie, H. and Wang, C.-H. (2007). Fabrication and characterization of PLGA/HAp composite scaffolds for delivery of BMP-2 plasmid DNA. Journal of Controlled Release *120*, 111–121.

Novoa-Carballal, R., Silva, C., Möller, S., Schnabelrauch, M., Reis, R.L., and Pashkuleva, I. (2014). Tunable nano-carriers from clicked glycosaminoglycan block copolymers. Journal of Materials Chemistry B *2*, 4177–4184.

Orive, G., Anitua, E., Pedraz, J.L., and Emerich, D.F. (2009). Biomaterials for promoting brain protection, repair and regeneration. Nature Reviews Neuroscience *10*, 682–692.

Pannier, A.K. and Shea, L.D. (2004). Controlled release systems for DNA delivery. Molecular Therapy *10*, 19–26.

Panyam, J. and Labhasetwar, V. (2003). Biodegradable nanoparticles for drug and gene delivery to cells and tissue. Advanced Drug Delivery Reviews *55*, 329–347.

Park, S.-Y., Kim, K.-H., Shin, S.-Y., Koo, K.-T., Lee, Y.-M., and Seol, Y.-J. (2013). Dual delivery of rhPDGF-BB and bone marrow mesenchymal stromal cells expressing the BMP2 gene enhance bone formation in a critical-sized defect model. Tissue Engineering Part A *19*, 2495–2505.

Pawar, V., Gohil, V., Tamboli, R., Giridhar, R., and Yadav, M.R. (2012). Cartilage targeted chemical delivery of naproxen and ibuprofen for the treatment of arthritis. Journal of Pharmacy & Bioallied Sciences *4*, S10.

Perez, R.A., Won, J.-E., Knowles, J.C., and Kim, H.-W. (2013). Naturally and synthetic smart composite biomaterials for tissue regeneration. Advanced Drug Delivery Reviews *65*, 471–496.

Pérez-Martínez, F.C., Guerra, J., Posadas, I., and Ceña, V. (2011). Barriers to non-viral vector-mediated gene delivery in the nervous system. Pharmaceutical Research *28*, 1843–1858.

Pilakka-Kanthikeel, S., Atluri, V.S.R., Sagar, V., Saxena, S.K., and Nair, M. (2013). Targeted brain derived neurotrophic factors (BDNF) delivery across the blood-brain barrier for neuro-protection using magnetic nano carriers: an in-vitro study. PLoS One *8*, e62241.

Place, L.W., Sekyi, M., and Kipper, M.J. (2014). Aggrecan-mimetic, glycosaminoglycan-containing nanoparticles for growth factor stabilization and delivery. Biomacromolecules *15*, 680–689.

Ponte, A.L., Marais, E., Gallay, N., Langonne, A., Delorme, B., Herault, O., Charbord, P., and Domenech, J. (2007). The in vitro migration capacity of human bone marrow mesenchymal stem cells: comparison of chemokine and growth factor chemotactic activities. Stem Cells *25*, 1737–1745.

Qiu, C., Coutinho, P., Frank, S., Franke, S., Law, L.-y., Martin, P., Green, C.R., and Becker, D.L. (2003). Targeting connexin43 expression accelerates the rate of wound repair. Current Biology *13*, 1697–1703.

Rea, J.C., Gibly, R.F., Barron, A.E., and Shea, L.D. (2009). Self-assembling peptide–lipoplexes for substrate-mediated gene delivery. Acta Biomaterialia *5*, 903–912.

Rice, J.J., Martino, M.M., De Laporte, L., Tortelli, F., Briquez, P.S., and Hubbell, J.A. (2013). Engineering the regenerative microenvironment with biomaterials. Advanced Healthcare Materials *2*, 57–71.

des Rieux, A., Ucakar, B., Mupendwa, B.P.K., Colau, D., Feron, O., Carmeliet, P., and Préat, V. (2011). 3D systems delivering VEGF to promote angiogenesis for tissue engineering. Journal of Controlled Release *150*, 272–278.

Roche, S., D'Ippolito, G., Gomez, L.A., Bouckenooghe, T., Lehmann, S., Montero-Menei, C.N., and Schiller, P.C. (2013). Comparative analysis of protein expression of three stem cell populations: models of cytokine delivery system in vivo. International Journal of Pharmaceutics *440*, 72–82.

Ruenraroengsak, P., Cook, J.M., and Florence, A.T. (2010). Nanosystem drug targeting: facing up to complex realities. Journal of Controlled Release *141*, 265–276.

Santo, V.E., Gomes, M.E., Mano, J.F., and Reis, R.L. (2012). Chitosan–chondroitin sulphate nanoparticles for controlled delivery of platelet lysates in bone regenerative medicine. Journal of Tissue Engineering and Regenerative Medicine *6*, s47–s59.

Santos, J.L., Oramas, E., Pêgo, A.P., Granja, P.L., and Tomás, H. (2009). Osteogenic differentiation of mesenchymal stem cells using PAMAM dendrimers as gene delivery vectors. Journal of Controlled Release *134*, 141–148.

Serrano, A.L., Baeza-Raja, B., Perdiguero, E., Jardí, M., and Muñoz-Cánoves, P. (2008). Interleukin-6 is an essential regulator of satellite cell-mediated skeletal muscle hypertrophy. Cell Metabolism *7*, 33–44.

Shah, N.J., Hyder, M.N., Quadir, M.A., Courchesne, N.-M.D., Seeherman, H.J., Nevins, M., Spector, M., and Hammond, P.T. (2014). Adaptive growth factor delivery from a polyelectrolyte coating promotes synergistic bone tissue repair and reconstruction. Proceedings of the National Academy of Sciences *111*, 12847–12852.

Shi, C., Chen, W., Zhao, Y., Chen, B., Xiao, Z., Wei, Z., Hou, X., Tang, J., Wang, Z., and Dai, J. (2011). Regeneration of full-thickness abdominal wall defects in rats using collagen scaffolds loaded with collagen-binding basic fibroblast growth factor. Biomaterials *32*, 753–759.

Sivak, W.N., White, J.D., Bliley, J.M., Tien, L.W., Liao, H.T., Kaplan, D.L., and Marra, K.G. (2014). Delivery of chondroitinase ABC and glial cell line-derived neurotrophic factor from silk fibroin conduits enhances peripheral nerve regeneration. Journal of Tissue Engineering and Regenerative Medicine. doi:10.1002/term.1970

Suh, J.S., Lee, J.Y., Choi, Y.J., You, H.K., Hong, S.-D., Chung, C.P., and Park, Y.J. (2014). Intracellular delivery of cell-penetrating peptide-transcriptional factor fusion protein and its role in selective osteogenesis. International Journal of Nanomedicine *9*, 1153.

Sy, J.C. and Davis, M.E. (2010). Delivering regenerative cues to the heart: cardiac drug delivery by microspheres and peptide nanofibers. Journal of Cardiovascular Translational Research *3*, 461–468.

Sy, J.C., Seshadri, G., Yang, S.C., Brown, M., Oh, T., Dikalov, S., Murthy, N., and Davis, M.E. (2008). Sustained release of a p38 inhibitor from non-inflammatory microspheres inhibits cardiac dysfunction. Nature Materials *7*, 863–868.

Tan, Q., Tang, H., Hu, J., Hu, Y., Zhou, X., Tao, Y., and Wu, Z. (2011). Controlled release of chitosan/heparin nanoparticle-delivered VEGF enhances regeneration of decellularized tissue-engineered scaffolds. International Journal of Nanomedicine *6*, 929–942.

Tessmar, J.K. and Göpferich, A.M. (2007). Matrices and scaffolds for protein delivery in tissue engineering. Advanced Drug Delivery Reviews *59*, 274–291.

Thorne, R.G. and Frey II, W.H. (2001). Delivery of neurotrophic factors to the central nervous system. Clinical Pharmacokinetics *40*, 907–946.

Tidball, J.G. (2005). Inflammatory processes in muscle injury and repair. American Journal of Physiology-Regulatory, Integrative and Comparative Physiology *288*, R345–R353.

Tros de Ilarduya, C., Sun, Y., and Düzgüneş, N. (2010). Gene delivery by lipoplexes and polyplexes. European Journal of Pharmaceutical Sciences *40*, 159–170.

Uchio, K., Graham, M., Dean, N.M., Rosenbaum, J., and Desmoulière, A. (2004). Down-regulation of connective tissue growth factor and type I collagen mRNA expression by connective tissue growth factor antisense oligonucleotide during experimental liver fibrosis. Wound Repair and Regeneration *12*, 60–66.

Uitto, J., Olsen, D.R., and Fazio, M.J. (1989). Extracellular matrix of the skin: 50 years of progress. Journal of Investigative Dermatology *92*, 61S–77S.

Vermonden, T., Censi, R., and Hennink, W.E. (2012). Hydrogels for protein delivery. Chemical Reviews *112*, 2853–2888.

Visser, L.C., Arnoczky, S.P., Caballero, O., Kern, A., Ratcliffe, A., and Gardner, K.L. (2009). Growth factor-rich plasma increases tendon cell proliferation and matrix synthesis on a synthetic scaffold: an in vitro study. Tissue Engineering Part A *16*, 1021–1029.

Vo, T.N., Kasper, F.K., and Mikos, A.G. (2012). Strategies for controlled delivery of growth factors and cells for bone regeneration. Advanced Drug Delivery Reviews *64*, 1292–1309.

Wang, N.X. and von Recum, H.A. (2011). Affinity-based drug delivery. Macromolecular Bioscience *11*, 321–332.

Wei, G. and Ma, P.X. (2008). Nanostructured biomaterials for regeneration. Advanced Functional Materials *18*, 3568–3582.

Wu, A.M., Lisowska, E., Duk, M., and Yang, Z. (2009). Lectins as tools in glycoconjugate research. Glycoconjugate Journal *26*, 899–913.

Yin, H., Moulton, H.M., Seow, Y., Boyd, C., Boutilier, J., Iverson, P., and Wood, M.J. (2008). Cell-penetrating peptide-conjugated antisense oligonucleotides restore systemic muscle and cardiac dystrophin expression and function. Human Molecular Genetics *17*, 3909–3918.

Yoo, J.-W., Irvine, D.J., Discher, D.E., and Mitragotri, S. (2011). Bio-inspired, bioengineered and biomimetic drug delivery carriers. Nature Reviews Drug Discovery *10*, 521–535.

Young, S., Wong, M., Tabata, Y., and Mikos, A.G. (2005). Gelatin as a delivery vehicle for the controlled release of bioactive molecules. Journal of Controlled Release *109*, 256–274.

Zhang, S., Xu, Y., Wang, B., Qiao, W., Liu, D., and Li, Z. (2004). Cationic compounds used in lipoplexes and polyplexes for gene delivery. Journal of Controlled Release *100*, 165–180.

Zhang, Y., Chan, H.F., and Leong, K.W. (2013). Advanced materials and processing for drug delivery: the past and the future. Advanced Drug Delivery Reviews *65*, 104–120.

Zhang, C., Chen, J., Feng, C., Shao, X., Liu, Q., Zhang, Q., Pang, Z., and Jiang, X. (2014). Intranasal nanoparticles of basic fibroblast growth factor for brain delivery to treat Alzheimer's disease. International Journal of Pharmaceutics *461*, 192–202.

Zhong, Y. and Bellamkonda, R.V. (2008). Biomaterials for the central nervous system. Journal of the Royal Society Interface *5*, 957–975.

Ziv-Polat, O., Shahar, A., Levy, I., Skaat, H., Neuman, S., Fregnan, F., Geuna, S., Grothe, C., Haastert-Talini, K., and Margel, S. (2014). The role of neurotrophic factors conjugated to iron oxide nanoparticles in peripheral nerve regeneration: in vitro studies. BioMed Research International *2014*, 10.

3

NANOMATERIALS FOR NEURAL REGENERATION

Melike Sever, Busra Mammadov, Mevhibe Gecer, Mustafa O. Guler, and Ayse B. Tekinay

Institute of Materials Science and Nanotechnology, National Nanotechnology Research Center (UNAM), Bilkent University, Ankara, Turkey

3.1 INTRODUCTION

3.1.1 Extracellular Matrix of Central Nervous System

The central nervous system (CNS) consists of a dense network of cells leaving a smaller volume for the extracellular matrix (ECM) components (10–20% for the brain unlike most other tissues (Cragg, 1979)). Before 1971, CNS was thought to be totally free of ECM and consisting just of a very dense network of cells (Tani and Ametani, 1971). Yet, this small amount is sufficient to modulate the survival, neurite outgrowth, and synaptic plasticity of neurons. Given its importance in modulation of cell function, neural ECM undergoes prominent changes during maturation of the nervous system. Embryonic and postnatal ECM is rather a loose network, while there occurs some condensation throughout adulthood starting from the postnatal period, according to specific needs of neural cells (Bandtlow and Zimmermann, 2000; Dityatev and Schachner, 2003). In fact, the balance among ECM components is so important that any imbalance leads to massive degeneration in the CNS.

Therapeutic Nanomaterials, First Edition. Edited by Mustafa O. Guler and Ayse B. Tekinay.

Being quite different from that of other tissues, ECM content also differs dramatically between specific locations in the CNS. Two different ECM types are deposited in the basement membrane, perineuronal nets (PNN) and gap filling interstitial matrix. Basement membrane (basal lamina) is the one that is more similar to ECM of other tissues in terms of its components. Collagen, entactin (laminin (LN)–nidogen complexes), fibronectin (FN), dystroglycan, and perlecan are the predominant proteins and proteoglycans (PGs) of basement membrane (Lau et al., 2013). This structure is associated with blood vessels and is important in the stabilization of the blood–brain barrier (Baeten and Akassoglou, 2011).

PNN are the dense network of matrix around cell bodies that is mainly responsible for maintenance of synaptic plasticity. Main components of PNN are hyaluronan, chondroitin sulfate proteoglycans (CSPGs), and tenascins (Bonneh-Barkay and Wiley, 2009; Lau et al., 2013). Being absent in the embryonic tissue, these structures form a barrier around the cells preventing the formation of new synaptic connections while preserving the ones already formed. Mammalian nervous system has a comparable plasticity that continues through postnatal development until maturation of the CNS. During this period, named as critical period, the nervous system continues to develop along with making new connections influenced by experience. Plasticity declines upon closure of the critical period, the time of which coincides with the formation of PNN indicating them as barriers to formation of new connections (Bonneh-Barkay and Wiley, 2009).

Being absent from PNN, small amounts of fiber-forming collagens and elastins along with FN and LN are present in the interstitial matrix. Sulfated PGs and hyaluronan along with tenascins are the main components of this ECM (Lau et al., 2013).

PGs are the predominant components of CNS ECM. PGs are classified into four main groups depending on the GAG chain: CSPGs, heparan sulfate proteoglycans (HSPGs), dermatan sulfate proteoglycans (DSPGs), and keratan sulfate proteoglycans (KSPGs). CSPGs and HSPGs are the main PGs of the CNS, CSPGs being the most abundant (Bonneh-Barkay and Wiley, 2009). Each group is also composed of a variety of PGs due to different protein cores carrying GAG chains. Lecticans are the most abundant CSPG type in the brain, which are classified into four subgroups. Neurocan and brevican are neural tissue-specific lecticans, while aggrecan and versican are common to ECM of other tissues (Yamaguchi, 2000). Besides sulfated GAGs attached to protein cores, hyaluronic acid (HA) is also abundant in the CNS tissues. HA is a large polymeric carbohydrate chain that is nonsulfated and does not contain a protein core. It binds to other PGs and forms a dense mesh-like ECM structure in PNN (Kwok et al., 2010).

Depending on the protein core and GAG chain, PGs in the CNS are permissive or nonpermissive to neurite extension. GAGs can exert their effects on cells either by inducing direct signaling through cell surface receptors (like HA achieves through CD44 receptor) or by their affinity to growth-active molecules. Affinity of HSPGs to growth factors is quite important in the modulation of neuronal function, especially through FGF signaling (improved upon syndecan binding to FGF (Walicke, 1988)). CSPGs can also bind to a range of molecules including semaphorin 5A, binding to which is found to change permissive nature of semaphorin to inhibitory (Kantor et al., 2004). Binding of CSPGs to other ECM proteins, such as tenascin and LN, blocks axon growth-promoting ability of these proteins and thus inhibits axon growth (Mckeon et al., 1995). While binding of CSPG to LN is inhibitory, binding by HSPG improves axonal elongation (Riopelle and Dow, 1990). CSPGs can also inhibit neural regeneration by limiting calcium availability by binding either to extracellular calcium or to calcium channels, thereby preventing the uptake by neurons (Sharma et al., 2012). Binding of CSPGs to PTPsigma receptor on neurons leads to phosphorylation of neurotrophin receptors TrkB and TrkC, possibly leading to a downregulation of responsiveness to neurotrophins (Faux et al., 2007). Binding of HSPGs to the same receptor in a competitive manner (as in the case of semaphorin 5A) abolishes CSPG inhibitory effect on neurotrophin responsiveness (Kantor et al., 2004). Inhibitory effects of CSPGs are either due to the protein core or sulfated GAG chain, depending on the CSPG type. For versican, neurocan, and phosphacan, protein core is indispensable in axonal inhibition, while GAG chain is responsible for inhibition by aggrecan and brevican. Chondroitinase treatment, which degrades the GAG chain, eliminates the inhibitory effect of aggrecan and brevican but not of the other CSPGs (Maeda and Noda, 1996; Margolis et al., 1996; Niederost et al., 1999; Schmalfeldt et al., 2000; Snow et al., 1990; Yamada et al., 1997).

Although limiting the regenerative capacity of CNS with their inherent inhibitory properties, CSPGs have indispensable role in neural development, both for cell migration and formation of neural connections. Neural crest cells avoid migration through local environments with CSPG abundance. The presence of CSPGs along their way is a trajectory for them to migrate to their defined positions (Kubota et al., 1999). Repulsive nature of CSPGs for growth cones is also quite important in axonal pathfinding and correct formation of neural networks during embryonic development as well as postnatal development in critical period. Treating developing nervous system with chondroitinase ABC leads to various axon pathfinding errors, verifying the importance of

CSPGs in the process (Chung et al., 2000). CSPGs are also important in adult CNS for proper stabilization of synaptic connections and prevention of improper axonal sprouting and formation of disorganized neural networks (Carulli et al., 2006).

Having a regulatory function on neural network formation during development and keeping connections made stable, CSPGs are quite important in healthy CNS. However, they should be kept in balance in order not to overinhibit neural functions. This balance becomes disorganized upon a trauma leading to production of reactive astrocytes. Astrocytes start to proliferate and deposit dense ECM in the injury site and fill the gap caused by the injury. This preserves the overall structure of the tissue, stabilizes blood–brain barrier, and limits the damage to injury site, thereby preventing secondary damage (Faulkner et al., 2004). Although being important for the overall health of the CNS, the ECM that is deposited hastily by astrocytes is responsible for the failure of neuronal regeneration. This de novo produced tissue is called "glial scar," and it is impermissive to any axonal sprouting due to the abundance of CSPGs. Hence, nearby cells that are healthy cannot enter into the area and thus cannot make new synaptic connections to tolerate the loss of the ones in the injury site. Furthermore, synaptic connections by surviving cells in the injury site cannot be made either due to the inhibitory effect on plasticity.

Besides CSPGs, myelin components are also inhibitory for neural regeneration after CNS injury. The major inhibitory myelin components can be listed as myelin-associated glycoprotein (MAG), Nogo-A, oligodendrocyte myelin glycoprotein (OMgP), and myelin lipid sulfatide (Winzeler et al., 2011; Yiu and He, 2006). Myelin inhibitors MAG, Nogo-A, and OMgP are found to exert their inhibitory effect through binding to Nogo-66 receptor 1 (NgR1) on neurons (Domeniconi et al., 2002; Fournier et al., 2001; Liu et al., 2002; Wang et al., 2002). The presence of NgR1 is essential for growth cone collapse upon encountering myelin inhibitors (Chivatakarn et al., 2007; Kim et al., 2004). Paired immunoglobulin-like receptor B (PirB) and its human homolog (LILRB2) are recently found as another receptor for these three myelin inhibitors, and blocking this receptor along with the NgR1 largely eliminates myelin inhibition of neurite outgrowth (Atwal et al., 2008).

The third type of CNS inhibitory molecules includes axon guidance molecules semaphorin and ephrin. Having a role in network stabilization by limiting neurite outgrowth, most semaphorins are inhibitory for neurons. In the glial scar, class 3 semaphorins (Sema3s) contribute to the inhibitory environment in a CSPG-dependent manner. Interfering CSPG–Sema3 interaction eliminates the inhibitory nature of this molecule *in vitro* (Pasterkamp and Verhaagen, 2006). Besides Sema3s, membrane-associated semaphorins like Sema4D are also upregulated in the glial scar and inhibit neurite outgrowth (Moreau-Fauvarque et al., 2003). Ephrins exert their inhibitory effect by binding to EphA and EphB receptor tyrosine kinase on neural cell surface. EphrinB3 is expressed in CNS myelin, ephrin B2 is upregulated upon injury by reactive astrocytes, and ephrinA5 expression is found to be increased in cortical lesion, and they all inhibit neurite outgrowth (Benson et al., 2005; Bundesen et al., 2003; Carmichael et al., 2005).

3.1.2 ECM of Peripheral Nervous System

The reaction of the nervous tissue to any injury leading to scar tissue formation acts as a barrier for regeneration in the CNS, while it supports regeneration in the peripheral nervous system (PNS). Different types of glial cells in the CNS and the PNS, astrocytes and Schwann cells, lead to the formation of completely different environments for axons struggling to heal from an injury. While astrocytes produce a growth-inhibitory environment consisting of chemorepulsive guidance molecules, Schwann cells produce a growth promoting environment by secretion of supportive ECM molecules like neurotrophins as well as dealing with the clearance of inhibitory myelin products after injury. A chain of events occurring after an injury in the peripheral nerve leads to the formation of a permissive environment for axonal regeneration. Wallerian degeneration, occurring in the distal stump of the injured nerve, leads to fragmentation of the damaged axon and associated myelin products. Fragmented products are then cleared by macrophages that are recruited to the injury site (Chen et al., 2007). Myelin-associated proteins are known to inhibit axonal elongation, and their clearance after injury in the PNS forms a permissive environment as opposed to that in the CNS. While the distal stump undergoes Wallerian degeneration, axonal regeneration starts from the proximal end. Schwann cells dedifferentiate and start to proliferate. They myelinate newly elongating axon, thus providing an electrical insulation to the regenerating axon (Fawcett and Keynes, 1990).

Permissive environment, axon guidance cues, and intrinsic growth capacity of peripheral nerves are crucial in successful regeneration of the PNS. Unlike their CNS counterpart, peripheral nerves have an intrinsic growth capacity that is activated upon injury. Cyclic adenosine monophosphate (cAMP) level is upregulated upon peripheral nerve injury, and it activates protein kinase A (Neumann et al., 2002; Qiu et al., 2002). Protein kinase A is thought to take role in cytoskeletal reorganization by inhibiting RhoA. RhoA is normally activated by the presence of myelin debris, and its activation leads to halting of cytoskeletal reorganization required for axonal elongation. Hence, its inhibition by protein kinase A makes axon's cytoskeletal dynamics favorable for regeneration (Snider et al., 2002). Besides, cAMP has a transcriptional role by activating cAMP response element binding protein (CREB), which in turn enables cell to overcome myelin-associated inhibition of neurite outgrowth (Gao et al., 2004). Upregulation of intracellular cAMP levels leads to an increase in cell surface localization of TrkB receptors also, which in turn increases axons' responsiveness to neurotrophins (Liu et al., 2011; Meyer-Franke et al., 1998).

Structurally different from that of the CNS, the PNS is abundant in basal lamina. Secreted by Schwann cells, basal lamina of the peripheral nerve tissue contains LN, type IV collagen, and HSPGs (Bunge et al., 1990). Being present in healthy peripheral nerve, basal lamina production is upregulated in the injured nerve (Chen et al., 2007). Besides its importance in myelination of nerves by Schwann cells, basal lamina components are also important for axonal elongation. LN is a positive regulator of neurite outgrowth, and it is overexpressed after peripheral nerve injury. Abolishing the expression of LN by Schwann cells impairs axon regeneration severely. Besides their role in neurite outgrowth, LN are also important for Schwann cells' proper differentiation and myelination of growing axons (Chen and Strickland, 2003). In addition to LN, NGF, BDNF, GDNF, IGF, CNTF, and FGF-2 are also upregulated in the injured peripheral nerve. Increased amount of these growth factors in the injury site provides better axonal elongation along with proper myelination by Schwann cells (Giger et al., 2010).

Myelin inhibition, an important problem for CNS regeneration, is not a big deal in the PNS. There are several routes by which peripheral nerves deal with this problem. Responsiveness of the axon to myelin products is thought to be different in the PNS when compared to the CNS, which is actually related to the intrinsic growth capacity of peripheral nerves. Intracellular cAMP levels, for example, determine whether MAG will act as a growth inhibitory molecule or not. Besides, lack of some myelin proteins in PNS myelin is thought to make PNS myelin less inhibitory through other major components such as MAG that it still consists (Chen et al., 2007). Clearance of myelin from the injury site is another important event, which does not take place in the CNS. It is rapidly cleared away from the injured peripheral nerve by hand-to-hand functioning of Schwann cells and macrophages. In the early stages after injury, Schwann cells are the main phagocytic cells in the injured nerve, and they start myelin clearance as early as 5 h after injury. They secrete phospholipase A2 that degrades phosphatidylcholine, which is abundant in myelin. Lysophosphatidylcholine is produced by hydrolysis of phosphatidylcholine, and it has a myelinolytic activity (Martini et al., 2008). Schwann cells also recruit macrophages to the injured peripheral nerve by secreting several chemokines including TNF-a, LIF, Il-1a, Il-1b, and Il-6 (Gaudet et al., 2011). Infiltration of macrophages to the nerve is essential in clearing myelin debris from the distal stump that undergoes Wallerian degeneration. Inhibitory myelin proteins are not cleared in the CNS due to the lack of macrophages in the lesion site, the infiltration of which is restricted by the blood–brain barrier.

Upon peripheral nerve injury, fibrinogen from the blood starts to infiltrate into the nerve where it is converted to fibrin. Accumulated fibrin prevents Schwann cell migration to the injury site and prevents remyelination (Akassoglou et al., 2002). Schwann cells upregulate production of tissue plasminogen activator (tPA) and activates fibrinolytic pathway thereby. Lysis of fibrin enables Schwann cell to enter the injured nerve and start myelination of the growing axon (Akassoglou et al., 2000).

3.1.3 Urgent Need for Materials to Induce Regeneration in Nervous Tissue

Due to the inherent lack of regeneration in the CNS, brain and spinal cord trauma makes quite a high number of people suffer from the lifelong effects. Ineffective therapies cause almost all survivors to continue their life with the aid of others.

Although the PNS is better than the CNS in terms of its regenerative potential, injuries leading to injuries to long nerve segments lead to disappointing therapeutic results. End-to-end bridging is not possible due to the tension that will lead to secondary injury, and harvesting long segments of the nerve from other sides of the body for autografting is also out of question. Commercially available nerve tubes, although providing some axonal elongation, are not always effective.

These clinical challenges make development of new therapies for nervous system injuries and disorders crucial. There are plenty of papers published in the area showing the efforts to meet this requirement. However, to be successful, one should understand the native tissue properties and requirements of the resident cells. For example, neural cells are electrically active cells, and conductive materials might support regeneration better. Besides, incorporating biological signals such as short peptides that bind to cell surface receptors and induce axonal elongation is found to lead promising results (Mammadov et al., 2012a). Biocompatibility of the material to be used is also essential, and long-term toxicity assessments should be carried out before planning a clinical use. Biodegradable materials are advantageous, if no lifelong therapy is desired. However, end products after degradation should also be nontoxic to avoid any harm to regenerating tissue de novo.

3.2 NANOMATERIALS FOR NEURAL REGENERATION

3.2.1 Physical Functionalization of Nanomaterials to Induce Neural Differentiation

During the development of the nervous system, ECM plays an important role to support neuronal maturation and neurite extension. Cell adhesion, proliferation, and differentiation are other important significant criteria to be considered when designing a nanomaterial for neural tissue engineering. The main goal of regenerative medicine is to enhance tissue regeneration and healing after injury or disease leading to degeneration of the tissue of interest. Guidance of cell behavior by the features of a material at cell–biomaterial interfaces has significant importance, and developing novel biomaterials with certain surface modifications to induce controlled cell function would be beneficial for improving the therapeutic potential of current regenerative medicine protocols. The biodegradability and biocompatibility are the most important features of synthetic biomaterials. Physical properties of a material should also be taken into account in addition to biological and chemical properties when designing biomaterials for tissue engineering applications. Physical properties including stiffness, dimensionality, substrate topography, and electrical conductivity are important parameters for scaffold functionalization to induce neural differentiation (Fig. 3.1).

3.2.2 Effects of Mechanical Stiffness on Cellular Behavior

Mechanical properties of the microenvironment are important regulators of cell properties including morphology (Bard and Hay, 1975) and motility (Halliday and Tomasek, 1995) of the cells as well as differentiation (Engler et al., 2006). When designing a scaffold for neural differentiation, the mechanical properties of the scaffold should be similar to that of brain tissue, which is below 1 kPa (Georges et al., 2006).

In mechanically tunable matrices and substrates, it was shown that PNS cells (Flanagan et al., 2002; Willits and Skornia, 2004) and PC12 cells (Gunn et al., 2005; Leach et al., 2007) displayed variable neurite outgrowth and branching according to the substrate stiffness. In the case of neural stem cells (NSCs), the influence of the mechanical properties of the substrate on cell behavior has been studied (Saha et al., 2008). According to this study, NSCs prefer to differentiate into neurons with intermediate substrate stiffness (500 Pa), while increased stiffness favors the differentiation into astrocyte lineage (Saha et al., 2008). Also, it was shown that while substrate

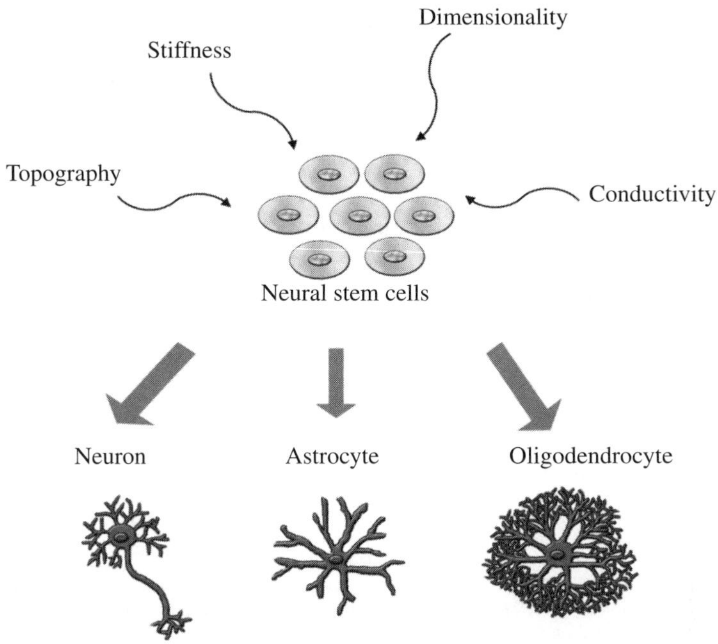

FIGURE 3.1 Physical properties to induce neural differentiation of neural stem cells (NSCs).

stiffness did not affect the number of neurons differentiated from NSCs, soft substrates promote neuronal maturity, and neurons differentiated on soft substrates displayed long neurites in the absence of added growth factors. In contrast, neurons showed a bipolar immature morphology on stiff substrates (Teixeira et al., 2009).

LN-coated methacrylamide chitosan (MAC) hydrogels with different stiffnesses were used to understand the contribution of substrate stiffness to neural stem/progenitor cell (NSPC) differentiation and proliferation. It was demonstrated that an optimal stiffness is present for both proliferation (3.5 kPa) and differentiation of NSPCs to neurons (<1 kPa) (Leipzig and Shoichet, 2009).

The effect of stiffness was also studied on human mesenchymal stem cells (hMSCs). This effect was investigated by using three-dimensional (3D) porous scaffolds generated by type I collagen (Col) and HA. Using different concentrations of 1-ethyl-3-(3-dimethylaminopropyl)carbodiimide (EDC) as a crosslinking agent, the elastic modulus of the 3D substrates was modified, and the stiffness was controlled in the range of 1–10 kPa. Results showed that mesenchymal stem cells (MSCs) were likely to differentiate into neurons with 1 kPa while they preferred glial lineage in the scaffold with 10 kPa stiffness verified with experiments focused on upregulation of neuronal mid and late protein markers after longer mechanical induction time for 21 days (Her et al., 2013).

In another study, spinal cord neurons were used to study cellular responses to substrate compliances. DNA crosslinked hydrogels with stiffness of approximately 100 Pa to 30 kPa were produced. DNA gel stiffnesses were modulated by changing the length of crosslinker, monomer concentration, and level of crosslinking. Although this variation in stiffness did not significantly affect the primary dendrite length, it was found that stiffer gels caused extending more primary dendrites and shorter axons with the reduction in focal adhesion kinase (FAK), suggesting the involvement of FAK in neuronal mechanosensing and neuritogenesis in response to stiffness. It provides cues for the design of scaffolds for neural tissue engineering depending on the mechanical properties (Jiang et al., 2008).

3.2.3 Effects of Dimensionality on Cellular Behavior

In vitro models are important tools to study cell behavior and fate in a highly controlled manner. Although two-dimensional (2D) cell cultures are commonly used in differentiation studies, 3D cell cultures are important *in vitro* models to fill the gap between 2D cell culture experiments and *in vivo* studies. Thus, use of 3D models is important to the study regeneration of neural tissue, which has a very low regeneration capacity. 3D models better mimic the ECM of neural cells in terms of providing support and enhanced diffusion of oxygen and nutrients.

There are fundamental differences between cells grown on a monolayer surface and in a 3D culture. Cells grown in 3D cultures have been shown to display longer neurite outgrowth and higher levels of survival as well as different patterns of differentiation when compared to 2D monolayers (Bellamkonda et al., 1995; Pardo and Honegger, 2000). Also, hippocampal neurons cultured on a 3D aragonite matrix displayed higher survival as compared to 2D counterparts with higher-density network formation (Peretz et al., 2007). In another study, 3D cell culture system with the functional motifs Arg–Gly–Asp (RGD), bone marrow homing peptide 1 (BMHP1), and BMHP2 was developed for the culture of adult NSCs. The proliferation and differentiation of NSCs were supported by these scaffolds through allowing a satisfactory supply of nutrients and oxygen (Cunha et al., 2011).

Stupp and coworkers generated a 3D network of nanofibers formed by self-assembly of peptide amphiphile (PA) molecules to encapsulate neural progenitor cells. They used neurite-promoting LN epitope IKVAV, and this artificial nanofiber system gave rise to rapid differentiation of neural progenitor cells into neurons while suppressing the astrocyte differentiation (Silva et al., 2004).

Also, LN-derived IKVAV-linked self-assembling peptide RADA16 was used to form a 3D matrix as a functional peptide-based scaffold. NSCs were encapsulated within this 3D peptide hydrogel to increase the reconstruction of the injured brain. The signal coming from the extended IKVAV sequence directs the adhesion of NSCs and leads to neuronal differentiation. In addition to *in vitro* studies, rat brain surgery model was generated to illustrate the damage in cerebral neocortex/neopallium loss. Injection of NSC-encapsulated 3D hydrogel filled the cavity caused by injury. They revealed enhanced survival of encapsulated NSCs and reduced glial scar formation with histological analysis. This 3D hydrogel model provided improvement in brain tissue regeneration 6 weeks after posttransplantation (Cheng et al., 2013).

3.2.4 Effects of Substrate Topography on Cell Behavior

The natural environment of the cells has complex topographical cues, which are different between different surfaces used for *in vitro* cultures. Nanotopography is an important tool to guide cell differentiation, and it can be modified in size and shape according to desired application. Interaction of the cells with nanotopographical features can induce different responses including changes in cell morphology (Dalby et al., 2002a), adhesion (Gallagher et al., 2002), proliferation (Dalby et al., 2002b), and gene regulation (Dalby et al., 2002c). It is important to understand that neurite growth along topographical patterns is important for tissue engineering applications in neurology.

Cecchini et al. analyzed time-lapse images to understand the differentiation dynamics of PC12 cells on nanopatterned biocompatible substrates. Depth gratings were produced on tissue-culture polystyrene substrates by nanoimprint lithography, and the results suggested that the substrate topography can act cooperatively with nerve growth factor (NGF) signaling to regulate differentiation and neurite outgrowth (Cecchini et al., 2007).

Embryonic stem cell (ESC) differentiation into specific cell lineages is one of the major concerns in regenerative medicine. In order to analyze the effect of different nanostructures on differentiation of ESC-derived neuronal precursors into neuronal lineage without biochemical factors, patterns in polydimethylsiloxane (PDMS) consisting of groove and pillar arrays of submicrometric lateral resolution were produced. It was found that neuronal differentiation was enhanced when neuronal precursors were cultured on nanopillars compared to flat PDMS substrates and neurites grew mostly on the top of the higher pillars (Migliorini et al., 2011).

Polyurethane acrylate nanoscale ridge/groove pattern arrays were generated to induce the differentiation of hESCs into a neuronal lineage without the use of any differentiation-inducing agents. The hESCs were found to differentiate into various neurons, but not into glial cells, such as astrocytes, as evidenced by immunofluorescence results (Lee et al., 2010).

In addition, nanotopographical control of neuronal polarity was studied, and the interaction between focal adhesions and topographical features was exploited to design a set of scaffolds made of cyclic olefin copolymer (COC), yielding control over neuronal polarity establishment and neurite pathfinding. By this way, specific neuronal polarity states were selectively favored by a set of biocompatible textured scaffolds, and it was demonstrated that this selection can be tailored through varying the topographical constraint applied to the maturation of focal adhesions during neuritogenesis (Ferrari et al., 2011).

3.2.5 Effects of Electrical Conductivity on Cell Behavior

Neurons are electrically excitable cells, and as a result of their inherent nature, they are capable of transmitting electrochemical signals throughout the nervous system, making them highly sensitive to electrical stimuli. The action potential generated at the synapse is the key component of neural communication, so electrical conductivity is an important physical property to enhance neural cell activity (Hodgkin and Huxley, 1952). Providing electrical conductivity to the scaffolds and electrical stimulation of the cells cultured on these scaffolds might be useful to improve synaptic connections.

One of the new aspects of scaffold generation is the design of conductive substrates, which provide external electrical stimuli to the stem cells, because these stimuli can affect the cell behavior including proliferation, migration, and differentiation (Chang et al., 2011; Huang et al., 2012).

Electrical conductivity can be provided to the scaffold through utilization of conductive polymers and carbon-based materials, which include carbon nanotubes (CNTs), graphite, and graphene (Kotov et al., 2009). Combining the conductivity of CNT with the alignment of poly(L-lactic acid) (PLLA) nanofibers supports and increases the neural differentiation of mouse embryonic stem cells (mESCs), which was detected by enhanced expression of mature neuronal markers such as Map-2 and NSE even in the absence of direct electrical stimulation (Kabiri et al., 2012).

The effect of electroactive fibers on responses of neuronal cells was also investigated (Bhang et al., 2012). In this study, electrically active electrospun fibers were generated from blends of poly(lactide-*co*-ε-caprolactone) and a conductive polymer, polyaniline (PAni). These electrically conductive fibers sustained stable electrical properties in cell culture conditions for up to 7 days and enhanced NGF-induced neurite outgrowth of PC12 cells verified with morphological changes and the expression of differentiation proteins.

3.3 CHEMICAL AND BIOLOGICAL FUNCTIONALIZATION OF NANOMATERIALS FOR NEURAL DIFFERENTIATION

By mimicking several unique characteristics of the natural environment of cells, synthetic materials for neural regeneration can be improved chemically and biologically. Especially bioactivation of materials can be achieved by addition of small chemical moieties to the scaffold particularly found in specific tissues or addition of biologically active molecules derived from natural ECM.

3.3.1 Effects of Biologically Active Molecules on Cell Behavior

Incorporation of the bioactive regions of ECM proteins that are used to direct and maintain the cell differentiation can be used to modify or reconstruct the scaffolds to be used as cellular environments (Ma et al., 2008). With the modification of these bioactive proteins, newly designed materials may be developed to enhance differentiation efficiency.

In order to control differentiation of NSCs for clinical applications, construction of new composite biomaterials to find an ideal combination of ECM proteins and growth factors is another approach. To induce NSC differentiation into specific cell types, gold surfaces modified by the photo-assisted patterning of an alkanethiol self-assembled monolayer (SAM) were used for immobilization of natural and synthetic matrices (FN, LN, poly(ethyleneimine) (PEI), PLL, RGD, IKVAV) along with growth factors such as epidermal growth factor (EGF), ciliary neurotrophic factor (CNTF), NGF, and neurotrophin-3 (NT-3). This study showed that immobilized LN/NGF, FN/NGF, Pro-F/NGF, FN/NT-3, and Pro-F/NT-3 on spots promoted more neuronal differentiation, while astroglial differentiation was promoted with FN/CNTF and Pro-F/CNTF among other combinations (Nakajima et al., 2007).

The fate of NSCs depends on the ECM–cell interactions, especially integrin interactions, which are directly connected with laminin proteins (Nakajima et al., 2007). In order to evaluate the impact of the laminin microenvironment on primary murine neurosphere culture, methylcellulose

(MC) scaffolds functionalized with laminin-1 (MC-x-LN1) were used as a scaffold. This study revealed that MC-x-LN1 induces NSC survival and maturation compared to that of MC only controls. Moreover, higher neuronal and oligodendrocytic differentiation was observed in MC-x-LN1 scaffold than in MC controls. Correspondingly, LN1 production and $\alpha_6\beta_1$ integrin levels were significantly extended on MC-x-LN1 than in MC controls (Stabenfeldt et al., 2010).

Aerogels were also used as a class of material, with tunable properties including chemical, physical, and surface properties. In order to evaluate the ability of polyurea-crosslinked silica aerogels (PCSA) coated with PLL (a synthetic polymer of L-lysine), basement membrane extract (BME) or LN1 was used for culture of DRG neurons. PCSA displayed the ability to induce directional differentiation and growth of DRGs. Furthermore, laminin was demonstrated to be a highly efficient surface material for attachment and growth of dorsal root ganglion (DRG) neurons on PCSA (Sabri et al., 2012).

Functional polymer scaffolds produced by covalent modification of PLLA nanofibers with laminin were also used to mimic bioactive properties of the ECM proteins by electrospinning technique. Cultured PC12 cells differentiated into neural cells in the presence of NGF more efficiently on these modified surfaces than those on the unblended PLLA. Using this modification technique, different types of biocompatible polymers with a significant potential in nerve tissue regeneration applications can be acquired (Koh et al., 2008). In another study, fibers made up of poly(l-lactic acid)-*co*-poly-(3-caprolactone) and collagen were fabricated by electrospinning to investigate human bone marrow-derived MSCs for neural differentiation on this nanofibrous scaffold. Neural induction of MSCs was carried out by culturing in the presence of β-mercaptoethanol, EGF, NGF, and BDNF. These results showed that the differentiation of MSCs into neurons was comparatively higher on Poly(l-lactic acid)-*co*-poly-(3-caprolactone)–collagen surfaces than those on Poly(l-lactic acid)-*co*-poly-(3-caprolactone) surface without collagen treatment. Moreover, the expression of neuronal markers neurofilament (NF) and nestin were observed within multipolar elongations by immunofluorescence (Prabhakaran et al., 2009).

3.3.2 Effects of Chemical Groups on Cellular Behavior

Chemical modification by specific terminal groups can modulate NSCs' response to different scaffolds including their differentiation. In the NSC extracellular environment, high abundance of specific chemical moieties is believed to direct neural differentiation, so substrates functionalized with chemical groups may be beneficial in controlling the behavior of NSCs.

Glass surfaces chemically modified with hydroxyl (—OH), methyl (—CH_3), amino (—NH_2), carboxylate (—COOH), mercapto (—SH), and sulfonate (—SO_3H) groups were used for culture of NSCs to investigate the commitment of stem cells to a specific neural subtype during differentiation. —SO_3H-modified surfaces were found to induce oligodendrocyte differentiation more. —NH_2 had a tendency to induce more neuronal differentiation and inhibit the oligodendrocyte differentiation of NSCs in comparison with —COOH (Ren et al., 2009).

Biomimetic polymeric scaffolds have recently become the focus point of research on nerve regeneration. Another method of chemical modification of polymeric scaffolds is synthesizing critical protein, acetylcholine, a neurotransmitter, and a neuromodulator in both the PNS and the CNS. In the PNS, acetylcholine binds to acetylcholine receptors on skeletal muscle fibers, and it opens the ligand-gated sodium channels. In the CNS, damages in the cholinergic system have been shown to be associated with memory deficit in diseases such as Alzheimer's disease. Acetylcholine-like biomimetic polymers including a bioactive unit mimicking the acetylcholine, a bioinert unit poly(ethylene glycol) monomethyl ether-glycidyl methacrylate (MePEG-GMA), and dimethylaminoethyl methacrylate (DMAEMA) were synthesized via click chemistry and free radical polymerization technique. Rat hippocampal neurons were used to observe the effect of synthesized polymer that has a significant effect on neuronal growth when 1:60 (mol/mol) ratio of MePEG-GMA to DMAEMA is used (Tu et al., 2011).

Other studies demonstrated that a substrate energy of adhesion (Murnane et al., 2002) generated by chemical heterogeneity, especially in the form of local gradients (Lamour et al., 2009), can be important for the guidance of neurite and neural regeneration on biosynthetic materials. In one study, SAMs of alkylsiloxanes, which had an extensive effect on generating surface energy distribution on glass, were used as a substrate. It was showed that chemical heterogeneity had a crucial impact on PC12 cell differentiation and adhesion within 48 h without NGF treatment (Lamour et al., 2010).

One of the promising techniques for nerve repair and regeneration is to support and induce axonal growth by application of nanometer-scale materials, particularly nanofibers and nanotubes mimicking the natural tubular structures such as microtubules, ion channels, and axons. Nanotubes can be fabricated from a variety of different materials such as carbon, silicon, synthetic polymers, DNA, lipids, and proteins. CNTs are novel materials that are able to regulate nerve cell behavior. They are composed of sheets of graphite rolled in a continuous cylinder and classified as single-walled CNTs (SWCNTs) or multiwalled CNTs (MWCNTs) depending on the number of layers formed (Harrison and Atala, 2007). CNTs can be modified with positively or negatively charged chemical groups such as $-NH_2$, $-COOH$, and $-SH$. The positively charged moieties, especially amine groups, may induce neural differentiation by electrostatic interaction with the negatively charged molecules localized on neural cell surface (Hu et al., 2004). PEG is also used to modify SWCNTs and found to reduce the toxicity of CNTs and increase the length of neurites extended by PC12 cells (Malarkey et al., 2008). However, the clinical use of CNTs remains controversial regarding the toxicity and the uncertainty of long-term effects for *in vivo* studies.

3.3.3 Effects of Biofunctionalization on Cellular Behavior Through ECM-Derived Short Peptides

The ECM-derived short peptides are promising candidates to be presented as functional domains on the scaffold surface for use in neural regeneration. These are biocompatible and biodegradable synthetic materials that interact with cell surface receptors similarly to native ECM proteins.

One of the most known ECM-derived peptides is IKVAV sequence derived from laminin. This sequence is a neurite-guiding integrin-binding epitope. Superporous P(HEMA-AEMA) hydrogels were constructed radically as a scaffold by copolymerization of 2-hydroxyethyl methacrylate with 2-aminoethyl methacrylate (AEMA) and ethylene dimethacrylate in the presence of ammonium oxalate crystals. Differentiation of rat MSCs was examined on P(HEMA-AEMA) hydrogels modified with laminin-derived IKVAV, as well as on PHEMA, P(HEMA-AEMA)-SH, and P(HEMA-AEMA)-TPy hydrogels. β-III tubulin, NF, and synaptophysin were expressed highly by fetal NSCs cultured on IKVAV-p(HEMA-AEMA) scaffolds (Kubinova et al., 2010).

Polyamide nanofibers are constituted of nanofibrillar surface that can mimic the spatial organization and porosity of the natural ECM. Primary cerebellar granule, cerebral cortical, hippocampal, motor, and DRG neurons were isolated and cultured on polyamide electrospun nanofibers modified with neuroactive tenascin C-derived peptides. These scaffolds were found to enhance neuronal adhesion, neuritogenesis, and neurite extension *in vitro* compared to unmodified polyamide scaffolds and PLL-coated glass cover slips (Ahmed et al., 2006).

A hydrogel scaffold produced by fiber-templating method and modified with laminin-derived oligopeptides, CDPGYIGSR and CQAASIKVAV, was used in another study to enhance neurite outgrowth and cell adhesion of primary neurons. The results of this study indicated that modification of poly(HEMA-*co*-AEMA) surface with peptides promoted extension of longer neurites when compared to the control P(HEMA-*co*-AEMA). Besides, there was no statistical difference between the neurite lengths of cells grown on modified hydrogels and positive control scaffold made of PLL/laminin (Yu and Shoichet, 2005).

In order to mimic the unique properties of natural ECM scaffolds, self-assembled peptide nanofiber system is one of the most versatile scaffolds due to their ease of modification for presenting bioactive epitopes found in ECM proteins or other neural differentiation inducer proteins. PA molecules contain an alkyl tail connected to hydrophobic amino acids and a hydrophilic bioactive epitope. In aqueous solutions, nanofibers form by self-assembly of individual molecules through charge neutralization and hydrogen bonding. Laminin epitope, IKVAV, presented on peptide nanofibers was used to promote differentiation of neural progenitor cells to neurons and for inducing neurite extension. These synthetic scaffolds provided selective differentiation into neurons very rapidly, while less astrocyte differentiation was observed (Silva et al., 2004). By using the self-assembly approach, PA nanofibers can be used to present multiple bioactive epitopes in one nanofiber system. Use of laminin-derived IKVAV-PAs along with heparan sulfate-mimicking PAs was found to act cooperatively in enhancing neural differentiation of PC12 cells. Moreover, these neural cells were shown to extend neurites even in the presence of inhibitory environment such as CSPGs on peptide nanofiber scaffolds (Mammadov et al., 2012a). Heparin-mimetic nanofibers were found to bind NGF with higher affinity than control peptide nanofibers. Also the results demonstrated that binding of growth factors to the scaffold did not lead to any loss of NGF bioactivity as demonstrated by PC12 differentiation (Mammadov et al., 2012b).

In peripheral nerve regeneration, PA nanofibers can also be used for the delivery of biological molecules such as sonic hedgehog (SHH) protein. In the animal model, SHH signaling was inhibited in the cavernous nerve in rats, and SHH protein was released by PA nanofiber hydrogels. SHH delivery enhanced the regeneration of the cavernous nerve, reduced penile apoptosis, and improved erectile function (Angeloni et al., 2011). In another study, the combination of the epitope-presenting PA molecules with type I collagen by a hybrid matrix approach sustained neuronal survival and control over dendrite and axon growth of Purkinje cells. While collagen matrix provided a desirable mechanical support, laminin-derived PA molecules were used to adjust the bioactivity concentration of the matrix (Sur et al., 2012).

3.4 CONCLUSION

The nervous system is the most complex system of the body due to the complex interactions of neurons with each other, which is hard to reestablish when disordered. Besides its complexity, the inherent fragile nature of neurons combined with the strictly inhibitory environment of the CNS makes regeneration even more difficult, if not impossible. Neurons lost due to a trauma or neurodegenerative disorder are not replenished, except very constrained areas where stem cells reside within the brain. Instead, such pathological conditions usually get even worse due to the proliferation of astrocytes and the inhibitory ECM that they deposit. Neurons that survive after the primary injury cannot make new connections in the scar tissue since they are not allowed to elongate axons. Their nerve terminals start to degenerate eventually leading to cell death resulting in secondary injury. Although the environment and the glial cells in the PNS are not inhibitory but supportive for regeneration, regeneration in PNS injuries is not without problems. Injuries that cause damage in long nerve segments are quite problematic. End-to-end suturing is not preferred to avoid the tension that will lead to secondary injury. Autografting is also not possible as harvesting long segments of nerves from other sides of the body is not possible. Commercially available nerve tubes are the only option for such injuries though they don't offer high success in clinical outcome.

Poor regenerative capacity of the nervous system along with the lack of successful therapies puts a pressure on clinicians and scientists to develop new approaches to relieve the burden of neural injuries and neurodegenerative disorders. Mimicking the healthy ECM of the nervous system by means of its mechanical, chemical, and biological properties is essential to offer neurons an environment where axonal regeneration and new synaptic connections can take place. A variety of materials from polymers to natural materials and synthetic self-assembled systems are currently being developed to fill the gap in the area. To find the optimal material for clinical success in neural regeneration, however, one should consider essential requirement of the material being biocompatible as well as it structural and functional properties' similarity to the natural tissue. Current literature on developing and applying such materials to induce or enhance neural regeneration is reviewed in here. To sum up, we can conclude that it is essential to make a smart design considering all the necessities of the tissue to be applied and currently there has been a lot of effort to generate such materials with clinical success due to their high demand.

REFERENCES

Ahmed, I., Liu, H.Y., Mamiya, P.C., Ponery, A.S., Babu, A.N., Weik, T., Schindler, M., and Meiners, S. (2006). Three-dimensional nanofibrillar surfaces covalently modified with tenascin-C-derived peptides enhance neuronal growth in vitro. J Biomed Mater Res A *76*, 851–860.

Akassoglou, K., Kombrinck, K.W., Degen, J.L., and Strickland, S. (2000). Tissue plasminogen activator-mediated fibrinolysis protects against axonal degeneration and demyelination after sciatic nerve injury. J Cell Biol *149*, 1157–1166.

Akassoglou, K., Yu, W.M., Akpinar, P., and Strickland, S. (2002). Fibrin inhibits peripheral nerve remyelination by regulating Schwann cell differentiation. Neuron *33*, 861–875.

Angeloni, N.L., Bond, C.W., Tang, Y., Harrington, D.A., Zhang, S., Stupp, S.I., McKenna, K.E., and Podlasek, C.A. (2011). Regeneration of the cavernous nerve by Sonic hedgehog using aligned peptide amphiphile nanofibers. Biomaterials *32*, 1091–1101.

Atwal, J.K., Pinkston-Gosse, J., Syken, J., Stawicki, S., Wu, Y., Shatz, C., and Tessier-Lavigne, M. (2008). PirB is a functional receptor for myelin inhibitors of axonal regeneration. Science *322*, 967–970.

Baeten, K.M. and Akassoglou, K. (2011). Extracellular matrix and matrix receptors in blood-brain barrier formation and stroke. Dev Neurobiol *71*, 1018–1039.

Bandtlow, C.E. and Zimmermann, D.R. (2000). Proteoglycans in the developing brain: new conceptual insights for old proteins. Physiol Rev *80*, 1267–1290.

Bard, J.B. and Hay, E.D. (1975). The behavior of fibroblasts from the developing avian cornea. Morphology and movement in situ and in vitro. J Cell Biol *67*, 400–418.

Bellamkonda, R., Ranieri, J.P., Bouche, N., and Aebischer, P. (1995). Hydrogel-based three-dimensional matrix for neural cells. J Biomed Mater Res *29*, 663–671.

Benson, M.D., Romero, M.I., Lush, M.E., Lu, Q.R., Henkemeyer, M., and Parada, L.F. (2005). Ephrin-B3 is a myelin-based inhibitor of neurite outgrowth. Proc Natl Acad Sci U S A *102*, 10694–10699.

Bhang, S.H., Jeong, S.I., Lee, T.J., Jun, I., Lee, Y.B., Kim, B.S., and Shin, H. (2012). Electroactive electrospun polyaniline/poly[(L-lactide)-*co*-(epsilon-caprolactone)] fibers for control of neural cell function. Macromol Biosci *12*, 402–411.

Bonneh-Barkay, D. and Wiley, C.A. (2009). Brain extracellular matrix in neurodegeneration. Brain Pathol *19*, 573–585.

Bundesen, L.Q., Scheel, T.A., Bregman, B.S., and Kromer, L.F. (2003). Ephrin-B2 and EphB2 regulation of astrocyte-meningeal fibroblast interactions in response to spinal cord lesions in adult rats. J Neurosci *23*, 7789–7800.

Bunge, M.B., Clark, M.B., Dean, A.C., Eldridge, C.F., and Bunge, R.P. (1990). Schwann-cell function depends upon axonal signals and basal lamina components. Ann N Y Acad Sci *580*, 281–287.

Carmichael, S.T., Archibeque, I., Luke, L., Nolan, T., Momiy, J., and Li, S.L. (2005). Growth-associated gene expression after stroke: evidence for a growth-promoting region in peri-infarct cortex. Exp Neurol *193*, 291–311.

Carulli, D., Rhodes, K.E., Brown, D.J., Bonnert, T.P., Pollack, S.J., Oliver, K., Strata, P., and Fawcett, J.W. (2006). Composition of perineuronal nets in the adult rat cerebellum and the cellular origin of their components. J Comp Neurol *494*, 559–577.

Cecchini, M., Bumma, G., Serresi, M., and Beltram, F. (2007). PC12 differentiation on biopolymer nanostructures. Nanotechnology *18*, 505103.

Chang, K.A., Kim, J.W., Kim, J.A., Lee, S., Kim, S., Suh, W.H., Kim, H.S., Kwon, S., Kim, S.J., and Suh, Y.H. (2011). Biphasic electrical currents stimulation promotes both proliferation and differentiation of fetal neural stem cells. PLoS One *6*, e18738.

Chen, Z.L. and Strickland, S. (2003). Laminin gamma 1 is critical for Schwann cell differentiation, axon myelination, and regeneration in the peripheral nerve. J Cell Biol *163*, 889–899.

Chen, Z.L., Yu, W.M., and Strickland, S. (2007). Peripheral regeneration. Annu Rev Neurosci *30*, 209–233.

Cheng, T.Y., Chen, M.H., Chang, W.H., Huang, M.Y., and Wang, T.W. (2013). Neural stem cells encapsulated in a functionalized self-assembling peptide hydrogel for brain tissue engineering. Biomaterials *34*, 2005–2016.

Chivatakarn, O., Kaneko, S., He, Z., Tessier-Lavigne, M., and Giger, R.J. (2007). The Nogo-66 receptor NgR1 is required only for the acute growth cone-collapsing but not the chronic growth-inhibitory actions of myelin inhibitors. J Neurosci *27*, 7117–7124.

Chung, K.Y., Taylor, J.S.H., Shum, D.K.Y., and Chan, S.O. (2000). Axon routing at the optic chiasm after enzymatic removal of chondroitin sulfate in mouse embryos. Development *127*, 2673–2683.

Cragg, B. (1979). Brain extracellular space fixed for electron microscopy. Neurosci Lett *15*, 301–306.

Cunha, C., Panseri, S., Villa, O., Silva, D., and Gelain, F. (2011). 3D culture of adult mouse neural stem cells within functionalized self-assembling peptide scaffolds. Int J Nanomedicine *6*, 943–955.

Dalby, M.J., Riehle, M.O., Johnstone, H., Affrossman, S., and Curtis, A.S. (2002a). In vitro reaction of endothelial cells to polymer demixed nanotopography. Biomaterials *23*, 2945–2954.

Dalby, M.J., Riehle, M.O., Johnstone, H.J., Affrossman, S., and Curtis, A.S. (2002b). Polymer-demixed nanotopography: control of fibroblast spreading and proliferation. Tissue Eng *8*, 1099–1108.

Dalby, M.J., Yarwood, S.J., Riehle, M.O., Johnstone, H.J.H., Affrossman, S., and Curtsi, A.S.G. (2002c). Increasing fibroblast response to materials using nanotopography: morphological and genetic measurements of cell response to 13-nm-high polymer demixed islands. Exp Cell Res *276*, 1–9.

Dityatev, A. and Schachner, M. (2003). Extracellular matrix molecules and synaptic plasticity. Nat Rev Neurosci *4*, 456–468.

Domeniconi, M., Cao, Z.U., Spencer, T., Sivasankaran, R., Wang, K.C., Nikulina, E., Kimura, N., Cai, H., Deng, K.W., Gao, Y., et al. (2002). Myelin-associated glycoprotein interacts with the Nogo66 receptor to inhibit neurite outgrowth. Neuron *35*, 283–290.

Engler, A.J., Sen, S., Sweeney, H.L., and Discher, D.E. (2006). Matrix elasticity directs stem cell lineage specification. Cell *126*, 677–689.

Faulkner, J.R., Herrmann, J.E., Woo, M.J., Tansey, K.E., Doan, N.B., and Sofroniew, M.V. (2004). Reactive astrocytes protect tissue and preserve function after spinal cord injury. J Neurosci *24*, 2143–2155.

Faux, C., Hawadle, M., Nixon, J., Wallace, A., Lee, S., Murray, S., and Stoker, A. (2007). PTPsigma binds and dephosphorylates neurotrophin receptors and can suppress NGF-dependent neurite outgrowth from sensory neurons. Biochim Biophys Acta *1773*, 1689–1700.

Fawcett, J.W. and Keynes, R.J. (1990). Peripheral nerve regeneration. Annu Rev Neurosci *13*, 43–60.

Ferrari, A., Cecchini, M., Dhawan, A., Micera, S., Tonazzini, I., Stabile, R., Pisignano, D., and Beltram, F. (2011). Nanotopographic control of neuronal polarity. Nano Lett *11*, 505–511.

Flanagan, L.A., Ju, Y.E., Marg, B., Osterfield, M., and Janmey, P.A. (2002). Neurite branching on deformable substrates. Neuroreport *13*, 2411–2415.

Fournier, A.E., GrandPre, T., and Strittmatter, S.M. (2001). Identification of a receptor mediating Nogo-66 inhibition of axonal regeneration. Nature *409*, 341–346.

Gallagher, J.O., McGhee, K.F., Wilkinson, C.D., and Riehle, M.O. (2002). Interaction of animal cells with ordered nanotopography. IEEE Trans Nanobiosc *1*, 24–28.

Gao, Y., Deng, K.W., Hou, J.W., Bryson, J.B., Barco, A., Nikulina, E., Spencer, T., Mellado, W., Kandel, E.R., and Filbin, M.T. (2004). Activated CREB is sufficient to overcome inhibitors in myelin and promote spinal axon regeneration in vivo. Neuron *44*, 609–621.

Gaudet, A.D., Popovich, P.G., and Ramer, M.S. (2011). Wallerian degeneration: gaining perspective on inflammatory events after peripheral nerve injury. J Neuroinflammation *8*, 110.

Georges, P.C., Miller, W.J., Meaney, D.F., Sawyer, E.S., and Janmey, P.A. (2006). Matrices with compliance comparable to that of brain tissue select neuronal over glial growth in mixed cortical cultures. Biophys J *90*, 3012–3018.

Giger, R.J., Hollis, E.R., and Tuszynski, M.H. (2010). Guidance molecules in axon regeneration. Cold Spring Harb Perspect Biol *2*, a001867.

Gunn, J.W., Turner, S.D., and Mann, B.K. (2005). Adhesive and mechanical properties of hydrogels influence neurite extension. J Biomed Mater Res A *72A*, 91–97.

Halliday, N.L. and Tomasek, J.J. (1995). Mechanical properties of the extracellular matrix influence fibronectin fibril assembly in vitro. Exp Cell Res *217*, 109–117.

Harrison, B.S. and Atala, A. (2007). Carbon nanotube applications for tissue engineering. Biomaterials *28*, 344–353.

Her, G.J., Wu, H.C., Chen, M.H., Chen, M.Y., Chang, S.C., and Wang, T.W. (2013). Control of three-dimensional substrate stiffness to manipulate mesenchymal stem cell fate toward neuronal or glial lineages. Acta Biomater *9*, 5170–5180.

Hodgkin, A.L. and Huxley, A.F. (1952). A quantitative description of membrane current and its application to conduction and excitation in nerve. J Physiol *117*, 500–544.

Hu, H., Ni, Y., Montana, V., Haddon, R.C., and Parpura, V. (2004). Chemically functionalized carbon nanotubes as substrates for neuronal growth. Nano Lett *4*, 507–511.

Huang, Y.J., Wu, H.C., Tai, N.H., and Wang, T.W. (2012). Carbon nanotube rope with electrical stimulation promotes the differentiation and maturity of neural stem cells. Small *8*, 2869–2877.

Jiang, F.X., Yurke, B., Firestein, B.L., and Langrana, N.A. (2008). Neurite outgrowth on a DNA crosslinked hydrogel with tunable stiffnesses. Ann Biomed Eng *36*, 1565–1579.

Kabiri, M., Soleimani, M., Shabani, I., Futrega, K., Ghaemi, N., Ahvaz, H.H., Elahi, E., and Doran, M.R. (2012). Neural differentiation of mouse embryonic stem cells on conductive nanofiber scaffolds. Biotechnol Lett *34*, 1357–1365.

Kantor, D.B., Chivatakarn, O., Peer, K.L., Oster, S.F., Inatani, M., Hansen, M.J., Flanagan, J.G., Yamaguchi, Y., Sretavan, D.W., Giger, R.J., et al. (2004). Semaphorin 5A is a bifunctional axon guidance cue regulated by heparan and chondroitin sulfate proteoglycans. Neuron *44*, 961–975.

Kim, J.E., Liu, B.P., Park, J.H., and Strittmatter, S.M. (2004). Nogo-66 receptor prevents raphespinal and rubrospinal axon regeneration and limits functional recovery from spinal cord injury. Neuron *44*, 439–451.

Koh, H.S., Yong, T., Chan, C.K., and Ramakrishna, S. (2008). Enhancement of neurite outgrowth using nano-structured scaffolds coupled with laminin. Biomaterials *29*, 3574–3582.

Kotov, N.A., Winter, J.O., Clements, I.P., Jan, E., Timko, B.P., Campidelli, S., Pathak, S., Mazzatenta, A., Lieber, C.M., Prato, M., et al. (2009). Nanomaterials for neural interfaces. Adv Mater *21*, 3970–4004.

Kubinova, S., Horak, D., Kozubenko, N., Vanecek, V., Proks, V., Price, J., Cocks, G., and Sykova, E. (2010). The use of superporous Ac-CGGASIKVAVS-OH-modified PHEMA scaffolds to promote cell adhesion and the differentiation of human fetal neural precursors. Biomaterials *31*, 5966–5975.

Kubota, Y., Morita, T., Kusakabe, M., Sakakura, T., and Ito, K. (1999). Spatial and temporal changes in chondroitin sulfate distribution in the sclerotome play an essential role in the formation of migration patterns of mouse neural crest cells. Dev Dyn *214*, 55–65.

Kwok, J.C., Carulli, D., and Fawcett, J.W. (2010). In vitro modeling of perineuronal nets: hyaluronan synthase and link protein are necessary for their formation and integrity. J Neurochem *114*, 1447–1459.

Lamour, G., Journiac, N., Souès, S., Bonneau, S., Nassoy, P., Hamraoui, A. (2009). Influence of surface energy distribution on neuritogenesis. Colloids Surf B Biointerfaces *72*(2), 208–218.

Lamour, G., Eftekhari-Bafrooei, A., Borguet, E., Soues, S., and Hamraoui, A. (2010). Neuronal adhesion and differentiation driven by nanoscale surface free-energy gradients. Biomaterials *31*, 3762–3771.

Lau, L.W., Cua, R., Keough, M.B., Haylock-Jacobs, S., and Yong, V.W. (2013). Pathophysiology of the brain extracellular matrix: a new target for remyelination. Nat Rev Neurosci *14*, 722–729.

Leach, J.B., Brown, X.Q., Jacot, J.G., DiMilla, P.A., and Wong, J.Y. (2007). Neurite outgrowth and branching of PC12 cells on very soft substrates sharply decreases below a threshold of substrate rigidity. J Neural Eng *4*, 26–34.

Lee, M.R., Kwon, K.W., Jung, H., Kim, H.N., Suh, K.Y., Kim, K., and Kim, K.S. (2010). Direct differentiation of human embryonic stem cells into selective neurons on nanoscale ridge/groove pattern arrays. Biomaterials *31*, 4360–4366.

Leipzig, N.D. and Shoichet, M.S. (2009). The effect of substrate stiffness on adult neural stem cell behavior. Biomaterials *30*, 6867–6878.

Liu, B.P., Fournier, A., GrandPre, T., and Strittmatter, S.M. (2002). Myelin-associated glycoprotein as a functional ligand for the Nogo-66 receptor. Science *297*, 1190–1193.

Liu, K., Tedeschi, A., Park, K.K., and He, Z.G. (2011). Neuronal intrinsic mechanisms of axon regeneration. Annu Rev Neurosci, 34, 131–152.

Ma, W., Tavakoli, T., Derby, E., Serebryakova, Y., Rao, M.S., and Mattson, M.P. (2008). Cell-extracellular matrix interactions regulate neural differentiation of human embryonic stem cells. BMC Dev Biol *8*, 90.

Maeda, N. and Noda, M. (1996). 6B4 proteoglycan/phosphacan is a repulsive substratum but promotes morphological differentiation of cortical neurons. Development *122*, 647–658.

Malarkey, E.B., Reyes, R.C., Zhao, B., Haddon, R.C., and Parpura, V. (2008). Water soluble single-walled carbon nanotubes inhibit stimulated endocytosis in neurons. Nano Lett *8*, 3538–3542.

Mammadov, B., Mammadov, R., Guler, M.O., and Tekinay, A.B. (2012a). Cooperative effect of heparan sulfate and laminin mimetic peptide nanofibers on the promotion of neurite outgrowth. Acta Biomater *8*, 2077–2086.

Mammadov, R., Mammadov, B., Guler, M.O., and Tekinay, A.B. (2012b). Growth factor binding on heparin mimetic peptide nanofibers. Biomacromolecules *13*, 3311–3319.

Margolis, R.K., Rauch, U., Maurel, P., and Margolis, R.U. (1996). Neurocan and phosphacan: two major nervous tissue-specific chondroitin sulfate proteoglycans. Perspect Dev Neurobiol *3*, 273–290.

Martini, R., Fischer, S., Lopez-Vales, R., and David, S. (2008). Interactions between Schwann cells and macrophages in injury and inherited demyelinating disease. Glia *56*, 1566–1577.

Mckeon, R.J., Hoke, A., and Silver, J. (1995). Injury-induced proteoglycans inhibit the potential for laminin-mediated axon growth on astrocytic scars. Exp Neurol *136*, 32–43.

Meyer-Franke, A., Wilkinson, G.A., Kruttgen, A., Hu, M., Munro, E., Hanson, M.G., Reichardt, L.F., and Barres, B.A. (1998). Depolarization and cAMP elevation rapidly recruit TrkB to the plasma membrane of CNS neurons. Neuron *21*, 681–693.

Migliorini, E., Grenci, G., Ban, J., Pozzato, A., Tormen, M., Lazzarino, M., Torre, V., and Ruaro, M.E. (2011). Acceleration of neuronal precursors differentiation induced by substrate nanotopography. Biotechnol Bioeng *108*, 2736–2746.

Moreau-Fauvarque, C., Kumanogoh, A., Camand, E., Jaillard, C., Barbin, G., Boquet, I., Love, C., Jones, E.Y., Kikutani, H., Lubetzki, C., et al. (2003). The transmembrane semaphorin Sema4D/CD100, an inhibitor of axonal growth, is expressed on oligodendrocytes and upregulated after CNS lesion. J Neurosci *23*, 9229–9239.

Murnane, A.C., Brown, K., and Keith, C.H. (2002). Preferential initiation of PC12 neurites in directions of changing substrate adhesivity. J Neurosci Res *67*(3), 321–328.

Nakajima, M., Ishimuro, T., Kato, K., Ko, I.K., Hirata, I., Arima, Y., and Iwata, H. (2007). Combinatorial protein display for the cell-based screening of biomaterials that direct neural stem cell differentiation. Biomaterials *28*, 1048–1060.

Neumann, S., Bradke, F., Tessier-Lavigne, M., and Basbaum, A.I. (2002). Regeneration of sensory axons within the injured spinal cord induced by intraganglionic cAMP elevation. Neuron *34*, 885–893.

Niederost, B.P., Zimmermann, D.R., Schwab, M.E., and Bandtlow, C.E. (1999). Bovine CNS myelin contains neurite growth-inhibitory activity associated with chondroitin sulfate proteoglycans. J Neurosci *19*, 8979–8989.

Pardo, B. and Honegger, P. (2000). Differentiation of rat striatal embryonic stem cells in vitro: monolayer culture vs. three-dimensional coculture with differentiated brain cells. J Neurosci Res *59*, 504–512.

Pasterkamp, R.J. and Verhaagen, J. (2006). Semaphorins in axon regeneration: developmental guidance molecules gone wrong? Philos Trans R Soc Lond B Biol Sci *361*, 1499–1511.

Peretz, H., Talpalar, A.E., Vago, R., and Baranes, D. (2007). Superior survival and durability of neurons and astrocytes on 3-dimensional aragonite biomatrices. Tissue Eng *13*, 461–472.

Prabhakaran, M.P., Venugopal, J.R., and Ramakrishna, S. (2009). Mesenchymal stem cell differentiation to neuronal cells on electrospun nanofibrous substrates for nerve tissue engineering. Biomaterials *30*, 4996–5003.

Qiu, J., Cai, C.M., Dai, H.N., McAtee, M., Hoffman, P.N., Bregman, B.S., and Filbin, M.T. (2002). Spinal axon regeneration induced by elevation of cyclic AMP. Neuron *34*, 895–903.

Ren, Y.J., Zhang, H., Huang, H., Wang, X.M., Zhou, Z.Y., Cui, F.Z., and An, Y.H. (2009). In vitro behavior of neural stem cells in response to different chemical functional groups. Biomaterials *30*, 1036–1044.

Riopelle, R.J. and Dow, K.E. (1990). Functional interactions of neuronal heparan-sulfate proteoglycans with laminin. Brain Res *525*, 92–100.

Sabri, F., Cole, J.A., Scarbrough, M.C., and Leventis, N. (2012). Investigation of polyurea-crosslinked silica aerogels as a neuronal scaffold: a pilot study. PLoS One *7*, e33242.

Saha, K., Keung, A.J., Irwin, E.F., Li, Y., Little, L., Schaffer, D.V., and Healy, K.E. (2008). Substrate modulus directs neural stem cell behavior. Biophys J *95*, 4426–4438.

Schmalfeldt, M., Bandtlow, C.E., Dours-Zimmermann, M.T., Winterhalter, K.H., and Zimmermann, D.R. (2000). Brain derived versican V2 is a potent inhibitor of axonal growth. J Cell Sci *113*, 807–816.

Sharma, K., Selzer, M.E., and Li, S. (2012). Scar-mediated inhibition and CSPG receptors in the CNS. Exp Neurol *237*, 370–378.

Silva, G.A., Czeisler, C., Niece, K.L., Beniash, E., Harrington, D.A., Kessler, J.A., and Stupp, S.I. (2004). Selective differentiation of neural progenitor cells by high-epitope density nanofibers. Science *303*, 1352–1355.

Snider, W.D., Zhou, F.Q., Zhong, J., and Markus, A. (2002). Signaling the pathway to regeneration. Neuron *35*, 13–16.

Snow, D.M., Lemmon, V., Carrino, D.A., Caplan, A.I., and Silver, J. (1990). Sulfated proteoglycans in astroglial barriers inhibit neurite outgrowth in vitro. Exp Neurol *109*, 111–130.

Stabenfeldt, S.E., Munglani, G., Garcia, A.J., and LaPlaca, M.C. (2010). Biomimetic microenvironment modulates neural stem cell survival, migration, and differentiation. Tissue Eng Part A *16*, 3747–3758.

Sur, S., Pashuck, E.T., Guler, M.O., Ito, M., Stupp, S.I., and Launey, T. (2012). A hybrid nanofiber matrix to control the survival and maturation of brain neurons. Biomaterials *33*, 545–555.

Tani, E. and Ametani, T. (1971). Extracellular distribution of ruthenium red-positive substance in the cerebral cortex. J Ultrastruct Res *34*, 1–14.

Teixeira, A.I., Ilkhanizadeh, S., Wigenius, J.A., Duckworth, J.K., Inganas, O., and Hermanson, O. (2009). The promotion of neuronal maturation on soft substrates. Biomaterials *30*, 4567–4572.

Tu, Q., Li, L., Zhang, Y.R., Wang, J.C., Liu, R., Li, M.L., Liu, W.M., Wang, X.Q., Ren, L., and Wang, J.Y. (2011). The effect of acetylcholine-like biomimetic polymers on neuronal growth. Biomaterials *32*, 3253–3264.

Walicke, P.A. (1988). Interactions between basic fibroblast growth factor (FGF) and glycosoaminoglycans in promoting neurite outgrowth. Exp Neurol *102*, 144–148.

Wang, K.C., Koprivica, V., Kim, J.A., Sivasankaran, R., Guo, Y., Neve, R.L., and He, Z.G. (2002). Oligodendrocyte-myelin glycoprotein is a Nogo receptor ligand that inhibits neurite outgrowth. Nature *417*, 941–944.

Willits, R.K. and Skornia, S.L. (2004). Effect of collagen gel stiffness on neurite extension. J Biomater Sci Polym Ed *15*, 1521–1531.

Winzeler, A.M., Mandemakers, W.J., Sun, M.Z., Stafford, M., Phillips, C.T., and Barres, B.A. (2011). The lipid sulfatide is a novel myelin-associated inhibitor of CNS axon outgrowth. J Neurosci *31*, 6481–6492.

Yamada, H., Fredette, B., Shitara, K., Hagihara, K., Miura, R., Ranscht, B., Stallcup, W.B., and Yamaguchi, Y. (1997). The brain chondroitin sulfate proteoglycan brevican associates with astrocytes ensheathing cerebellar glomeruli and inhibits neurite outgrowth from granule neurons. J Neurosci *17*, 7784–7795.

Yamaguchi, Y. (2000). Lecticans: organizers of the brain extracellular matrix. Cell Mol Life Sci *57*, 276–289.

Yiu, G. and He, Z.G. (2006). Glial inhibition of CNS axon regeneration. Nat Rev Neurosci *7*, 617–627.

Yu, T.T. and Shoichet, M.S. (2005). Guided cell adhesion and outgrowth in peptide-modified channels for neural tissue engineering. Biomaterials *26*, 1507–1514.

4

THERAPEUTIC NANOMATERIALS FOR CARTILAGE REGENERATION

ELIF ARSLAN, SEHER USTUN YAYLACI, MUSTAFA O. GULER, AND AYSE B. TEKINAY

Institute of Materials Science and Nanotechnology, National Nanotechnology Research Center (UNAM), Bilkent University, Ankara, Turkey

4.1 INTRODUCTION

Cartilage tissue is present throughout the human body and is mainly responsible for load bearing, lubrication, and articulation (Fox et al., 2009; Han et al., 2011). It is categorized under three types: hyaline cartilage, elastic cartilage, and fibrocartilage. The tissue is avascular, aneural, and alymphatic, which causes its low capacity for self-repair following damage (Mahmoudifar and Doran, 2012). Cartilage is secreted by cartilage-specific cells, called chondrocytes, that comprise 3–5% of adult cartilage (Taipaleenmaki, 2010). During embryonic development, mesenchymal stem cells (MSCs) surround the cartilage in the deep region of the perichondrium and subsequently differentiate into chondrocytes (Slomianka, 2009). Besides its cellular components, cartilage also has an extracellular matrix (ECM) that is largely responsible for the load-bearing function of the tissue (Taipaleenmaki, 2010). The primary extracellular constituents of cartilage are the fibrillar collagen network and proteoglycans (Han et al., 2011). Specifically, type II collagen is the primary collagen fibril found in the cartilaginous ECM. Other

Therapeutic Nanomaterials, First Edition. Edited by Mustafa O. Guler and Ayse B. Tekinay.

collagens found in cartilage are the types III, VI, IX, X, XI, XII, and XIV, all of which contribute to the mature extracellular microenvironment of the tissue (Eyre, 2002). Proteoglycans are also a major element of the cartilage ECM, with aggrecan, the glycosaminoglycans (GAGs) chondroitin sulfate and keratan sulfate, hyaluronan, and the glycoprotein lubricin (PRG4) occurring commonly in the tissue matrix (Lin et al., 2005). The load-bearing, articulative, and lubricant properties of cartilage also depend heavily on the water content of the tissue. Water is highly abundant in cartilage and comprises 80% of its total weight. Significant amounts of water are integrated within the collagen matrix in interfibrillar spaces (Lu et al., 2006). The interfibrillar water forms a gelatinous structure and assists in spreading the pressure gradient across the tissue, thus relieving the solid matrix from excessive mechanical stress (Mow et al., 1980, 1992). The ability of healthy cartilage tissue to endure significant weight loads is attributed to two basic mechanisms of water: the frictional resistance of water flow and the pressurization of water within the matrix (Fox et al., 2009).

The joints of the human body are composed of articular cartilage (AC), synovial membranes, ligaments, subchondral bone, and menisci (Clouet et al., 2009). AC is the main focus of this chapter and indeed most cartilage regeneration studies, since it is commonly affected by diseases and traumatic injuries that necessitate its therapeutic regeneration. AC is a heterogeneous type of cartilage and contains different types of material organizations and matrix components across its structure. It is composed of four tissue layers that are horizontally stacked across the tissue surface and can be categorized with respect to their ECM composition and cellular morphology (Stroebel, 2007). The superficial zone forms 10–20% of AC tissue by volume, consists of flattened chondrocytes and tightly packed collagens (mainly types II and IX), and is predominantly responsible for the protection of the underlying layers from shear stresses, as well as providing tensile strength to the tissue (Fox et al., 2009). The middle (transitional) zone forms 40–60% of cartilage tissue; is composed of round chondrocytes, thicker collagen fibrils, and proteoglycans; and assists in resisting compressive forces (Fox et al., 2009). The deep zone forms 30% of the tissue and contains the thickest collagen fibers observed in cartilage. It is rich in proteoglycans and largely responsible for creating the resistance of cartilage against compressive forces (Fox et al., 2009). The last layer, called the tidemark, is located between the deep zone and the calcified part of cartilage; it displays a high proteoglycan content that also allows it to resist compression. In addition to the layers of cartilage proper, there is an additional calcified layer between the bone and cartilage. This intermediary layer integrates the two tissue types and contains hypertrophic chondrocytes.

While structural differences can be used to classify cartilage into four strata, compositional, collagen-based, and positional differences also exist within the cells and matrix elements of a given cartilage layer. Three types of cartilage ECM are recognized: pericellular, territorial, and interterritorial (Fox et al., 2009). The pericellular matrix is a thin layer that is mostly composed of proteoglycans, noncollagenous proteins, and glycoproteins and is responsible for supporting chondrocytes. It is also suggested that this layer may play a role in establishing the signal transduction networks required for the proper functioning of cartilage tissue (Eggli et al., 1985). The pericellular matrix is encompassed by the territorial matrix, which is composed mainly of a network of thin collagen fibrils that spread throughout the cartilage tissue. It has been suggested that this matrix is responsible for the resistance of cartilage against mechanical stresses and that it bears the majority of the mechanical load on the tissue (Engel et al., 1969). The interterritorial region is the largest among the three regions of the cartilage ECM and contributes the most to the biochemical composition and function of the AC. This matrix has abundant proteoglycan content in the form of collagen bundles, which are arranged in parallel in the superficial zone, perpendicular on the surface of the deep zone, and in intermediate angles for middle zone collagens (Mow and Guo, 2002).

The physiological metabolism of cartilage tissue is as crucial as its structural and biocompositional features, and its investigation is vital for the determination of processes underlying the diseases, defects, and repair mechanisms of cartilage. Cartilage metabolism is mainly anaerobic, since it is an avascular and alymphatic tissue. Nutrition is provided by diffusion from the synovial fluid, as the cartilage matrix contains pores of approximately 6 nm in diameter and is therefore limited in its capacity for material exchange (Maroudas, 1968; Mow et al., 1992). Chondrocytes are the main effectors of the metabolic activities of the tissue, despite their scarcity (~3–5% of cartilage tissue). They are also responsible for the synthesis, maintenance, and repair of the ECM (Buckwalter and Mankin, 1997). Several factors and molecules surround the chondrocytes and affect their ECM synthesis and degradation. For example, proinflammatory cytokines play a role in regulating the catabolic and anabolic metabolisms of chondrocytes and affect their capacity to synthesize and degrade ECM molecules (Buckwalter and Mankin, 1997). Growth factors, on the other hand, are involved in the regulation of proteoglycan synthesis: Although there is a dearth of information about the mechanisms and cellular pathways behind their effects, several growth factors are known to be important for cartilage proteoglycan synthesis. These include transforming growth factor beta-1 (TGF-β1), insulin-like growth factor, interleukin-1, and tumor necrosis factor-α (Fox et al., 2009).

The homeostasis of cartilage tissue contributes to its structural stability and depends on the healthy turnover of its ECM components; thus, deficiencies in turnover mechanisms commonly lead to disease. Enzymes responsible for the degradation of ECM molecules include metalloproteinases and cathepsins (Dejica et al., 2008; Woessner, 1991). The physiological metabolism of cartilage is also modified according to its stage of growth. Chondrocytes rapidly synthesize ECM components and expand the new tissue during embryonic development, which necessitates high rates of activity. However, the chondrocytes of adult cartilage tissues are at a steady state and function only to replace degraded or damaged matrix materials (Buckwalter, 1995).

There is a constant cross-communication between chondrocytes and their ECM, through which various functions of the tissue are performed. Chondrocytes are responsible for the turnover of the ECM, while the ECM facilitates nutrient exchange and determines which molecules can pass through the matrix to reach the chondrocytes. This function of the ECM depends on its composition and organization, especially on its proteoglycan components (Buckwalter and Mankin, 1997). The ECM itself also imparts signals in a more direct manner by transferring mechanical signals (e.g., joint loading) to the chondrocytes (Buckwalter, 1995).

Overall, a thorough characterization of the biomechanical function of cartilage is necessary to understand the nature of the tissue and the mechanisms behind its degeneration and regeneration. As such, it is unsurprising that cartilage tissue regeneration is one of the most comprehensively investigated topics in tissue engineering, especially in light of the fact that osteochondral defects and joint diseases are severely debilitating disorders with no known cures. Osteoarthritis (OA) in particular is the most common joint disease in the world and primarily affects the population aged over 60 (Arden and Nevitt, 2006; Clouet et al., 2009). OA has a detrimental effect on several regions of cartilage tissue, including AC, the synovium, and subchondral bone. It often manifests as a decrease in the thickness of AC, the growth of bone on the joint margins, and the alteration of the molecular components of the synovial fluid (Clouet et al., 2009). Although many treatment methods have been developed for OA, the complete amelioration of its effects has so far been outside the reach of modern medicine. In clinics, regular therapies for OA include weight reduction, physical therapy, drug treatments (e.g., pain relievers), and the administration of chondroprotective molecules; more severe conditions may necessitate surgical interventions such as arthroplasty and osteotomy (Clouet et al., 2009). These treatments are only able to alleviate the symptoms of OA and cannot reverse the condition itself; as such, significant advances in regenerative medicine and tissue engineering are necessary for the development of prospective clinical treatments and therapeutics for osteochondral diseases.

4.2 CURRENT TREATMENT METHODS FOR CARTILAGE INJURIES

Defects in AC can result from traumatic injury or pathological degeneration. The capacity of cartilage to repair injury is limited, and when not treated, such defects can progress to degenerative arthritis. Various repair strategies are currently used in the treatment of cartilage defects. Arthroscopic techniques, such as lavages and debridements, aim to clean out the defect site and alleviate pain associated with the inflammation of the cartilage tissue (Hunziker, 2002). These procedures are considered to be palliative, as they only provide short-term relief by removing unstable cartilage flaps and preventing inflammation; they do not restore the function of the damaged site (Shannon et al., 2001). An arthroscopic lavage involves the washing of the joint space for the removal of blood or loose debris, while debridement is the removal of unstable cartilage flaps. The main targets of these procedures are patients with very small cartilage defects (Lewis et al., 2006).

Other arthroscopic techniques aim to utilize intrinsic repair capacity of cartilage and include drilling and microfractures. Both of these techniques consist of the penetration and subsequent bleeding of subchondral bone to induce fibrocartilage formation. They rely on the formation of a blood clot at the defect site by stimulating the bone marrow, resulting in the deposition of a mechanically weak fibrocartilage tissue at the defect site. The microfracture technique is used for defects smaller than 2.5 cm^2 and involves the removal of the unstable cartilage for the creation of a well-defined defect environment that is surrounded by healthy cartilage (Bhosale and Richardson, 2008). In this procedure, a microscopic awl is used to penetrate subchondral bone without damaging its integrity. The defect is then filled with clot material from the subchondral bone, which contains MSCs that will oversee the subsequent repair of cartilage tissue. The outcome of the treatment depends mainly on how well the patient can avoid overexerting the cartilage tissue with excessive weight loads. Clinical data suggest that 75% of patients display short-term clinical improvements; however, functional deterioration of the defect site begins within 24 months in 48–80% of cases (Gobbi et al., 2014). Low numbers of MSCs, or their dilution by synovial fluid, can result in a low regeneration capacity and early deterioration following the procedure. Adequate regeneration therefore depends on the optimization of the number of cells extracted from the subchondral bone, which is performed by controlling the depth and numbers of the holes produced on the cartilage surface.

Grafting is the preferred way of treating larger cartilage defects and consists of the removal of healthy cartilage tissue from nonweight-bearing areas of the knee, followed by the transplantation of the removed tissue into the defect site (Görtz and Bugbee, 2007). Multiple plugs may be required depending on the depth and the extent of the defect site. Grafts are self-secured to the defect site without using any adhesives; their success depends heavily on whether draft plugs from nonload-bearing regions are able to withstand the stresses applied to the weight-bearing parts of the tissue. Another concern about the treatment of cartilage injuries through osteochondral grafting is the negative effect the procedure has on the viability of cells at the margins of the grafts. Cells in these regions are more prone to be affected by stress, and the resulting cell death may lead to the degeneration of the tissue over time (Boscainos et al., 2008). In addition, the dead space between the graft and the native tissue affects the integrity of repaired cartilage.

A new generation of tissue repair strategies involving the external administration of autologous or donor cells into the defect site has been developed to replace damaged cartilage with biomechanically stronger and functional hyaline cartilage. Autologous chondrocyte implantation (ACI) is one such strategy and utilizes the patient's own chondrocytes. It is conducted in three stages: Chondrocytes are first arthroscopically extracted from nonload-bearing parts of cartilage, subsequently expanded in an *in vitro* environment until sufficient numbers of cells are obtained, and finally reimplanted back into the defect site and secured with a periosteal graft taken from the tibia (Brittberg et al., 1994). The use of the patient's own chondrocytes prevents immune responses, and the isolation of chondrocytes can be performed through a minor biopsy operation, which reduces potential complications at the donor site. However, the operations involved are complex, and the procedure necessitates a longer recovery time to allow for the maturation of new cartilage tissue (Dehne et al., 2009). Graft hypertrophy is one of the most pronounced postsurgery complications of this technique.

Second-generation ACI use restorable collagen matrices instead of periosteal flaps. A combination matrix of collagens I and III has been shown to reduce graft hypertrophy and the number and depth of incisions necessary for the operation (Tuan, 2007). However, the use of artificial scaffolds increases the risk of immune reaction. Technical constraints during the culture and transplantation of cells, such as nonhomogeneous distribution of chondrocytes, leakage of the transplanted cell suspension due to gravity, and the risk of dedifferentiation of chondrocytes, limit the usefulness of this technique for the production of functional hyaline cartilage and create large variations in the success of the operation (Vavken and Samartzis, 2010).

These difficulties have led to the development of a third generation of ACI, composed of biomechanically stable and functional tissue matrices that integrate better into host tissues (Bartlett, 2005). Use of scaffold-based techniques has some major advantages over scaffold-free techniques: Scaffolds can better assume the three-dimensional (3D) structure of the defect, provide a suitable substrate for cellular adhesion, ensure the homogeneous distribution of cells, and prevent leakage during transplantation. Matrix-induced autologous chondrocyte implantation (MACI) is an example of scaffold-based techniques in clinical use (Kon et al., 2011, 2013). Unlike second-generation ACI methods, cells are seeded onto 3D porcine-derived type I/III collagen matrices and cultured for an additional 3 days prior to implantation. These matrices are designed in a way that one side is porous and allows the infiltration of MSCs, while the other side exhibits high mechanical strength and has a low-friction surface that allows easier integration into the chondral cavity. The membrane patch is secured by a fibrin glue during implantation, eliminating the need for tedious watertight sutures. Materials used in MACI include collagen type I/III, hyaluronan (Hyalograft®-C, HYAFF®11; Fidia Advanced Biopolymers, Abano Terme, Italy) and fibrin (Tissucol, Baxter, Austria) scaffolds (Iwasa et al., 2008).

In vitro studies have demonstrated that the use of bilayer collagen membranes presents an adequate environment for cell attachment and chondrogenic differentiation (Gigante et al., 2007). However, there are no clinical findings that clearly prove the superiority of MACI over existing techniques (Memon and Quinlan, 2012): Indeed, no differences were found in the arthroscopic appearances, histological observations, and rates of tissue hypertrophy between collagen I/III-based ACI-C and bilayer collagen-based MACI grafts used in the treatment of symptomatic cartilage defect cases (Bartlett et al., 2005). MACI is nonetheless an attractive treatment method, as it provides a means of arthroscopic implantation that reduces invasiveness and operation times. Nonetheless, chondrocytes tend to lose their differentiated states during clonal expansion, which adversely affects their ability to synthesize functional cartilage ECM and results in large variances in the outcomes of ACI and MACI treatments. Three days of culture in collagen matrices may result in the formation of only an immature tissue environment, leading to the redifferentiation of cells and a corresponding decrease in their matrix synthesis capability (Schulze-Tanzil, 2009). Longer culture periods may permit chondrocytes to synthesize their own matrix, resulting in more mature and stable tissue replacements.

4.3 TISSUE ENGINEERING EFFORTS

Existing treatment techniques are insufficient for the long-term repair of cartilage and need to be complemented with engineered systems containing cartilage-mimetic material architectures, inductive signals, or stem cells to produce reproducible, safe, and functional tissue constructs. There are four main approaches for cartilage regeneration: cultured cell implantation, engineered tissue construct implantation, scaffoldless tissue regeneration, and guided tissue regeneration. Cartilage growth, development, and repair are dependent on both biomechanical and biological signals. *In vitro* culture environments that attempt to induce cartilage repair or recapitulate the structural and biochemical elements of cartilage environment can therefore increase the success of implantation. Scaffold materials can directly alter the sensory landscape of the cells they contain, in addition to providing mechanical cues necessary for chondrogenic differentiation. As such, a great variety of factors should be considered in tandem for the design of optimal cartilage scaffolds. In a general sense, the material should allow the transport of nutrients and waste products, facilitate the migration and development of cells, possess an intrinsic biocompatibility, and provide the biochemical signals necessary for the survival and maintenance of chondrocytes.

Scaffold surface characteristics have a direct role on controlling the adhesion, migration, and differentiation of cells (Ma et al., 2005). Material properties such as scaffold morphology, hydrophilicity, and surface charge should therefore be considered for the development of functional scaffolds that can interact with the surrounding environment. An additional factor in the design of cartilage scaffolds is the time of implantation, that is, whether the construct is set to be implanted immediately or following an *in vitro* culture period. Constructs implanted immediately should be stable in the physical environment of cartilage and protect the seeded cells until neotissue formation is completed. However, this kind of structural integrity and durability is not particularly necessary for cell-laden constructs matured in *in vitro* environments, since newly formed tissue itself may be able to withstand the native, high-shear cartilage environment. The material should also be able to fill the defect site and integrate with the surrounding native tissue for the implantation to succeed. In addition, degradation rate of scaffolds in native tissue should be in concert with the formation rate of neotissue, as faster degradation results in shape retention issues, while slow degradation intervenes with the ingrowth of neotissue.

Material source (natural or synthetic), physical properties, biofunctionality, and tissue integration capability can affect the choice and design parameters of a scaffold. A wide range of natural and synthetic polymers have been investigated in cartilage tissue engineering efforts (Chung and Burdick, 2008; Kim et al., 2012; Makris et al., 2014). Natural materials are preferred due to their low costs and similarity to the cartilage ECM, which allows their natural degradation, facilitates their interaction with cells through functional groups, and ensures that the scaffold will be biocompatible. Agarose, alginate, hyaluronic acid, fibrin glue, chitosan, type I and II collagen, and reconstituted tissue matrices are among the natural materials used in cartilage tissue engineering (Zhao et al., 2013). However, natural polymers lack the mechanical properties necessary to withstand the high-stress environment of cartilage and tend to undergo rapid degradation following implantation. Synthetic polymers, in contrast, provide tailorable physical and chemical properties through different synthesis methods and can be processed in different sizes and shapes. In addition, they present low immunogenic responses and toxicity and tend to have lower batch-to-batch variances compared to natural materials.

4.3.1 Natural Polymers

Agarose and alginate are both derived from algae and provide a biocompatible 3D environment that has been shown to preserve the rounded morphology of chondrocytes. Alginate and agarose are continuous hydrogel-based materials that can transmit applied forces to encapsulated chondrocytes and allow the uniform distribution of seeded cells; this property has led to their use in studies investigating effects of dynamic loading on cell behavior (Chahine et al., 2009; Grogan et al., 2012; Henrionnet et al., 2012; Mauck et al., 2015). However, alginate is not degraded rapidly in the body and interferes with growth of neotissue. Covalent crosslinking of alginate can be performed to tune the mechanical properties, degradation kinetics, and swelling ratio of alginate (Kuo and Ma, 2001). Photocrosslinkable alginate systems allow noninvasive implantation of scaffolds that can fill defects of any size and geometry. In a recent study, alginate was modified with 2-aminoethyl methacrylate (AEMA) and crosslinked under 365 nm UV light with the help of a photoinitiator, creating hydrogels with adjustable degradation kinetics based on the rate of methacrylation (Jeon et al., 2009).

Collagen types I and II are the most abundant proteins in the native cartilage ECM and have been used in the development of bioactive scaffolds. Collagen-derived materials bear inherent biological cues that facilitate cell attachment and can be remodeled by cells during their development. Like other natural materials, collagen has to be processed to decrease its antigenicity before use. This process removes immunogenic components within collagen and crosslinks the remaining atellopeptides by aldehyde or carbodiimide chemistry. It can easily be processed into various physical configurations, such as tubes, sheets, fleeces, and sponges; however, gels and fibers are the forms most commonly studied in cartilage tissue engineering (Parenteau-Bareil et al., 2010). The predominant collagen of cartilage is type II collagen, and chondrocytes seeded in type II collagen scaffolds were previously demonstrated to maintain their phenotype. However poor availability and undesirable mechanical properties of type II collagen limit its use in tissue engineering studies.

Hyaluronan is a major GAG component of cartilage and synovial fluid and helps in the lubrication of joints. It is a nonsulfated GAG and is involved in cell differentiation, ECM organization, and cell motility (Parenteau-Bareil et al., 2010). Hyaluronan can be derived from various animal tissues or produced by microbial fermentation. Hyaluronan can be injected to fill defects in any shape and is therefore suited for less invasive applications. However, hyaluronan is quickly degraded *in vivo* by cell-secreted hyaluronidases. As with agarose and alginate, hyaluronan and hyaluronic acid can be crosslinked to produce porous solid platforms with higher resistances to degradation (Jin et al., 2010). Aqueous solutions of hyaluronic acid can be crosslinked through covalent crosslinking or photocrosslinking. HYAFF®'7 and HYAFF®11 are the ethyl and benzyl esters of hyaluronan, respectively, and remain intact in the body for around 2 months prior to their degradation by the hydrolysis of their ester bonds (Grigolo et al., 2002).

Fibrin is a protein involved in the blood clotting process and used as an adhesive in surgery. It is composed of fibrinogen monomers. The fibrinogen molecule has two sets of three polypeptide chains and is solidified through the binding of thrombin. The stability and mechanical integrity of the resulting fibrin gel are dependent on pH and the concentrations of fibrinogen and calcium ions. Fibrin gels can be derived from the patient's own blood and subsequently utilized as autologous scaffold. Because of its complete biodegradability and injectable forms, it is popular in various tissue engineering applications. However, it needs to be blended with different materials to enhance its mechanical properties.

Chitosan is the deacetylated form of chitin, one of the most abundant polysaccharides in nature. This polysaccharide has an intrinsic antibacterial ability, triggers mild immunological reactions, and has a structure similar to GAG molecules. Its highly cationic nature and resemblance to GAGs allow it to bind growth factors and adhesion proteins. Chitosan is degraded by deacetylation, and the host tissue progressively metabolizes the sugar units detached from its structure. The degradation rate of chitosan scaffolds can be modified during processing by altering the number of acetyl units. Similar to fibrin, it can be injected into target tissues due to its unique temperature-dependent gelation property: It is liquid at room temperature and gel at physiological temperature. Generally, chitosan is combined with other materials, such as poly(lactic-*co*-glycolic acid) (PLGA) and hyaluronan, to improve chondrocyte attachment, proliferation, and matrix synthesis (Jayakumar et al., 2010).

4.3.2 Synthetic Polymers

Synthetic polymers show predictable chemical and physical properties and can be customized easily to meet specific requirements, including degradation rates, mechanical properties, and biological activities. In addition, synthetic polymers are generally cheaper than natural polymers, can be produced in large quantities, and have longer shelf lives (Gunatillake and Adhikari, 2003; Kim and Mooney, 1998).

Short-chain saturated aliphatic polyesters, including poly(glycolic acid) (PGA), poly(lactic acid) (PLA), and PLGA copolymers, are commonly used for scaffold materials because of their biodegradability and the US Food and Drug Administration (FDA) approval for their clinical use. Their degradation products are monomeric glycolic acids or lactic acids and can be resorbed *in vivo*. However, these products alter the local pH, resulting in insufficient tissue ingrowth and triggering inflammatory responses. Total degradation takes place within 24 months and accompanies the loss of mechanical durability. These polymers can be produced in different shapes and forms, but for cartilage tissue engineering purposes, they are generally utilized as nonwoven meshes or feltlike forms (Gunatillake and Adhikari, 2003).

PGA is a hydrophilic polymer and enables the attachment of chondrocytes onto its surface. However, its acidic degradation products adversely affect cell proliferation, and polymers implanted into the defect site undergo rapid degradation, losing 50% of their mass within 2 weeks (Middleton and Tipton, 2000). For these reasons, applications of PGA are limited. PLA is similar to PGA, although its extra methyl group renders it more hydrophobic. PLA also degrades slower than PGA, allowing the deposition of a replacement ECM in the defect site before the total loss of its mechanical integrity. PLGA is a copolymer composed of PGA and PLA monomers. The overall material properties are dependent on the ratio of each polymer (Maher et al., 2011).

PCL is another polymer approved by the US FDA. It has a longer degradation time compared to PGA/PLA/PLGA; however, it also has poor wettability characteristics and cellular interaction potential (Middleton and Tipton, 2000).

4.3.3 Composite Materials

Single-phase homogeneous scaffolds cannot fully replicate the structure and function of cartilage or withstand the mechanical stresses at the defect site until the formation of new tissue. The disadvantages associated with natural and synthetic materials can be overcome by combining two or more polymers into a single material. The use of multiple polymers can improve the interaction of the scaffold material with native tissue and therefore enhance cartilage regeneration. Inert polymers in particular are coated or blended with biofunctional natural polymers in order to enhance cellular attachment, proliferation, and matrix synthesis. For example, it has been reported that filling the empty fraction of PLGA with chondrocytes in fibrin glue creates a material that exhibits homogeneous cell distribution and better infiltration capacity during neotissue formation (Ameer et al., 2002). In another study, the immobilization of hyaluronic acid on PLGA scaffolds was shown to enhance chondrocyte attachment and differentiation and also to prevent the dedifferentiation of chondrocytes (Yoo et al., 2005). The bioactivity of scaffolds can be improved by the incorporation of bioactive molecules, including growth factors, adhesion proteins, and short peptide sequences into their structure. The most important growth factors used in cartilage regeneration and tissue engineering approaches are the transforming growth factor β (TGF-β) family members, particularly TGF-β1 and TGF-β3. Various *in vitro* and *in vivo* studies proved that the use of TGF-β assists in the maintenance of the chondrocyte phenotype and stimulates the synthesis of a collagen II-rich ECM (Blunk and Sieminski, 2002; van der Kraan et al., 1992). Other growth factors used for the enhancement of cartilage regeneration are insulin-like growth factor I (IGF-I), basic fibroblast growth factor (bFGF) (FGF-2), bone morphogenetic proteins (BMPs), and hedgehog (hh) and wingless (Wnt) proteins.

In situ delivery of growth factors involves the covalent or noncovalent binding of the factors to the scaffold. Noncovalent binding is achieved by the physical entrapment of the growth factor within the scaffolding material. In one study, bFGF was immobilized onto the surface of PLGA after carbon dioxide plasma treatment, which provided the formation of ionic binding sites for positively charged FGF and greatly enhanced its binding to the polymer matrix (Shen et al., 2008). Covalent binding allows the prolonged release, slow degradation, and cellular internalization of the growth factor. Functional groups that are incorporated to polymers by physical/chemical binding or copolymerization can be used for the conjugation of growth factors onto the scaffold material (Cabanas-dan et al., 2014; Lee et al., 2011).

Many research efforts focus on replicating tissue function in terms of both the structural architecture and mechanical and biological functions. Self-assembling peptides are used extensively to provide short amino acid-based biological signals to cells. Unlike long chains, the use of short chains does not sterically hinder active domains and increases the stability of the structure (Hartgerink et al., 2001). In one study, researchers designed peptide amphiphile (PA) molecules that display a TGF-β-binding epitope to sequester endogenously released TGF-β. When implanted with MSCs, TGF-β-binding PA significantly enhanced the recovery of microfracture-treated cartilage defects without the addition of exogenous growth factors (Shah et al., 2010). As an alternative approach, the incorporation of cell-interacting sequences to the scaffold material is also used to elicit specific cellular responses. Methacrylated hyaluronic acid hydrogels have been functionalized with N-cadherin-mimetic peptides to induce chondrogenic differentiation of MSCs, and results showed that the conjugation of N-cadherin-mimetic peptides promoted chondrogenesis and cartilage-specific ECM production of MSCs (Bian et al., 2013).

4.3.4 Physical Stimuli

Conventional cell culture methods lack the biochemical and structural characteristics of the ECM. As such, it is hard to obtain highly organized, layered, and differentiated constructs. Nutrient diffusion and waste transport issues, especially the accumulation of anabolic products, result in fluctuations in the metabolic state of the medium and affect the homogeneity of scaffolds in static cultures (Dunkelman et al., 1995). However, mechanical stimulation can be used during the conditioning of cell-seeded constructs in *ex vivo* environments to improve the mechanical and functional properties of resulting constructs (Vunjak-Novakovic et al., 1999). Cartilage is a tissue that responds to mechanical stimuli, and its maintenance and development are dependent on strong mechanical forces. It has been shown that ECM synthesized by chondrocytes under dynamic conditions display better compositions and zonal organization (Vunjak-Novakovic et al., 1999). Bioreactors have been developed to apply mechanical loading regimens and create reproducible and homogeneous tissue constructs. Functional tissue engineering efforts have consequently focused on the modulation of dynamic compression, shear stress, and hydrostatic pressure through bioreactors, with the aim of producing cartilage-mimetic constructs under a well-defined set of mechanical conditions (Mabvuure et al., 2012). Compressive loading is the main mechanical stimulus in the native cartilage environment, especially in articulating joints. In joints, the surface of the cartilage on one side physically compresses the opposing cartilage in a continuous manner. The most commonly used bioreactors are compression bioreactors, which generally use platens to contact the construct surface. Dynamic compression, in which the loading is cyclical, has better outcomes than static compression in terms of ECM synthesis and chondrocyte proliferation rate. The combination of dynamic loading and growth factor application on agarose constructs has demonstrated that the two stimuli have a synergistic effect on the ECM synthesis (Mauck et al., 2003).

The other type of loading exerted on chondrocytes in their native environment is hydrostatic pressure. As mentioned previously, cartilage is a highly hydrated tissue, and mature cartilage is comprised primarily of water (70–80% by weight). During joint function, stress imparted on cartilage is distributed on cartilage surface homogeneously by water entrapped in the tissue; this effect is enhanced by the small effective pore size of cartilage tissue. As stress continues to be applied on the joint surface, water is expelled from the tissue and synovial fluid transmits the mechanical stress to water, thus decreasing friction and dissipating energy. During its normal function, cartilage is exposed to 3–10 MPa of stress at a frequency of around 1 Hz. Tissue engineering efforts using 0.1 and 15 MPa and 0.05 and 1 Hz pressures and frequencies have produced positive results for chondrocyte phenotype maintenance (Athanasiou et al., 2013). However, the application of constant hydrostatic pressure over long periods of time results in low cellular viability and matrix synthesis (Elder and Athanasiou, 2009). For that reason, in addition to pressure and frequency, the duration of hydrostatic pressure needs to be optimized for each construct or explant system.

4.4 CLINICAL THERAPEUTICS FOR CARTILAGE REGENERATION

Age-related problems spearhead the list of cartilage-related joint disorders. Although more common in the elderly, joint problems can also be seen in younger individuals due to traumatic accidents or obesity (Bosnakovski et al., 2006; Clouet et al., 2009). Cartilage injuries are hard to treat due to the low self-regeneration capacity of the tissue and the tendency of such injuries to deteriorate over time. Current treatments against OA incorporate both pharmacological and alternative treatments. Although severe cases cannot be treated with alternative methods alone, nonpharmacological approaches such as weight reduction, patient education and self-management, referral to a physical therapist, aerobics, muscle strengthening, walking aids, thermal modalities, transcutaneous electrical nerve stimulation, and acupuncture may assist in the recuperation process of arthritic patients (Hochberg et al., 2012). Pharmacological therapies, however, are common and indeed necessary for more advanced forms of OA. Two types of drugs—slow acting and fast acting—are prescribed for OA treatment (Table 4.1). In addition, several drug candidates are still under development. Novel analgesic and anti-inflammatory drugs, for example, aim to efficiently reduce joint pain while eliminating the adverse effects of currently prescribed drugs. Moreover, a class of molecules called disease-modifying osteoarthritis drugs (DMOADs) are currently under investigation for their ability to reduce tissue deformation. Finally, chondroprotective drugs such as growth factors and hormones are also under development for potential use in OA therapy (Clouet et al., 2009) (Table 4.2).

4.5 CONCLUSIONS AND FUTURE PERSPECTIVES

The unique structure and indispensable function of cartilage tissue render it difficult to develop new therapeutic approaches for osteochondral defects and diseases. Even though several clinical strategies are currently used in the treatment of such injuries, no method is so far capable of facilitating the long-term repair of cartilage tissue. At the cellular level, however, the application of stem cells onto defect sites appears to be promising, as MSCs have a high differentiation capacity and can secrete cartilage-specific ECM materials under the right conditions. However, difficulties in cell isolation, stem cell heterogeneity, and potential hypertrophy of the defect site and limited reproducibility in academic studies limit the use of stem cells in clinical applications. Tissue engineering strategies also hold great promise for the development of novel therapeutics and treatments, although few have made it into clinical use. Recent efforts have focused on improving current scaffold designs through the use of multipurpose materials that incorporate physical and biochemical cues to improve cell adhesion, survival, and maintenance. In addition, mechanical properties are as important as biochemical cues for the development of functional scaffolds

TABLE 4.1 Current Pharmacological Drugs Used in Clinics

	Drug	Function	Side Effects	References
Fast-Acting Drugs	Acetaminophen (paracetamol)	Analgesic	High doses of acetaminophen (up to 4 g/day) can trigger adverse hepatic events in patients with hepatic insufficiency	Hochberg et al. (2012)
	Nonsteroidal anti-inflammatory drugs (NSAIDs) Cyclooxygenase-2 (Cox-2) inhibitors	Analgesic	Gastrointestinal, renal, and cardiovascular toxicity	Ayral (2001), Gerwin et al. (2006), Glass (2006), Michael et al. (2010)
	Glucocorticoids (intra-articular injection)	Analgesic	Adverse metabolic effects	Michael et al. (2010)
	Opioids	Analgesic	Wide range of adverse effects, such as gastrointestinal problems (nausea, vomiting, and constipation), alteration in cognitive function, dependence, and respiratory depression	Glass (2006)

	Drug	Origin	Function	References
Slow-Acting Drugs	Glucosamine	Natural precursor of GAGs; stimulates GAG and collagen synthesis by chondrocytes	Reduces the rate of joint space narrowing	Derfoul et al. (2007), Pavelká et al. (2002)
	Chondroitin sulfate	Major component of cartilaginous ECM	Decreases the activity of catabolic enzymes in osteoarthritic cartilage and stimulates the synthesis of GAGs and collagens	McAlindon et al. (2000)
	S-Adenosyl methionine (SAM)	Small nonprotein metabolite—a coenzyme involved in methyl group transfers between enzymes	Increases the synthesis of GAGs in articular chondrocytes, suggesting that it may be able to aid in the repair of damaged cartilage through this mechanism. Oral administration of SAM induces a significant decrease in pain and improvements in joint function, comparable to that of NSAID	Najm et al. (2004)
	Hyaluronic acid (HA)	Polysaccharide ubiquitously found in ECMs	Used as viscosupplementation through intra-articular injections, it restores the viscoelastic and tribologic properties of the synovial fluid. Also shown to be chondroprotective, as it is able to stimulate the production of TIMPs in chondrocytes	Brockmeier and Shaffer (2006), Simon and Jackson (2006), Wang et al. (2004)

TABLE 4.2 Potential Future Therapeutics in Preclinical and Clinical Trials

	Drug Type	Function	References
Analgesic and Anti-Inflammatory Drugs	COX/LOX inhibitor	Anti-inflammatory activity with no detrimental effects on blood pressure and good gastrointestinal tolerability and safety	Clouet et al. (2009), Qvist et al. (2008), Steinmeyer and Konttinen (2006)
	Cyclooxygenase-inhibiting nitric oxide donators (CINODs)		
	NSAIDs		
	Transient receptor potential vanilloid subfamily 1 receptor agonist (TRPV1)		
	Serotonin–norepinephrine reuptake inhibitor		
	Bradykinin B2 receptor antagonist		
Disease-Modifying Osteoarthritis Drugs (DMOADs)	Matrix metalloproteinases (MMPs)	Slowing of OA-associated cartilage degeneration by targeting catabolic enzymes or cytokine-activated signaling cascades	Clouet et al. (2009), Pelletier and Martel-Pelletier (2007), Qvist et al. (2008)
	ADAMTS-5		
	Cathepsin K		
	IL-1-converting enzyme (ICE)		Chevalier et al. (2005)
	p38 MAP kinase	Synthetic inhibitors of various signaling pathways implicated in the physiopathology of OA, such as MAP kinase	Pelletier and Martel-Pelletier (2007), Qvist et al. (2008)
	NF-κB		
	Osteoprotegerin	Bone resorption	Tat et al. (2009)

Chondroprotective Drugs	TGF-β FGF (e.g., FGF-18) BMPs (BMP-2 and BMP-7)	Growth factors involved in cartilage repair	Blaney Davidson et al. (2007), Moore et al. (2005), Hayashi et al. (2008)
Novel Scaffolds	Collagen I/II-based hydrogels	Biodegradable three-dimensional matrices incorporating both biochemical and physical clues. Clinical use limited but still present, including cases such as	
	Polymer-based hydrogels	• Bone marrow MSCs incorporated in a collagen gel transplantation in an athlete with a grade IV cartilage defect	Kuroda et al. (2007)
	Peptide-based nanofiber hydrogels	• A poly(ethylene glycol) diacrylate (PEGDA) hydrogel used in a pilot clinical study alongside microfracture surgery in focal defects of 15 patients	Johnstone et al. (2013), Nguyen et al. (2011), Shah et al. (2010)

capable of facilitating the regeneration of cartilage, and nanomaterials can be designed to provide the physical cues necessary for the maintenance of chondrocytes. Native ECM-like and fiber-reinforced hydrogels are good material candidates for scaffold-incorporated approaches.

New design strategies also concentrate on the anatomical and structural complexity of cartilage tissue for advanced treatments. The heterogeneity of cartilage defects and diseases is another obstacle that is faced in the treatment of these injuries, and novel tissue engineering approaches can be optimized depending on whether the defect in question is osteochondral or solely chondral, leading to the creation of therapeutic platforms that are specific to a given type and severity of cartilage injury (potential future therapeutics in preclinical and clinical trials is summarized in Table 4.2). Overall, new strategies in the field of cartilage regeneration focus on the unique biochemical and physical properties of native cartilage to design novel tissue constructs that are decorated with cartilage-specific signals and display suitable anatomical geometries and mechanical properties for the treatment of large tissue defects.

REFERENCES

Ameer, G.A., Mahmood, T.A., and Langer, R. (2002). A biodegradable composite scaffold for cell transplantation. J. Orthop. Res. *20*, 16–19.

Arden, N. and Nevitt, M.C. (2006). Osteoarthritis: epidemiology. Best Pract. Res. Clin. Rheumatol. *20*, 3–25.

Athanasiou, K.A., Darling, E.M., Hu, J.C., DuRaine, G.D., and Reddi, A.H. (2013). Articular Cartilage (CRC Press, Boca Raton, FL).

Ayral, X. (2001). Injections in the treatment of osteoarthritis. Best Pract. Res. Clin. Rheumatol. *15*, 609–626.

Bartlett, W., Skinner, J.A, Gooding, C.R., Carrington, R.W.J., Flanagan, A.M., Briggs, T.W.R., and Bentley, G. (2005). Autologous chondrocyte implantation versus matrix-induced autologous chondrocyte implantation for osteochondral defects of the knee: a prospective, randomised study. J. Bone Joint Surg. Br. *87*, 640–645.

Bhosale, A.M. and Richardson, J.B. (2008). Articular cartilage: structure, injuries and review of management. Br. Med. Bull. *87*, 77–95.

Bian, L., Guvendiren, M., Mauck, R.L., and Burdick, J.A. (2013). Hydrogels that mimic developmentally relevant matrix and N-cadherin interactions enhance MSC chondrogenesis. Proc. Natl. Acad. Sci. U.S.A. *110*, 10117–10122.

Blaney Davidson, E., Van der Kraan, P., and van Den Berg, W. (2007). TGF-β and osteoarthritis. Osteoarthr. Cartil. *15*, 597–604.

Blunk, T. and Sieminski, A. (2002). Differential effects of growth factors on tissue-engineered cartilage. Tissue Eng.. *8*, 73–84.

Boscainos, P.J., Gross, A.E., and Kellett, C.F. (2008). Articular defects of the knee. 4, 1–13.

Bosnakovski, D., Mizuno, M., Kim, G., Takagi, S., Okumura, M., and Fujinaga, T. (2006). Chondrogenic differentiation of bovine bone marrow mesenchymal stem cells (MSCs) in different hydrogels: influence of collagen type II extracellular matrix on MSC chondrogenesis. Biotechnol. Bioeng. *93*, 1152–1163.

Brittberg, M., Lindahl, A., Nilsson, A., Ohlsson, C., Isaksson, O., and Peterson, L. (1994). Treatment of deep cartilage defects in the knee with autologous chondrocyte transplantation. N. Engl. J. Med. *331*, 889–895.

Brockmeier, S.F. and Shaffer, B.S. (2006). Viscosupplementation therapy for osteoarthritis. Sports Med. Arthrosc. Rev. *14*, 155–162.

Buckwalter, J.A. (1995). Activity vs. rest in the treatment of bone, soft tissue and joint injuries. Iowa Orthop. J. *15*, 29–42.

Buckwalter, J.A. and Mankin, H. (1997). Articular cartilage: degeneration and osteoarthritis, repair, regeneration, and transplantation. Instr. Course Lect. *47*, 487–504.

Cabanas-dan, J., Huskens, J., and Jonkheijm, P. (2014). Chemical strategies for the presentation and delivery of growth factors. J. Mater. Chem. B *2*, 2381–2394.

Chahine, N.O., Albro, M.B., Lima, E.G., Wei, V.I., Dubois, C.R., Hung, C.T., and Ateshian, G.A. (2009). Effect of dynamic loading on the transport of solutes into agarose hydrogels. Biophys. J. *97*, 968–975.

Chevalier, X., Giraudeau, B., Conrozier, T., Marliere, J., Kiefer, P., and Goupille, P. (2005). Safety study of intraarticular injection of interleukin 1 receptor antagonist in patients with painful knee osteoarthritis: a multicenter study. J. Rheumatol. *32*, 1317–1323.

Chung, C. and Burdick, J.A. (2008). Engineering cartilage tissue. Adv. Drug Deliv. Rev. *60*, 243–262.

Clouet, J., Vinatier, C., Merceron, C., Pot-vaucel, M., Maugars, Y., Weiss, P., Grimandi, G.L., and Guicheux, J.R.M. (2009). From osteoarthritis treatments to future regenerative therapies for cartilage. Drug Discov. Today *14*, 913–925.

Dehne, T., Karlsson, C., Ringe, J., Sittinger, M., and Lindahl, A. (2009). Chondrogenic differentiation potential of osteoarthritic chondrocytes and their possible use in matrix-associated autologous chondrocyte transplantation. Arthritis Res. Ther. *11*, R133.

Dejica, V.M., Mort, J.S., Laverty, S., Percival, M.D., Antoniou, J., Zukor, D.J., and Poole, A.R. (2008). Cleavage of type II collagen by cathepsin K in human osteoarthritic cartilage. Am. J. Pathol. *173*, 161–169.

Derfoul, A., Miyoshi, A.D., Freeman, D.E., and Tuan, R.S. (2007). Glucosamine promotes chondrogenic phenotype in both chondrocytes and mesenchymal stem cells and inhibits MMP-13 expression and matrix degradation. Osteoarthr. Cartil. *15*, 646–655.

Dunkelman, N.S., Zimber, M.P., Lebaron, R.G., Pavelec, R., Kwan, M., and Purchio, A.F. (1995). Cartilage production by rabbit articular chondrocytes on polyglycolic acid scaffolds in a closed bioreactor system. Biotechnol. Bioeng. *46*, 299–305.

Eggli, P.S., Herrmann, W., Hunziker, E.B., and Schenk, R.K. (1985). Matrix compartments in the growth plate of the proximal tibia of rats. Anat. Rec. *211*, 246–257.

Elder, B.D. and Athanasiou, K.A. (2009). Hydrostatic pressure in articular cartilage tissue engineering: from chondrocytes to tissue regeneration. Tissue Eng. Part B Rev. *15*, 43–53.

Engel, A., Larsson, T., and Skandiakoncernen, (1969). Aging of connective and skeletal tissue. Symposium October 1–3, 1968 (Stockholm, Nordiska bokhandelns).

Eyre, D. (2002). Collagen of articular cartilage. Arthritis Res. *4*, 30–35.

Fox, A.J.S., Bedi, A., and Rodeo, S.A. (2009). The basic science of articular cartilage: structure, composition, and function. Sports Health *1*, 461–468.

Gerwin, N., Hops, C., and Lucke, A. (2006). Intraarticular drug delivery in osteoarthritis. Adv. Drug Deliv. Rev. *58*, 226–242.

Gigante, A., Bevilacqua, C., Ricevuto, A., Mattioli-Belmonte, M., and Greco, F. (2007). Membrane-seeded autologous chondrocytes: cell viability and characterization at surgery. Knee Surg. Sports Traumatol. Arthrosc. *15*, 88–92.

Glass, G.G. (2006). Osteoarthritis. Dis. Mon. *52*, 343–362.

Gobbi, A., Karnatzikos, G., and Kumar, A. (2014). Long-term results after microfracture treatment for full-thickness knee chondral lesions in athletes. Knee Surg. Sports Traumatol. Arthrosc. 22, 1986–1996.

Görtz, S. and Bugbee, W.D. (2007). Allografts in articular cartilage repair. Instr. Course Lect. *56*, 469–480.

Grigolo, B., Lisignoli, G., Piacentini, A., Fiorini, M., Gobbi, P., Mazzotti, G., Duca, M., Pavesio, A., and Facchini, A. (2002). Evidence for redifferentiation of human chondrocytes grown on a hyaluronan-based biomaterial (HYAFF s 11): molecular, immunohistochemical and ultrastructural analysis. Biomaterials *23*, 1187–1195.

Grogan, S.P., Ph, D., Sovani, S., Pauli, C., Chen, J., Hartmann, A., Colwell, C.W., Lotz, M.K., and Lima, D.D.D. (2012). Effects of perfusion and dynamic loading on human neocartilage formation in alginate hydrogels. Tissue Eng. Part A *18*, 1784–1792.

Gunatillake, P.A. and Adhikari, R. (2003). Biodegradable synthetic polymers for tissue engineering. Eur. Cell. Mater. *5*, 1–16.

Han, L., Grodzinsky, A., and Ortiz, C. (2011). Nanomechanics of the cartilage extracellular matrix. Annu. Rev. Mater. Res. *41*, 133–168.

Hartgerink, J.D., Beniash, E., and Stupp, S.I. (2001). Self-assembly and mineralization of peptide-amphiphile nanofibers. Science *294*, 1684–1688.

Hayashi, M., Muneta, T., Ju, Y.-J., Mochizuki, T., and Sekiya, I. (2008). Weekly intra-articular injections of bone morphogenetic protein-7 inhibits osteoarthritis progression. Arthritis Res. Ther. *10*, R118.

Henrionnet, C., Wang, Y., Roeder, E., Gambier, N., Galois, L., Mainard, D., Bensoussan, D., Gillet, P., and Pinzano, A. (2012). Effect of dynamic loading on MSCs chondrogenic differentiation in 3-D alginate culture. Biomed. Mater. Eng. *22*, 209–218.

Hochberg, M. C., Altman, R. D., April, K. T., Benkhalti, M., Guyatt, G., McGowan, J., and Tugwell, P. (2012). American College of Rheumatology 2012 recommendations for the use of nonpharmacologic and pharmacologic therapies in osteoarthritis of the hand, hip, and knee. Arthritis Care Res. *64*, 465–474.

Hunziker, E.B. (2002). Articular cartilage repair: basic science and clinical progress. A review of the current status and prospects. Osteoarthr. Cartil. *10*, 432–463.

Iwasa, J., Engebretsen, L., Shima, Y., and Ochi, M. (2008). Clinical application of scaffolds for cartilage tissue engineering. Knee Surg. Sports Traumatol. Arthrosc. *17*, 561–577.

Jayakumar, R., Prabaharan, M., Nair, S.V., and Tamura, H. (2010). Novel chitin and chitosan nanofibers in biomedical applications. Biotechnol. Adv. *28*, 142–150.

Jeon, O., Bouhadir, K.H., Mansour, J.M., and Alsberg, E. (2009). Photocrosslinked alginate hydrogels with tunable biodegradation rates and mechanical properties. Biomaterials *30*, 2724–2734.

Jin, R., Moreira Teixeira, L.S., Dijkstra, P.J., van Blitterswijk, C.A., Karperien, M., and Feijen, J. (2010). Enzymatically-crosslinked injectable hydrogels based on biomimetic dextran-hyaluronic acid conjugates for cartilage tissue engineering. Biomaterials *31*, 3103–3113.

Johnstone, B., Alini, M., Cucchiarini, M., Dodge, G.R., Eglin, D., Guilak, F., Madry, H., Mata, A., Mauck, R.L., and Semino, C.E. (2013). Tissue engineering for articular cartilage repair—the state of the art. Eur. Cell. Mater. *25*, e67.

Kim, B.-S. and Mooney, D.J. (1998). Development of biocompatible synthetic extracellular matrices for tissue engineering. Trends Biotechnol. *16*, 224–230.

Kim, S.H., Park, D.Y., and Min, B.-H. (2012). A new era of cartilage repair using cell therapy and tissue engineering: turning current clinical limitations into new ideas. Tissue Eng. Regen. Med. *9*, 240–248.

Kon, E., Filardo, G., Condello, V., Collarile, M., Di Martino, A., Zorzi, C., and Marcacci, M. (2011). Second-generation autologous chondrocyte implantation: results in patients older than 40 years. Am. J. Sports Med. *39*, 1668–1675.

Kon, E., Filardo, G., Matteo, B., Di Perdisa, F., and Marcacci, M. (2013). Matrix assisted autologous chondrocyte transplantation for cartilage treatment a systematic review. Bone Joint Res. *2*, 18–25.

Kuo, C.K. and Ma, P.X. (2001). Ionically crosslinked alginate hydrogels as scaffolds for tissue engineering: part 1. Structure, gelation rate and mechanical properties. Biomaterials *22*, 511–521.

Kuroda, R., Ishida, K., Matsumoto, T., Akisue, T., Fujioka, H., Mizuno, K., Ohgushi, H., Wakitani, S., and Kurosaka, M. (2007). Treatment of a full-thickness articular cartilage defect in the femoral condyle of an athlete with autologous bone-marrow stromal cells. Osteoarthritis Cartilage *15*, 226–231.

Lee, K., Silva, E.A., and Mooney, D.J. (2011). Growth factor delivery-based tissue engineering: general approaches and a review of recent developments. J. R. Soc. Interface *8*, 153–170.

Lewis, P.B., McCarty, L.P., Kang, R.W., and Cole, B.J. (2006). Basic science and treatment options for articular cartilage injuries. J. Orthop. Sports Phys. Ther. *36*, 717–727.

Lin, Y., Luo, E., Chen, X., Liu, L., Qiao, J., Yan, Z., Li, Z., Tang, W., Zheng, X., and Tian, W. (2005). Molecular and cellular characterization during chondrogenic differentiation of adipose tissue-derived stromal cells in vitro and cartilage formation in vivo. J. Cell. Mol. Med. *9*, 929–939.

Lu, Y., Parker, K.H., and Wang, W. (2006). Effects of osmotic pressure in the extracellular matrix on tissue deformation. Philos. Trans. R. Soc., A *364*, 1407–1422.

Ma, Z., Ph, D., Kotaki, M., and Inai, R. (2005). Potential of nanofiber matrix as tissue-engineering scaffolds. Tissue Eng. *11*, 101–109.

Mabvuure, N., Hindocha, S., and Khan, W.S. (2012). The role of bioreactors in cartilage tissue engineering. Curr. Stem Cell Res. Ther. *7*, 287–292.

Maher, S.A., Ph, D., and Lowman, A.M. (2011). Hydrogels for the repair of articular cartilage defects. Tissue Eng. Part B Rev. *17*, 281–299.

Mahmoudifar, N. and Doran, P.M. (2012). Chondrogenesis and cartilage tissue engineering: the longer road to technology development. Trends Biotechnol. *30*, 166–176.

Makris, E.A., Gomoll, A.H., Malizos, K.N., Hu, J.C., and Athanasiou, K.A. (2015). Repair and tissue engineering techniques for articular cartilage. Nat Rev Rheumatol *11*, 21–34.

Maroudas, A. (1968). Physicochemical properties of cartilage in the light of ion exchange theory. Biophys. J. *8*, 575–595.

Mauck, R.L., Nicoll, S.B., Seyhan, S.L., Ateshian, G.A., and Hung, C.T. (2003). Synergistic action of growth factors and dynamic loading for articular cartilage tissue engineering. Tissue Eng. *9*, 597–611.

Mauck, R.L., Soltz, M.A., Wang, C.C.B., Wong, D.D., Chao, P.G., and Ateshian, G.A. (2015). Functional tissue engineering of articular cartilage through dynamic loading of chondrocyte-seeded agarose gels. J. Biomech. Eng. *122*(3), 252–260.

McAlindon, T.E., LaValley, M.P., Gulin, J.P., and Felson, D.T. (2000). Glucosamine and chondroitin for treatment of osteoarthritis: a systematic quality assessment and meta-analysis. JAMA *283*, 1469–1475.

Memon, A.R. and Quinlan, J.F. (2012). Surgical treatment of articular cartilage defects in the knee: are we winning? Adv. Orthop. *2012*, 528423.

Michael, J.W.-P., Schlüter-Brust, K.U., and Eysel, P. (2010). The epidemiology, etiology, diagnosis, and treatment of osteoarthritis of the knee. Dtsch. Arztebl. Int. *107*, 152.

Middleton, J.C. and Tipton, A.J. (2000). Synthetic biodegradable polymers as orthopedic devices. Biomaterials *21*, 2335–2346.

Moore, E., Bendele, A., Thompson, D., Littau, A., Waggie, K., Reardon, B., and Ellsworth, J. (2005). Fibroblast growth factor-18 stimulates chondrogenesis and cartilage repair in a rat model of injury-induced osteoarthritis. Osteoarthr. Cartil. *13*, 623–631.

Mow, V.C. and Guo, X.E. (2002). Mechano-electrochemical properties of articular cartilage: their inhomogeneities and anisotropies. Annu. Rev. Biomed. Eng. *4*, 175–209.

Mow, V.C., Kuei, S.C., Lai, W.M., and Armstrong, C.G. (1980). Biphasic creep and stress relaxation of articular cartilage in compression: theory and experiments. J. Biomech. Eng. *102*, 73–84.

Mow, V.C., Ratcliffe, A., and Robin Poole, A. (1992). Cartilage and diarthrodial joints as paradigms for hierarchical materials and structures. Biomaterials *13*, 67–97.

Najm, W., Reinsch, S., Hoehler, F., Tobis, J., and Harvey, P. (2004). S-Adenosyl methionine (SAMe) versus celecoxib for the treatment of osteoarthritis symptoms: a double-blind cross-over trial. [ISRCTN36233495]. BMC Musculoskelet. Disord. *5*, 6.

Nguyen, L.H., Kudva, A.K., Guckert, N.L., Linse, K.D., and Roy, K. (2011). Unique biomaterial compositions direct bone marrow stem cells into specific chondrocytic phenotypes corresponding to the various zones of articular cartilage. Biomaterials *32*, 1327–1338.

Parenteau-Bareil, R., Gauvin, R., and Berthod, F. (2010). Collagen-based biomaterials for tissue engineering applications. Materials (Basel). *3*, 1863–1887.

Pavelká, K., Gatterová, J., Olejarová, M., Machacek, S., Giacovelli, G., and Rovati, L.C. (2002). Glucosamine sulfate use and delay of progression of knee osteoarthritis: a 3-year, randomized, placebo-controlled, double-blind study. Arch. Intern. Med. *162*, 2113–2123.

Pelletier, J. and Martel-Pelletier, J. (2007). DMOAD developments: present and future. Bull. Hosp. Jt. Dis. *65*, 242.

Qvist, P., Bay-Jensen, A.-C., Christiansen, C., Dam, E.B., Pastoureau, P., and Karsdal, M.A. (2008). The disease modifying osteoarthritis drug (DMOAD): is it in the horizon? Pharmacol. Res. *58*, 1–7.

Schulze-Tanzil, G. (2009). Activation and dedifferentiation of chondrocytes: implications in cartilage injury and repair. Ann. Anat. *191*, 325–338.

Shah, R.N., Shah, N.A., Lim, M.M.D.R., Hsieh, C., Nuber, G., and Stupp, S.I. (2010). Supramolecular design of self-assembling nanofibers for cartilage regeneration. Proc. Natl. Acad. Sci. U. S. A. *107*, 3293–3298.

Shannon, F., Devitt, A., Poynton, A., Fitzpatrick, P., and Walsh, M. (2001). Short-term benefit of arthroscopic washout in degenerative arthritis of the knee. Int. Orthop. *25*, 242–245.

Shen, H., Hu, X., Bei, J., and Wang, S. (2008). The immobilization of basic fibroblast growth factor on plasma-treated poly(lactide-co-glycolide). Biomaterials *29*, 2388–2399.

Simon, T.M. and Jackson, D.W. (2006). Articular cartilage: injury pathways and treatment options. Sports Med. Arthrosc. Rev. *14*, 146–154.

Slomianka, L. (2009). Blue Histology—Skeletal Tissues—Cartilage. http://www.lab.anhb.uwa.edu.au/mb140/corepages/bone/bone.htm (accessed on October 28, 2015).

Steinmeyer, J.R. and Konttinen, Y.T. (2006). Oral treatment options for degenerative joint disease-presence and future. Adv. Drug Deliv. Rev. *58*, 168–211.

Stroebel, S. (2007). Engineering of cartilage tissue constructs in a 3-dimensional perfusion bioreactor culture system under controlled oxygen tension. In Naturwissenschaftlichen (Basel, Universität Basel), pp. 134.

Taipaleenmaki, H. (2010). Factors regulating chondrogenic differentiation. In Medical Biochemistry and Genetics (Turku, Finland University of Turku).

Tat, S.K., Pelletier, J.-P., Velasco, C.R., Padrines, M., and Martel-Pelletier, J. (2009). New perspective in osteoarthritis: the OPG and RANKL system as a potential therapeutic target? Keio J. Med. *58*, 29–40.

Tuan, R.S. (2007). A second-generation autologous chondrocyte implantation approach to the treatment of focal articular cartilage defects. Arthritis Res. Ther. *9*, 109.

Van der Kraan, P., Vitters, E., and van den Berg, W. (1992). Differential effect of transforming growth factor beta on freshly isolated and cultured articular chondrocytes. J. Rheumatol. *19*, 140–145.

Vavken, P. and Samartzis, D. (2010). Effectiveness of autologous chondrocyte implantation in cartilage repair of the knee: a systematic review of controlled trials. Osteoarthr. Cartil. *18*, 857–863.

Vunjak-Novakovic, G., Martin, I., Obradovic, B., Treppo, S., Grodzinsky, A.J., Langer, R., and Freed, L.E. (1999). Bioreactor cultivation conditions modulate the composition and mechanical properties of tissue-engineered cartilage. J. Orthop. Res. *17*, 130–138.

Wang, C.-T., Lin, J., Chang, C.-J., Lin, Y.-T., and Hou, S.-M. (2004). Therapeutic effects of hyaluronic acid on osteoarthritis of the knee. A meta-analysis of randomized controlled trials. J. Bone Joint Surg. Am. *86*, 538–545.

Woessner, J.F. (1991). Matrix metalloproteinases and their inhibitors in connective tissue remodeling. FASEB J. *5*, 2145–2154.

Yoo, H.S., Lee, E.A., Yoon, J.J., and Park, T.G. (2005). Hyaluronic acid modified biodegradable scaffolds for cartilage tissue engineering. Biomaterials *26*, 1925–1933.

Zhao, W., Jin, X., Cong, Y., Liu, Y., and Fu, J. (2013). Degradable natural polymer hydrogels for articular cartilage tissue engineering. J. Chem. Technol. Biotechnol. *88*, 327–339.

5

WOUND HEALING APPLICATIONS OF NANOMATERIALS

BERNA SENTURK, GOZDE UZUNALLI, RASHAD MAMMADOV, MUSTAFA O. GULER, AND AYSE B. TEKINAY

Institute of Materials Science and Nanotechnology, National Nanotechnology Research Center (UNAM), Bilkent University, Ankara, Turkey

5.1 INTRODUCTION

The prevalence of diabetes and obesity, as well as the increase in the percent of the aging population, has resulted in a sharp rise in the costs of wound care, creating a major and so far unmet need for effective treatments against wounds requiring long-term medical attention. Chronic skin ulcers caused by diabetes mellitus, pressure, or venous stasis affect about 6.5 million individuals worldwide (Jones et al., 2002), while 15% of diabetic patients suffer from foot ulcers (DFUs)—a leading cause of limb amputations (Brem and Tomic-Canic, 2007). In addition to chronic and traumatic wounds, burn injuries also represent a leading cause of death: The World Health Organization (WHO) reported a total of 11 million severe burn cases in 2004, and 265,000 deaths per year are caused by burns (WHO, April, 2014).

Although nanomaterials are promising candidates for assisting the wound regeneration process, they nonetheless are a recent area of research, and considerable effort is required to develop nanomaterial-based treatments

Therapeutic Nanomaterials, First Edition. Edited by Mustafa O. Guler and Ayse B. Tekinay.

to the point that they surpass the current medical standard of care. The clinical challenges in using nanomaterials for wound healing include safety issues related to the materials, the requirement for new materials to induce vascularization more rapidly than their conventional counterparts, and issues related to the adequate promotion of cell attachment to the wound bed. It is hoped that the development of a new generation of biomaterials will overcome the clinical challenges and provide a more systematic and effective approach for wound management.

A better insight into wound injuries and the mechanism of their recovery may allow the design of more effective wound dressings, for example, by combining the hormones and growth factors required for cell migration and proliferation with scaffold materials that imitate the structure of the healthy extracellular matrix (ECM). However, a deeper understanding of the wound physiology is necessary to achieve this goal or to appreciate the current level of progress in artificial wound regeneration. As such, this chapter first provides an overview of skin structure and its biological response to the tissue loss in wound healing process and subsequently reviews the classical and next-generation scaffolds and materials that are used to minimize scar formation while accelerating the healing process. We also provide a snapshot of the natural and synthetic biomaterials currently in clinical use in this field.

5.1.1 The Structure of Healthy Mammalian Skin

By far the largest organ of the mammalian body, skin, encompasses the whole surface of the organism as a physical barrier against external factors such as infection, dehydration, and UV radiation. In addition to physical protection, it is also responsible for thermoregulation, sensory detection, and endocrine function (e.g., vitamin D synthesis, conversion of prohormones into hormones). The skin is assisted in its function by a variety of specialized structures, such as hair follicles, sweat glands, and sebaceous glands, collectively called skin appendages. As a result of its underlying epidermal stem cell population, it also displays an exceptional capacity for self-renewal under normal physiological conditions (Menon, 2002; Yildirimer et al., 2012).

The skin consists of two distinct layers: an outermost covering called epidermis and an inner region called dermis. The demarcation of skin layers is an issue of contention, and three layers (epidermis, dermis, and hypodermis) are sometimes proposed; however, the present chapter considers the skin to consist of the two aforementioned layers.

The epidermis is a thin (from 0.04 to 1.6 mm) and avascular layer that consists mostly of specialized epithelial cells, such as keratinocytes, and nonepithelial cells, such as melanocytes, Langerhans cells, and Merkel cells. The keratinocytes of the stratified epidermis are wholly undifferentiated at the bottom of the epidermis, where they form a basal layer. These cells gradually move upward as they differentiate, eventually reaching the uppermost layers as squamous cells (stratum spinosum), granular cells (stratum granulosum), and cornified cells (stratum corneum). The epidermis is anchored to the dermis through hemidesmosomic connections to an interlayeric basement membrane. This membrane, which contains laminins, type IV collagen, perlecan, heparan sulfate proteoglycans, fibronectin, and nidogens, provides the interchange of nutrients between the two layers (Markeson et al., 2015; Menon, 2002). ECM components, such as collagen, elastin, proteoglycans, and glycosaminoglycans (GAG), and cellular structures, such as vessels, nerves, adipose cells, muscle fibers, glands, and hair follicles, form the inner layer of the skin, called the dermis. This region of connective tissue ranges from 1 to 4 mm in thickness and is composed of three major types of cells: fibroblasts, macrophages, and adipocytes (Markeson et al., 2015).

5.1.2 The Mechanisms of Wound Healing

Wound healing occurs in nearly all tissue types, and the order of events during the healing process is similar across different tissues. The regeneration of wounds is an interactive, complex, and dynamic process and starts immediately in response to tissue damage, which triggers the activation of a series of inter- and intracellular signaling pathways that ultimately culminate in the recovery of the anatomic continuity of the tissue. Following the initial step of blood clot formation, inflammatory pathways are activated for the secretion of specific soluble factors, and progenitor and stem cells undergo major metabolic changes to produce the cell populations required for the regeneration process (Gurtner et al., 2008). The tissue repair process is rapid, and the injured tissue is initially replaced with a poorly organized ECM that lacks skin appendages and is dissimilar to the tightly wound basket-weave structure of unwounded tissue. This "stopgap" approach results in the formation of a scar, and the injured tissue is not fully restored to its previous functionality following the regeneration process (Bayat et al., 2003).

Unlike the wound repair mechanisms of adult mammals, lower vertebrates (e.g., salamanders) are capable of total regeneration in organs such as limbs and eyes (Brockes et al., 2001). It is also known that mammalian embryos are able to undergo scarless healing (Ferguson et al., 1996). A suppressed or immature immune system appears to be a key necessity for regeneration to occur without scar tissue formation, though postnatal defects of liver and bone tissues have been reported to partially heal without scar formation (Gurtner et al., 2008).

The formation of scar tissue, although suboptimal in the long run, is evolutionarily advantageous: The ability to rapidly exclude opportunistic pathogens and partially restore tissue functionality outweighs the potential long-term disadvantages of scars. Nonetheless, the disadvantages inherent to the incomplete regeneration of wounded tissues are reflected in health-care costs, especially in patients with metabolic disorders like diabetes or obesity, as well as those suffering from old age. Due to the limited human capacity for recovery, even otherwise healthy individuals may face life-long disabilities if they receive severe injuries at a young age. However, a better understanding of the natural healing process and its limitations may allow the restoration of the original tissue structure and function in major wound or burn injuries, improving the life quality of patients and decreasing the economic burden by preventing the formation of scars. The process of wound repair has three distinct but spatially overlapping stages: inflammation, proliferation, and maturation.

5.1.2.1 Inflammation Inflammation is a crucial step that protects the organism against opportunistic pathogens and prepares the region of injury for the rest of the repair process. It is characterized by discomfort, loss of movement, edema, erythema, and heat. This phase proceeds for about 3–5 days in acute wounds, while it is extended in chronic wounds (Gardner et al., 2001).

The healing process begins immediately following injury and is initiated by the activation of the coagulation cascade, which prevents the loss of blood from the injury site. A platelet plug is formed at the wound bed, creating a provisional scaffold for migrating cells and serving as a physical barrier to protect the wound from the external environment. The formation of the clot is achieved through the aggregation of fibrin with fibronectin, vitronectin, and thrombospondin-1 (TSP-1) (Martin, 1997). The activated platelets, which trigger the release of growth factors such as epidermal growth factor (EGF), platelet-derived growth factor (PDGF) (Lynch et al., 1987), and transforming growth factor β (TGF-β) 1 and 2 (O'Kane and Ferguson, 1997), are also necessary in this step. Platelet cells also attach to the fibronectin matrix and release ECM proteins to further assist in the production of the blood clot.

Growth factors released by activated platelets, especially PDGF and TGF-β, are responsible for triggering the subsequent stage of inflammation. Neutrophils also arrive at the wound site in response to platelet degranulation, bacterial degradation products, and fibrin proteolysis. These cells are responsible for sterilizing the wound by killing opportunistic bacteria and secrete the cytokines that activate the resident populations of fibroblasts and keratinocytes (Gillitzer and Goebeler, 2001). Monocytes are also recruited to the wound site and are activated as macrophages. In the absence of excessive infection, neutrophils are phagocytosed by macrophages in response to the interaction between specific ECM proteins and integrin receptors. Another task of macrophages is the expression of important growth factors such as colony-stimulating factor-1 (which is responsible for the survival of monocytes and macrophages), PDGF, tumor necrosis factor-α (TNF-α), insulin-like growth factor (IGF), interleukin-1 (IL-1), and some TGF family members. The lack of macrophage infiltration appears to result in impaired wound healing, while the treatment of scars with a TGF-β1 inhibitor was shown to reduce inflammation and improve the dermal architecture of the scar tissue (Shah et al., 1995).

In addition to neutrophils and macrophages, mast cells and lymphocytes are also crucial for the wound healing process. The production of interleukin-4 by mast cells triggers the proliferation of fibroblasts, which in turn initiate the proliferation phase. Lymphocytes function both as immune effector cells and as growth factor sources; their secretions are associated with the remodeling process.

5.1.2.2 Proliferation The formation of new tissue is the second stage of wound healing, involving the deposition of a temporary ECM and the priming of the vascular network for the formation of new blood vessels. This step allows the isolation of the wound from external factors, further preparing the tissue environment and its constituent cells for the subsequent ECM remodeling process.

5.1.2.2.1 Reepithelialization Wound injuries compromise the integrity of the epidermal cover, and the injured dermis is therefore susceptible to damage from the external environment. Although a temporary barrier is provided by the blood clot, a more permanent protection is necessary for the later steps of wound repair to occur. Reepithelialization involves the formation of this protective layer and occurs through the migration, proliferation, and differentiation of local keratinocytes and epithelial stem cells from hair follicles or sweat glands (Lau et al., 2009; Martin, 1997; Miller et al., 1998). Two processes are required for this step: the dissolution of cell–cell and cell–basement membrane connections (Singer and Clark, 1999) and the activation of signaling pathways needed for the release of various growth factors and cytokines (Werner and Grose, 2003).

The integrity of uninjured epidermis is maintained by the hemidesmosomal links between keratinocytes and the basement membrane, as well as the desmosomal bridges that connect one keratinocyte to another. Wound injuries disrupt this structure and alter cell membrane permeability to ions by activating membrane-associated kinases, resulting in cell polarization through reorganization of intercellular filaments (Martin, 1997). However, keratinocytes in healthy tissue are still anchored to each other and the surrounding matrix, and their migration requires enzymatic cleavage of desmosomes and hemidesmosomes by collagenase and elastase. In addition to enzymatic cleavage, it is also known that the activation and relocalization of protein kinase Cα are important for the dissociation of desmosomal connections (Wallis et al., 2000).

Following migration, keratinocytes establish new epithelial layers at the wound edge by forming focal adhesions with the local integrin network. This process of dissociation and reanchoring is called as "keratinocyte shuffling" and allows keratinocytes to spread over the fibronectin-rich provisional matrix with the help of a gradient of cytokines and growth factors, such as IL-1, EGF, and TGF-α (Clark et al., 1982; Jiang et al., 1993). Keratinocyte shuffling is considered complete when each keratinocyte reforms its connections with its neighbors and reestablishes the intercellular hemidesmosome network (Jacinto et al., 2001).

The entire wound healing process is orchestrated by a complex cocktail of growth factors. EGF is one of the most important growth factors for the reepithelialization process, inducing the autophosphorylation of its receptor (EGF receptor (EGFR)) to promote cell proliferation (Martin, 1997) and migration (Rheinwald and Green, 1977). In addition, it has been demonstrated that a direct relationship exists between EGFR expression and epidermal thickness in wounded tissues (Stoscheck et al., 1992). Another important group is the fibroblast growth factor (FGF) family. It is known that FGF-2 is important for wound reepithelialization and keratinocyte motility (Bhora et al., 1995), while two other members of this family, FGF-7 (Marchese et al., 1995) and FGF-10 (Beer et al., 1997), are called keratinocyte growth factors and reported to be upregulated in skin wounds.

5.1.2.2.2 Angiogenesis Wound healing necessitates the coordinated activity of a large number of specialized cells, which must be supplied with oxygen and nutrients through a complex network of blood vessels. As such, angiogenesis in wound healing is a dynamic and multistep process, and several characteristics distinguish it from other forms of blood vessel generation. First, wound angiogenesis is unique in that it is activated by the immune response, although it is also regulated through the activity of endothelial cells, angiogenic growth factors, and ECM molecules. New blood vessels can be formed both by the sprouting of preexisting capillaries, which is called angiogenesis, and by the de novo production of bone marrow-derived endothelial precursors (EPCs), which is called vasculogenesis (Carmeliet, 2003; Eming et al., 2007a; Phelps and Garcia, 2009). Both occur during wound healing, and the processes are supported by one another.

The most important angiogenic factors are vascular endothelial growth factor (VEGF) and basic fibroblast growth factor (bFGF or FGF-2). Although a variety of cells can produce these factors, macrophages in particular are responsible for secreting FGF-2 and VEGF in response to the hypoxic environment of the wound bed (Greenhalgh, 1998). It is also known that damaged endothelial cells secrete FGF-2, while VEGF is released by keratinocytes (Martin, 1997). New vascular formation requires the degradation of the old ECM in order to create space for new tubules, and endothelial cells also perform this function by secreting a plasminogen activator to combine plasmin and procollagenase into active collagenase. The degradation of the basement membrane allows production of new vessels at the wound bed, and angiogenic growth factor secretion ceases when enough oxygen is supplied to the wound area (Bishop, 2008).

5.1.2.2.3 Granulation Tissue Formation The formation of granulation tissue, or new stroma, is a transition step between the proliferation and remodeling phases. This tissue invades into the fibrin/fibronectin-based provisional matrix about 4 days after injury. Its granular appearance is a result of the large numbers of capillaries, macrophages, fibroblasts, and collagen bundles present within its matrix (Singer and Clark, 1999).

This tissue is formed as a result of macrophage activity. In particular, macrophage-secreted growth factors, such as platelet-derived growth factor 4 (PDGF-4) (Heldin and Westermark, 1999) and TGF-β1 (Beck et al., 1991), stimulate the resident fibroblast population into secreting a new ECM that contains collagen type III, fibronectin, GAG, proteoglycans, and hyaluronic acid (Eckes et al., 2010). Over time, some fibroblasts differentiate into α-smooth muscle actin (α-SMA)-expressing myofibroblast cells (Darby et al., 1990). These cells form the contractile apparatus of the wound site and are characterized by large actin bundles, as well as by α-SMA. The contractile ability of these cells introduces a mechanotransduction element into the wound ECM. In addition, myofibroblasts can use the transmembrane connections between their intracellular actin and extracellular fibronectin domains (a system collectively called the fibronexus) to transmit their contractions across the tissue ECM (Gabbiani, 2003). Through their collective contractions, myofibroblasts may therefore physically move the wounded tissue area. This, in conjunction with the reepithelialization process, allows the closure of the wound.

5.1.2.3 Tissue Remodeling Remodeling or resolution is the last phase of the healing process, starting at day 21 and lasting for a year or more after injury. It is the phase of regression in which the capillary- and cell-rich granulation zone is replaced with a more mature wound tissue. Myofibroblasts, endothelial cells, and macrophages undergo apoptosis during this stage, and the new tissue only contains a collagen-rich ECM and a sparse population of cells (Greenhalgh, 1998). In addition, although a large number of capillaries are required to supply the metabolic demands of rapidly proliferating, migrating, and differentiating cells during the tissue repair process, these blood vessels are no longer required by this point. As such, the wound vascular density decreases to normal during this phase.

Another major activity at this stage is the production of a new collagen matrix. Collagen type III, which consists of small parallel bundles, is the dominant component of the ECM during the early stages of wound repair. As the wound matures, collagen III is replaced by a stronger type of collagen, collagen I, which forms thin fibrils and is the collagen type normally found in healthy skin (Lovvorn et al., 1999). This replacement is called as collagen deposition and under the control of matrix metalloproteinases (MMPs), which are secreted by endothelial cells, fibroblasts, and macrophages (Levenson et al., 1965). In tandem with the controlled degradation and redeposition of ECM proteins, the maintenance of tissue homeostasis in the skin is provided through cytokine-regulated mesenchymal–epidermal interactions (Szabowski et al., 2000).

Although the wound is considered healed at the end of the remodeling phase, damaged tissue nonetheless does not immediately regain its former strength. About 30% of the skin's breaking strength is lost following wounding and repair, which is attributed to the presence of a stronger collagen network in healthy tissue, laden with intermolecular crosslinks (Levenson et al., 1965).

5.1.3 Repair Process in Chronic Wounds

Factors such as type I diabetes, neuropathies, stress, and aging can interfere the natural healing process, resulting in the formation of long-term injuries called *chronic wounds*. These injuries are generally classified under three groups: diabetic ulcers, arterial and venous ulcers, and pressure ulcers. In all three types, the wound healing process is initiated through the formation of a blood clot, as is typical for the repair of normal skin injuries. However, the inflammation period is prolonged, and the subsequent repair of the damaged tissue area is delayed in chronic wounds (Eming et al., 2007b).

In addition, the production/degradation balance seen in natural wound healing is heavily disrupted in favor of degradation in chronic wounds, and the wound area is saturated with an elevated concentration of proteases that hinder the activity of ECM components, growth factors, and/or their receptors (Mast and Schultz, 1996).

Venous ulcers generally occur due to the incompetence of venous valves, which results in the pooling of blood within the vein and increases the pressure it experiences. This increased pressure allows blood proteins, especially fibrin, to leak into the extravascular space, preventing regular blood flow within the vein and creating a sink for growth factors required for wound healing (Walker, 1999). These effects, in turn, decrease collagen production within the wound area (Pardes et al., 1995). Arterial ulcers, in contrast, typically occur as result of narrowed or ischemic arteries. As the arterial blood flow is blocked, the wound area cannot be supplied with an adequate amount of oxygen, nutrients, and growth factors, which delays the wound healing process (Bonham, 2003).

Diabetes is another major cause of chronic wounds, as the neural and vascular pathologies experienced by diabetic patients may often impair the wound healing process. Hyperglycemia triggers the formation of advanced glycation end products, which have been reported to facilitate the formation of abnormal granulation tissue (Goldin et al., 2006). In addition, it is known that the metabolism of nitric oxide, a major factor in angiogenesis, is altered in diabetic patients (Linden et al., 2008). MMPs, a group of proteins involved in ECM degradation, also show increased activity in diabetic patients: Although the remodeling of the wound area is necessary for the repair process, these enzymes must be regulated through tissue inhibitors of metalloproteinases (TIMPs), and the balance between these two factors is compromised in diabetic conditions (Lobmann et al., 2002).

Pressure ulcers, or bedsores, occur when shearing force and/or pressure is applied to the soft tissues over long periods of time, especially in patients with limited mobility. Over time, shear force and pressure trigger reperfusion injury, ischemia, and, eventually, tissue necrosis (Defloor, 1999).

5.2 APPLICATIONS OF NANOMATERIALS FOR THE ENHANCEMENT OF WOUND HEALING PROCESS

The development of biodegradable and nontoxic materials for the enhancement of the wound healing process is an active area of research, and advances in nanotechnology have contributed greatly to the design efficiency of such materials. Nanobiomaterials allow the design of wound dressings that not only create a suitable biomolecular environment for the regeneration process to occur but also protect against infections at the wound site and facilitate the controlled release of bioactive materials such as growth factors.

Natural and synthetic polymers, such as polysaccharides (e.g., alginate, chitin, chitosan, heparin, chondroitin), proteoglycans, and proteins (e.g., collagen, gelatin, fibrin, keratin, silk fibroin, eggshell membrane) are commonly used in wound management, as those nanomaterials display low toxicities and can be naturally degraded over time in the body. The large surface-to-volume ratios inherent to nanoscale materials is another major advantage and allows a small volume of material to contain a large amount of therapeutic cargo. Advances in the design of soft materials, such as hydrogels and bioscaffolds, have revolutionized the field of wound management research, as these materials can be engineered to provide both the biophysical environment and biochemical signals necessary for the wound regeneration process.

Wound dressings protect the damaged tissue against opportunistic pathogens and assist the recovery of dermal and epidermal tissues. The choice of dressing material depends on the cause and type of the wound, with both natural and artificial dressings being available.

5.2.1 Artificial Skin

An "artificial skin" that can adequately replace damaged skin tissue is the ideal type of wound dressing, and considerable effort has been spent on developing such a material. Natural skin transplantation is the first choice for clinicians and surgeons to replace damaged skin in severe burn injuries; though only large and deep wounds warrant this procedure (skin grafting is generally not necessary for first- and second-degree burns, which heal with little to no scarring). On the other hand, the wound bed in the third-degree burns must be covered as quickly as possible with artificial or natural grafts (Centanni et al., 2011). Donor areas for natural skin transplantation include the chest, thighs, buttock, abdomen, or behind the ear. As the donor region must contain healthy skin, this procedure is not recommended for elderly patients with pressure sores and people with diabetic or other chronic ulcers. The treatment of these patients therefore necessitates the development of tissue engineering techniques for the artificial replacement of the skin.

Artificial tissues generally incorporate three main elements: a cell type, a differentiation-inducing substance, and a material matrix. Skin substitutes might be biological (e.g., xenografts, allografts, autografts, and amniotic substitutes) or synthetic (Halim et al., 2010). Xenografts are skin substitutes harvested from animals and serve as a temporary insulating layer during early stages of wound healing in humans. Allotransplantation, which refers to the transplant of substitute skin from the same species, allows a potential means of rapid intervention in burn wound management. Cadaveric skin allografts are one of the most common biological substitutes worldwide. There are two main strategies to preserve cadaveric skin allografts: cryopreservation and glycerol preservation. Glycerol has antibacterial and antiviral effects and is more cost efficient for the long-term storage and long-distance transport compared to cryopreserved skin. Amniotic substitutes derived from placenta of selected and screened donors are rich in collagen and various growth factors that support the healing process, thus improving wound closure rates and reducing scar formation (Halim et al., 2010).

Although these naturally derived skin substitutes are widely used in clinics, they cannot facilitate the complete regeneration of the skin due to limited donor sites, risk of infection, slow healing rates, and, in the case of autografts, the requirement to create new wounds to acquire the required skin tissue. As such, there is a substantial need for tissue-engineered skin constructs that support the complete regeneration of wound injuries. The development of artificial skin tissue may also enhance the scope of regenerative medicine, and techniques used in artificial skin production may be expanded for the generation of more complex structures, such as artificial organs. Additionally, advances in the design of therapeutic agents will no doubt also lead to a greater understanding of the pathophysiology of the wound healing process.

5.2.2 Natural Nanomaterials for Wound Healing

Natural materials are required to exhibit certain characteristics to be considered suitable for use as wound dressings. Foremost important among these criteria is the ability to serve as an adequate matrix scaffold for the cell types involved in the wound healing process (including stem cells, macrophages, fibroblasts, and/or epithelial cells). Also important is the ability to incorporate vital molecular signaling elements such as growth factors and signaling molecules. These functions allow the nanomaterial dressing to reduce inflammation, scar formation, and infection; and they may also play a role in various other processes. A large number of natural materials, such as collagen, gelatin, laminin, and chitin/chitosan, have been used as electrospun scaffolds for tissue engineering.

5.2.2.1 Collagen Collagen is the major protein of the ECM and acts as a structural scaffold in various tissues (Lodish et al., 2000). Types I and III are the main types of collagen found in skin tissue. Collagen subunits pack together to form long, thin fibrils, which typically assume a triple helical structure. Triple helices of type I collagen consist of two $\alpha 1$(I) chains and one $\alpha 2$(I) chain, while type III collagen is composed of three $\alpha 1$ chains. Collagen fibers in the skin are mainly synthesized by fibroblasts and myofibroblasts. Collagen deposition occurs at the wound site following the third day postinjury and involves the replacement of type III collagen with type I. The replacement process gradually increases the stiffness and tensile strength of the regenerating tissue, as the tensile strength of type I collagen is greater than that of type III collagen.

It is known that collagen not only serves a structural support but also contributes to matrix deposition and cellular differentiation and migration during the wound healing process. A number of collagen dressings have been reported in the literature, often incorporating additional materials such as alginate (to control material efflux from the scaffold) and silver nanoparticles (AgNPs) (to provide antimicrobial effects). Some of these designs are commercially available and frequently used in surgical or clinical procedures (Fitzgerald and Steinberg, 2009).

Apligraf (Graftskin) (Organogenesis/Novartis, Canton, MA) is a wound dressing product that contains allogeneic neonatal foreskin fibroblasts and keratinocytes in bovine collagen gel; it is used in the treatment of chronic foot and venous leg ulcers. *Cellular-allogeneic* OrCel (Ortec International, New York, NY) contains both fibroblasts and keratinocytes from neonatal foreskin, cultured in a type I collagen sponge. It is used for treatment of split-thickness donor sites in patients with burn and surgical wounds in epidermolysis bullosa (Supp and Boyce, 2005). Cultured skin substitutes, composed of collagen–GAG substrates containing autologous fibroblasts and keratinocytes, facilitate permanent wound closure in burn injuries and congenital nevus and chronic wounds (Boyce et al., 1993).

The porous structure, high permeability, hydrophilicity, stability, and cell-supporting nature of collagen make this material a popular choice for wound dressings. In addition, the development of new material combinations of collagens and engineered skin substitutes may further enhance the utility and versatility of wound dressings in clinical use.

5.2.2.2 Laminin Laminin is a widely expressed protein that contributes to the formation of the ECM, particularly the basement membrane. Laminins comprise a family of at least 15 large trimeric basement membrane proteins, each of which is composed of one α chain, one β chain, and one γ chain and may display a distinct, tissue-specific biological role (Bosman and Stamenkovic, 2003). Laminin is critical for the cell–material interactions that occur during the wound healing process, and displays the ability to promote cell proliferation and attachment. Defects on laminin expression have been correlated with delayed or impaired wound closure. Junctional epidermolysis bullosa (JEB) is an inherited disease affecting laminin and collagen expression and characterized by fragility of the skin and mucous membranes. It has been shown that laminin 5 is highly expressed in migrating keratinocytes in the wound bed (Larjava et al., 2000). Due to its promotion of cell proliferation and attachment, laminin-derived peptides are also used as wound dressing (Hashimoto et al., 2004; Malinda et al., 2008). In the peptide-based approach, topical application of laminin-derived peptides was shown to increase reepithelialization and granulation tissue formation in the early stages of wound healing in rats (Malinda et al., 2008). Similarly application of peptide conjugated scaffolds to the wound area, for example, the PPFLMLLKGSTR peptide derived from LG3 domain of α3 laminin, increased rates of skin reepithelialization (Min et al., 2010).

5.2.2.3 Chitin and Chitosan Chitin and chitosan are natural polysaccharides that display suitable biological and physicochemical characteristics for use as wound dressings. Their main advantages are oxygen permeability, biodegradability, and biocompatibility of these materials, as well as easily processability into hydrogels, fibers, membranes, scaffolds, and sponges. Chitin is an inexpensive material obtained from the exoskeleton of invertebrates. Chitosan is composed of randomly distributed β-(1-4)-linked D-glucosamine and *N*-acetyl-D-glucosamine units. After depolymerization, *N*-acetyl-D-glucosamine units induce fibroblast proliferation, critical for collagen deposition, and stimulate natural hyaluronic acid synthesis, which leads to faster wound healing and scar prevention.

Chitin–nanosilver composite scaffolds were found to possess excellent antibacterial activity against bacterial infection during the wound healing process (Jayakumar et al., 2010). Chitin and chitosan can also be used as slow-release drug delivery vehicles for growth factors to further accelerate the wound healing process (Dai et al., 2011).

5.2.2.4 Alginate-Based Nanomaterials Alginates are unbranched polysaccharides extracted from brown algae. Alginate-based wound dressings display hemostatic properties and are commonly used in bleeding wounds and burns. Materials such as calcium alginate, calcium sodium alginate, collagen/alginate, and gelatin/alginate are highly absorbent natural fiber dressings and can be obtained at minimal cost from processed algae. The high absorption capability of alginate facilitates the creation of a moist environment that promotes the formation of granulation tissue during the wound healing process (Mogoşanu and Grumezescu, 2014). The high water absorption capacity of alginate also helps to reduce blood loss. Chemical modifications with calcium and zinc can also be performed to improve the efficiency of alginate over conventional cotton wound dressings.

5.2.3 Synthetic Nanomaterials for Wound Healing

Synthetic materials allow greater control over properties such as porosity, controlled release capacity, biodegradability, and structural versatility, in addition to avoiding issues inherent to natural materials such as batch-to-batch variance and viral, bacterial, or other types of biological contamination. As such, they have been commonly used in the design of novel regenerative scaffolds that aim to provide optimum conditions for regeneration with high reproducibility. Increased control over material structure and composition is a major advantage for synthetic skin substitutes, the ultimate goal of which is to produce an ideal skin substitute that provides an effective and scar-free wound healing.

5.2.3.1 Carbon Nanotubes Carbon nanotubes (CNTs) are currently under consideration for clinical applications regarding wound healing, although there are serious concerns over the toxicity of these materials. CNT-based antiseptic bandages were produced by combining povidone-iodine with single-walled carbon nanotubes (SWNTs) (Simmons and Linhardt, 2012), and it has been shown that the presence of the multiwalled carbon nanotubes (MWCNTs) positively affects the wound healing process by decreasing the expression of genes that interfere with the function of the cytoskeleton, such as actin stress fibers (Zhang et al., 2011).

5.2.3.2 Hydrogels Hydrogels promote granulation, epithelialization, and autolytic debridement at the wound site by providing a moist environment that is conductive to the regeneration process. Commercially available hydrogel dressings are generally found in the form of membranes and sheets. Studies involving *in situ* formation of hydrogel wound dressings based on gelatin, oxidized alginate, and borax suggest that the application of hydrogels enhance the wound healing process in rat models (Balakrishnan et al., 2005).

Full-thickness wounds dressed with a combination of chemically cross-linked GAG hydrogels and the commercial wound dressing Tegaderm™ displayed a significant increase in reepithelialization at days 5 and 7 compared to wounds treated with Tegaderm alone, suggesting that modified hydrogels can improve the efficiency of treatments based on conventional wound healing agents (Kirker et al., 2002). Hydrogels can also be improved through incorporation of drugs or growth factors.

5.2.3.3 Antimicrobial AgNPs The antimicrobial properties of silver have been known and utilized for centuries, and the material has lately been used in wound dressings for the same purpose. AgNPs with diameters below 100 nm display increased antimicrobial activity due to their larger combined surface areas. Several methods exist for the incorporation of AgNPs into wound dressings. The antibacterial properties of silver are well known; however, the safety of using AgNPs and their fate within the body or the environment should be addressed.

5.2.4 Wound Dressings Containing Growth Factors

Growth factors are soluble secreted proteins capable of affecting a variety of cellular processes involved in tissue regeneration. However, their clinical applications are limited due to the absence of robust delivery systems and biomaterial carriers in addition to the risks associated with carcinogenicity of these materials. Nonetheless, growth factors have been found to speed the healing of acute and chronic wounds in humans, and some are approved for clinical use by the FDA. The PDGF analog REGRANEX™, for example, is currently used for the treatment of leg and foot ulcers in diabetic patients (Chan et al., 2006). Endothelial growth factor (EGF) is another major factor in wound regeneration and has been reported to play a role in keratinocyte migration, fibroblast proliferation and differentiation, and granulation tissue formation (Yildirimer et al., 2012). In addition to its rapid breakdown within the wound environment, EGF has a mitogenic effect upon cells (Yildirimer et al., 2012).

Among VEGF family composed of VEGF-A, VEGF-B, VEGF-C, VEGF-D, and placenta growth factor, in particular VEGF-A is highly expressed in acute wounds and promotes angiogenesis. The main initiator of expression of VEGF is hypoxic environment. It has been showed that VEGF-A improves reepithelialization of diabetic wounds via enhancement of vessel formation (Galiano et al., 2004).

bFGF is another main growth factor responsible for the induction of angiogenesis, which is the most critical step during wound repair. It has been shown that the treatment of rabbit ears with exogenous bFGF accelerates wound healing compared to nontreated control; however, no increase was observed in capillary density (Komori et al., 2005).

Although topical application of growth factors is used for the treatment of chronic wounds, these materials are released in a noncontrolled manner if not encapsulated within a controlled delivery matrix. Slow release is required for the effective delivery of these polypeptides to the wound site, and matrix-encapsulated growth factors will better enhance the wound regeneration process.

5.2.5 Biomimetic Materials

Recent advances in material sciences and polymer chemistry have resulted in the development of biomimetic peptide fibers, an entire new class of nanomaterials that can be used to assist in wound regeneration. The ECM is an enormously complex network of peptides, proteoglycans, growth factors, and soluble proteins, and its components play critical roles at each and every step of the wound healing process. The dysfunction of any of these components may result in delayed wound healing, as is observed in the chronic wounds of hyperglycemic patients. Development of biomimetic scaffolds that restore or replace the roles of ECM components may therefore allow the treatment of slow-healing dermal injuries that are currently difficult to treat using conventional approaches.

A bioengineered fibronectin matrix-mimetic substrate was shown to induce cell migration and stimulate high rates of cell proliferation (Roy et al., 2011). Similarly, collagen-mimetic peptides that assume a nanofibrous triple-helix formation were effectively used to enhance the thrombosis process (Kumar et al., 2014). Mesenchymal stem cell (MSC)-seeded biomimetic collagen hydrogels significantly enhanced angiogenesis compared to MSC injection alone (Rustad et al., 2012).

The main fabrication techniques to engineer such materials mimicking the nanofibrous ECM proteins are electrospinning, self-assembly, and phase separation. Development of those techniques allows the design and fabrication of biomimetic scaffolds that offer tremendous potential applications in wound healing of the skin. The development of novel biomimetic scaffolds capable of promoting and directing new vascular network growth would eliminate a critical problem in tissue engineering, serving as an efficient solution to the problems associated with chronic wound healing. Future research is expected to yield more information on the unique properties of biomimetic scaffolds.

5.2.6 Current Challenges in the Design of Nanomaterials for Chronic Wound Management

The WHO reports that 347 million people suffer from diabetes worldwide and expects the disease to be the seventh leading cause of death in 2030. The incidence of diabetes has been on the rise and may soon become one of the foremost public health challenges that the world will face in the decades ahead. Factors such as age, obesity, malnutrition, and vascular disease may contribute to wound infection and delayed wound healing in diabetic patients. The metabolic effects of diabetes, combined with reduced blood flow and neuropathies (nerve damage) associated with the disease, increases the chance of foot ulcers and the severity of wound infections, potentially leading to limb amputation and even death. The design of smart scaffolds as wound dressing is therefore necessary to increase the quality of life in patients suffering from diabetic ulcers and similar chronic wounds.

Due to the variations between the pathophysiologies of wound and burn injuries, it is difficult to develop an artificial dressing material that can facilitate optimal healing of all types of skin damage. The initial assessment of chronic wounds is therefore essential before deciding on wound dressing choice, which should be made by taking a full account of the wound and dressing characteristics. The choice of dressing should be based on wound size, depth, severity of infection, and location. Patients with severe infection should use an appropriate antibiotic therapy; additionally, the wound can be treated with silver and iodine products (Moura et al., 2013, Table 5.1).

5.2.6.1 Nanomaterials Used in the Treatment of Diabetic Wounds

Increase in the number of patients with delayed diabetic wound healing has made developing appropriate wound dressings for diabetic wounds an urgent necessity. Decreased reepithelialization and growth factor expression, a dysfunctional ECM, and impaired angiogenesis are the main major problems that interfere with the recovery of diabetic wounds.

Studies involving the local application of growth factors such as human epidermal growth factor (hEGF) have shown that treatment with high concentrations of hEGF is effective in promoting wound healing in diabetic foot injuries (Tsang et al., 2003). It has been shown that recombinant hEGF immobilized in a copolymer of poly(ε-caprolactone) (PCL) and poly(ethylene glycol) (PEG) may increase the proliferation and phenotypic expression of keratinocytes (Choi et al., 2008). Although materials such as collagen and hydrogels have been used to facilitate the healing of diabetic wounds, dressings that can encapsulate growth factors or other materials that induce angiogenesis are more favorable for this purpose.

TABLE 5.1 Commercially Available Wound Dressings and Their Recommended Regions of Application

Product Name	Type	Usage	Reference
AQUACEL®	Hydrofiber	Chronic wounds	Cohn et al. (2004)
AQUACEL Ag	Hydrofiber containing silver	Leg ulcers, superficial pressure ulcers, partial-thickness burns	Barnea et al. (2010)
REGRANEX® (FDA approved)	Gel containing platelet-derived growth factor (PDGF)	Chronic and traumatic wound	Chan et al. (2006)
OASIS®	Derived from porcine small intestinal submucosa (SIS)	Partial- and full-thickness wounds, pressure ulcers, venous ulcers, chronic vascular ulcers, diabetic ulcers, trauma wound	Mostow et al. (2005)
SANTYL® (FDA approved)	Ointment, biological enzymatic debriding agent	Chronic dermal ulcers, severe burns	Hansbrough et al. (1995)
IODOFLEX®	Cadexomer containing silver	Chronic wounds	Bianchi (2001)

The encapsulation of growth factors in a polymer matrix not only protects the growth factor against enzymatic degradation within the body but also allows its release into a localized area at a controllable rate. As such, current strategies for the treatment of diabetic wounds mainly involve administration of materials that provide controlled release of growth factors. Since decreased angiogenesis contributes to impaired tissue repair, VEGF is also commonly incorporated into scaffolds intended for use in the treatment of diabetic wounds (Galiano et al., 2004). In addition to delivery of growth factors, engineered biomimetic materials such as heparan sulfate have shown an enhanced wound healing prowess compared to unloaded scaffolds (Garcia-Filipe et al., 2007).

5.2.6.2 Nanomaterials Used for the Treatment of Peripheral Arterial Disease Peripheral arterial disease (PAD) is an atherosclerotic disease that is commonly seen in the elderly population. Diabetic PAD patients experience a gradual reduction in the delivery of oxygen- and nutrient-rich blood to the legs, which results in severe infections, osteomyelitis, and a higher risk of lower leg amputations (Lipsky et al., 2012; Reiber et al., 1999). PAD patients who also present with foot ulcerations tend to display other characteristics, such as neuropathic ulcers (Prompers et al., 2008). These ulcers are commonly located at the pressure points on the foot, mostly at the tips of the toes.

The lack of suitable natural materials and limitations of synthetic materials for treatment of PAD patients necessitate discovery of new routes for effective clinical therapy. Combinations of proper dressings with endothelial progenitor cells (EPC) are one important and exciting advance in the field of tissue engineering.

5.3 PEPTIDE NANOFIBER GELS FOR WOUND HEALING

Numerous materials for wound healing were mentioned in previous sections. Salient features of these materials are designed in an attempt to provide a desired niche for robust wound healing process. Fibrous structures of these materials serve as a scaffold for incoming cells. Growth factors or other bioactive molecules can be encapsulated or conjugated into materials to provide immigration of cells with necessary biological signals. Cell adhesion epitopes can be included in the material to facilitate cell attachment at damaged site. Mechanical properties of the materials should be cognate to the tissue they will function in. Besides these materials should be biodegradable and biocompatible. However, to gather all these properties in desired quality in one material is a formidable task. Obstacles here include synthesis of the fibrous materials with fibers of nanoscale size, conjugation of the bioactive elements onto them, and controlling their stiffness. On the other hand, nanofibrous materials formed by self-assembly of peptide amphiphile (PA) molecules have necessary features in remarkable quality. Nanofibers with a few nanometers in diameter and micron length can be obtained easily by this way. Their peptide part allows versatility in peptide epitopes and also provides easy way of addition of other chemical functional groups. Their structures are also comprised of biodegradable and biocompatible elements.

5.3.1 Relevance of Nanofibrous Structure of Peptide Gels for Wound Healing

5.3.1.1 Bioactive Epitopes ECM proteins such as laminin and fibronectin induce signaling pathways in cells through their specific epitopes. Induction of these signaling pathways is very critical for various cellular processes such as proliferation, differentiation, as well as organization of the cells in tissues. ECM proteins are evolved to carry various domains charged with different functions, which allows them to operate in different contexts. Our body is adapted to perform in complex environments with scarce resources and has high levels of sophistication, which allows it to respond appropriately to an enormous variety of environmental signals. Since resources to build its structure are scarce, economy principle imposed itself on the design of our body. Accordingly resources to build our intricate body are distributed so that it can respond to millions of adverse conditions. However, when it comes to triggering specific signaling pathway involved in certain repair mechanism, we only need the appropriate epitopes of specific proteins. Scale of response provided by our body in natural environment may not be optimal in each case since distribution of resources occurs according to the distribution of signals in natural environment where mammalian body has evolved. Materials that present these epitopes with artificially high density unlike to the natural ECM proteins would be more valuable for providing the optimal response to certain disease conditions. For example, the peptide nanofibers carrying IKVAV epitopes showed better capability of inducing neuronal differentiation than laminin proteins from which IKVAV epitopes derived (Silva et al., 2004). Here, PA design allowed to achieve presentation of IKVAV epitopes on them with enormously high density. PA nanofibers are also amenable to change concentration of epitopes for optimal arrangement of signals on nanofibers for cells (Shah et al., 2010).

5.3.1.2 Topographical Cues Topography of the extracellular environment influences cell shape and behavior extensively. Size and type of patterns on substrates affect cell proliferation, migration, and differentiation (Kim et al., 2012). Basement membrane of epithelial tissue has a complex 3D texture in the nanometer range (Kim et al., 2012). Nanofibrous architecture also has been shown to facilitate cell attachment through enhanced protein absorption (Woo et al., 2003). PA form 3D nanofiber network (Fig. 5.1), which resembles the architecture of natural ECM that contact with epithelial and dermal cells. Pore size of network also can be controlled via changing concentration of PA molecules, which provide suitable topographical cues for wound healing applications.

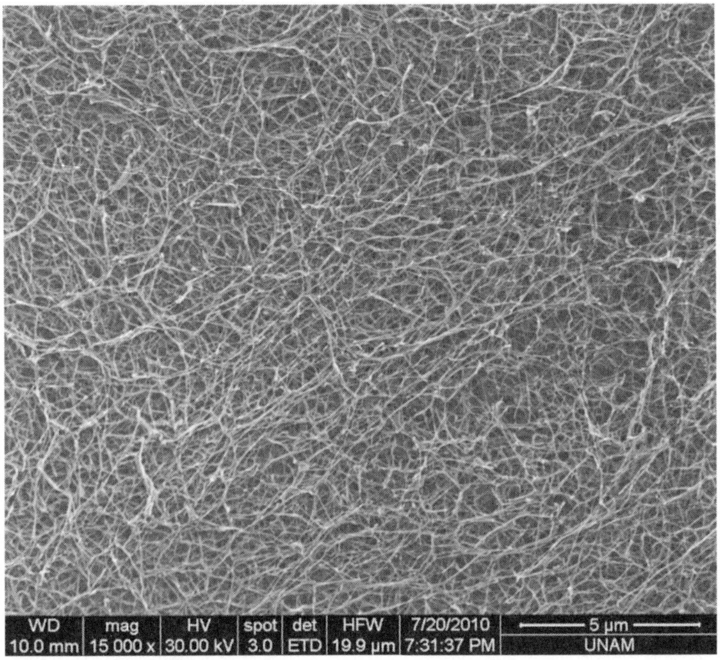

FIGURE 5.1 Three-dimensional nanofibrous network formed by peptide amphiphile nanofibers.

5.3.2 Engineered PA Nanofiber Gels for Wound Healing and Insights into Various Designs

5.3.2.1 Growth Factor Binding An enormously complex growth factor network takes role in wound healing mechanism (Werner and Grose, 2003). These growth factors regulate physiological processes such as angiogenesis (e.g., VEGF and FGF-2) and reepithelialization (e.g., EGF, HGF, and TGF-β), which are critical for repair process (Gurtner et al., 2008). Conventional approaches in tissue engineering focus on encapsulation of growth factors in scaffold systems to provide signals for cells. Burst release of growth factors from these materials became a big challenge with these designs since it causes aberrant regeneration besides inefficient expenditure of expensive growth factors. In order to overcome this, growth factors were conjugated to materials either noncovalently or covalently. Interaction of growth factors with the scaffold materials slows the release of growth factors that protect them from degradation and provides a signaling niche for cells inside the material. In order to conjugate growth factors to materials, the latter can be decorated with various binding sites. Indeed in natural ECM, heparan sulfates perform this function through providing binding sites for various heparin-binding growth factors. Heparin inclusion in scaffold systems improved its interaction with growth factors and bioactivity such as angiogenic performance (Rajangam et al., 2006). Functional groups from heparin improved growth factor binding capability of otherwise poorly performing material (Freeman et al., 2008). Other ECM proteins also have been proposed to have such binding sites (Martino and Hubbell, 2010). Some growth factors have domains allowing them to bind to various ECM proteins (Martino et al., 2014). Other growth factors can be engineered to carry that same domain to make them affine to ECM proteins (Martino et al., 2014).

PA nanofibers are versatile systems in terms of conjugation of binding sites for growth factors. Recently, we have designed a PA system carrying a specific arrangement of functional groups (—COOH, —OH, and $-SO_3$) on nanofibers (Mammadov et al., 2011). This nanofiber system showed better growth factor binding, angiogenesis, and neurogenesis than control PA nanofibers (Mammadov et al., 2011, 2012a, 2012b). They showed significant capability to bind growth factors taking role in wound healing such as HGF, VEGF, and FGF-2 (Mammadov et al., 2012b). Hence, this nanofiber system might accelerate wound healing mechanisms such as reepithelialization and angiogenesis. Besides facilitating the performance of the exogenously delivered growth factors, these materials might also bind to endogenously secreted growth factors and minimize requirement for the growth factors in wound healing. In addition to chemical functional groups, peptide sequences could be developed for facilitating binding of other growth factors to the nanofiber system. By using phage display system, Shah et al. discovered the peptide sequence for the optimal TGF-β binding to peptide nanofibers (Shah et al., 2010). The peptide nanofibers carrying this sequence increased performance of the TGF-β and eventually enhanced *in vivo* cartilage regeneration.

Besides designing the PA to bind to growth factors, they can also be engineered to carry bioactive domains of the growth factors. Receptor-binding regions of the growth factors are generally short peptide sequences, which can be attached into the PA. In this way, the cost of using whole growth factors can be eliminated, and the unwanted effects of other parts of growth factors can be reduced. Also, aggregation of the receptor-binding domains on nanofibers might have synergistic effect on induction of signaling pathways. For example, VEGF-mimetic peptide incorporated onto PAs and presented on nanofibers induced *in vivo* angiogenesis stronger than same peptide delivered in soluble form (Webber et al., 2011). Receptor-binding domains of other important growth factors in wound healing can be engineered onto PA to induce various signaling pathways.

5.3.2.2 Epitopes for Cell Adhesion One of the salient features of the wound healing mechanism is infiltration of damaged site by immense amounts of cells. The cell adhesion peptide epitopes derived from natural ECM elements (e.g., RGD or YIGSR) have been incorporated into materials in different studies to facilitate cell adhesion in damaged area. RGD incorporated into an interpenetrating network of gelatin increased cellularity and ECM organization when compared to RGD lacking material (Waldeck et al., 2007). These short peptide sequences can be easily conjugated into the PA nanofibers to facilitate cell attachment. Besides signaling purpose, this will also increase availability of the cells in material and also in damaged area, which will increase probability of their interaction with the growth factors tethered into the material. Materials with such multiple components would have synergistic effects on wound healing. Besides using peptide sequences such as RGD, which have target receptors in many cell types, specific peptide sequences can also be used to make materials bind to specific cell types (e.g., REDV sequence binds only to endothelial cells). We can change epitope density on nanofibers and obtain optimal designs for cell adhesion through dilution of adhesion epitope carrying PA molecule among control PA.

5.3.2.3 Adjustment of Mechanical Properties Mechanosensation of cells and tissues recently has gained increased interest after several studies showing the effect of substrate stiffness on cellular behavior. MSC have been shown to differentiate into neurons, muscle, and bone according to the stiffness of underlying substrate (Engler et al., 2006). The skin is also a mechanosensitive tissue (Evans et al., 2013). Keratinocytes have been shown to proliferate and migrate better on stiff substrates while differentiate better on soft substrates (Wang et al., 2012). These studies suggest that stiffness of the materials designed with purpose of wound healing will influence this very process. Better control on material stiffness will provide flexibility in the design of materials according to purpose. Various ways have been suggested to control the stiffness of PA nanofiber gels. Storage modulus increases with concentration of PA molecules (Dagdas et al., 2011). Ca^{2+}-mediated gelation produced gels with higher stiffness values than pH-mediated gelation, probably owing to interfiber crosslinking with Ca^{2+} ion (Dagdas et al., 2011). Increase in Ca^{2+} concentration caused increase in gel stiffness until PA neutralization has been reached (Stendahl et al., 2006). Also, PA gel moduli can be changed by type and valency of metal ions used to induce gel formation (Stendahl et al., 2006). PA molecules functionalized with 3,4-dihydroxy-L-phenylalanine (DOPA) residues showed increase in storage moduli in the order of magnitude when compared to ordinary PA gel (Ceylan et al., 2013). Fe^{3+}-mediated crosslinking of DOPA molecules allowed reversible gel formation and also gained self-healing property to gels. Besides the aforementioned ways, PA molecules can also be combined with polymer molecules to give them stronger mechanical properties. To conclude, PA nanofiber materials allow control of stiffness for wound healing purposes.

REFERENCES

Balakrishnan, B., Mohanty, M., Umashankar, P.R., and Jayakrishnan, A. (2005). Evaluation of an in situ forming hydrogel wound dressing based on oxidized alginate and gelatin. Biomaterials *26*, 6335–6342.

Barnea, Y., Weiss, J., and Gur, E. (2010). A review of the applications of the hydrofiber dressing with silver (Aquacel Ag) in wound care. Ther Clin Risk Manag *6*, 21–27.

Bayat, A., McGrouther, D.A., and Ferguson, M.W.J. (2003). Skin scarring. Br Med J *326*, 88–92.

Beck, L.S., Deguzman, L., Lee, W.P., Xu, Y., McFatridge, L.A., and Amento, E.P. (1991). TGF-beta 1 accelerates wound healing: reversal of steroid-impaired healing in rats and rabbits. Growth Factors *5*, 295–304.

Beer, H.D., Florence, C., Dammeier, J., McGuire, L., Werner, S., and Duan, D.R. (1997). Mouse fibroblast growth factor 10: cDNA cloning, protein characterization, and regulation of mRNA expression. Oncogene *15*, 2211–2218.

Bhora, F.Y., Dunkin, B.J., Batzri, S., Aly, H.M., Bass, B.L., Sidawy, A.N., and Harmon, J.W. (1995). Effect of growth factors on cell proliferation and epithelialization in human skin. J Surg Res *59*, 236–244.

Bianchi, J. (2001). Iodoflex and Iodosorb in the treatment of venous leg ulcers. Br J Nurs *10*, 342–346.

Bishop, A. (2008). Role of oxygen in wound healing. J Wound Care *17*, 399–402.

Bonham, P.A. (2003). Assessment and management of patients with venous, arterial, and diabetic/neuropathic lower extremity wounds. AACN Clin Issues *14*, 442–456; quiz 548–450.

Bosman, F.T. and Stamenkovic, I. (2003). Functional structure and composition of the extracellular matrix. J Pathol *200*, 423–428.

Boyce, S.T., Greenhalgh, D.G., Kagan, R.J., Housinger, T., Sorrell, J.M., Childress, C.P., Rieman, M., and Warden, G.D. (1993). Skin anatomy and antigen expression after burn wound closure with composite grafts of cultured skin cells and biopolymers. Plast Reconstr Surg *91*, 632–641.

Brem, H. and Tomic-Canic, M. (2007). Cellular and molecular basis of wound healing in diabetes. J Clin Invest *117*, 1219–1222.

Brockes, J.P., Kumar, A., and Velloso, C.P. (2001). Regeneration as an evolutionary variable. J Anat *199*, 3–11.

Carmeliet, P. (2003). Angiogenesis in health and disease. Nat Med *9*, 653–660.

Centanni, J.M., Straseski, J.A., Wicks, A., Hank, J.A., Rasmussen, C.A., Lokuta, M.A., Schurr, M.J., Foster, K.N., Faucher, L.D., Caruso, D.M., et al. (2011). StrataGraft skin substitute is well-tolerated and is not acutely immunogenic in patients with traumatic wounds: results from a prospective, randomized, controlled dose escalation trial. Ann Surg *253*, 672–683.

Ceylan, H., Urel, M., Erkal, T.S., Tekinay, A.B., Dana, A., and Guler, M.O. (2013). Mussel inspired dynamic cross-linking of self-healing peptide nanofiber network. Adv Funct Mater *23*, 2081–2090.

Chan, R.K., Liu, P.H., Pietramaggiori, G., Ibrahim, S.I., Hechtman, H.B., and Orgill, D.P. (2006). Effect of recombinant platelet-derived growth factor (Regranex) on wound closure in genetically diabetic mice. J Burn Care Res *27*, 202–205.

Choi, J.S., Leong, K.W., and Yoo, H.S. (2008). In vivo wound healing of diabetic ulcers using electrospun nanofibers immobilized with human epidermal growth factor (EGF). Biomaterials *29*, 587–596.

Clark, R.A., Lanigan, J.M., DellaPelle, P., Manseau, E., Dvorak, H.F., and Colvin, R.B. (1982). Fibronectin and fibrin provide a provisional matrix for epidermal cell migration during wound reepithelization. J Invest Dermatol *79*, 264–269.

Cohn, S.M., Lopez, P.P., Brown, M., Namias, N., Jackowski, J., Li, P., Mishkin, D., Lopez, J.M., and Group, U.o.M.W.S. (2004). Open surgical wounds: how does Aquacel compare with wet-to-dry gauze? J Wound Care *13*, 10–12.

Dagdas, Y.S., Tombuloglu, A., Tekinay, A.B., Dana, A., and Guler, M.O. (2011). Interfiber interactions alter the stiffness of gels formed by supramolecular self-assembled nanofibers. Soft Matter *7*, 3524–3532.

Dai, T., Tanaka, M., Huang, Y.Y., and Hamblin, M.R. (2011). Chitosan preparations for wounds and burns: antimicrobial and wound-healing effects. Expert Rev Anti Infect Ther *9*, 857–879.

Darby, I., Skalli, O., and Gabbiani, G. (1990). Alpha-smooth muscle actin is transiently expressed by myofibroblasts during experimental wound healing. Lab Invest *63*, 21–29.

Defloor, T. (1999). The risk of pressure sores: a conceptual scheme. J Clin Nurs *8*, 206–216.

Eckes, B., Nischt, R., and Krieg, T. (2010). Cell-matrix interactions in dermal repair and scarring. Fibrogenesis Tissue Repair *3*, 4.

Eming, S.A., Brachvogel, B., Odorisio, T., and Koch, M. (2007a). Regulation of angiogenesis: wound healing as a model. Prog Histochem Cytochem *42*, 115–170.

Eming, S.A., Krieg, T., and Davidson, J.M. (2007b). Inflammation in wound repair: molecular and cellular mechanisms. J Invest Dermatol *127*, 514–525.

Engler, A.J., Sen, S., Sweeney, H.L., and Discher, D.E. (2006). Matrix elasticity directs stem cell lineage specification. Cell *126*, 677–689.

Evans, N.D., Oreffo, R.O., Healy, E., Thurner, P.J., and Man, Y.H. (2013). Epithelial mechanobiology, skin wound healing, and the stem cell niche. J Mech Behav Biomed Mater *28*, 397–409.

Ferguson, M.W., Whitby, D.J., Shah, M., Armstrong, J., Siebert, J.W., and Longaker, M.T. (1996). Scar formation: the spectral nature of fetal and adult wound repair. Plast Reconstr Surg *97*, 854–860.

Fitzgerald, R.H. and Steinberg, J. (2009). Collagen in wound healing: are we onto something new or just repeating the past? Foot Ankle Online J 2, 3.

Freeman, I., Kedem, A., and Cohen, S. (2008). The effect of sulfation of alginate hydrogels on the specific binding and controlled release of heparin-binding proteins. Biomaterials *29*, 3260–3268.

Gabbiani, G. (2003). The myofibroblast in wound healing and fibrocontractive diseases. J Pathol *200*, 500–503.

Galiano, R.D., Tepper, O.M., Pelo, C.R., Bhatt, K.A., Callaghan, M., Bastidas, N., Bunting, S., Steinmetz, H.G., and Gurtner, G.C. (2004). Topical vascular endothelial growth factor accelerates diabetic wound healing through increased angiogenesis and by mobilizing and recruiting bone marrow-derived cells. Am J Pathol *164*, 1935–1947.

Garcia-Filipe, S., Barbier-Chassefiere, V., Alexakis, C., Huet, E., Ledoux, D., Kerros, M.E., Petit, E., Barritault, D., Caruelle, J.P., and Kern, P. (2007). RGTA OTR4120, a heparan sulfate mimetic, is a possible long-term active agent to heal burned skin. J Biomed Mater Res A *80*, 75–84.

Gardner, S.E., Frantz, R.A., Troia, C., Eastman, S., MacDonald, M., Buresh, K., and Healy, D. (2001). A tool to assess clinical signs and symptoms of localized infection in chronic wounds: development and reliability. Ostomy Wound Manage *47*, 40–47.

Gillitzer, R. and Goebeler, M. (2001). Chemokines in cutaneous wound healing. J Leukoc Biol *69*, 513–521.

Goldin, A., Beckman, J.A., Schmidt, A.M., and Creager, M.A. (2006). Advanced glycation end products: sparking the development of diabetic vascular injury. Circulation *114*, 597–605.

Greenhalgh, D.G. (1998). The role of apoptosis in wound healing. Int J Biochem Cell Biol *30*, 1019–1030.

Gurtner, G.C., Werner, S., Barrandon, Y., and Longaker, M.T. (2008). Wound repair and regeneration. Nature *453*, 314–321.

Halim, A.S., Khoo, T.L., and Mohd Yussof, S.J. (2010). Biologic and synthetic skin substitutes: an overview. Indian J Plast Surg *43*, S23–S28.

Hansbrough, J.F., Achauer, B., Dawson, J., Himel, H., Luterman, A., Slater, H., Levenson, S., Salzberg, C.A., Hansbrough, W.B., and Doré, C. (1995).

Wound healing in partial-thickness burn wounds treated with collagenase ointment versus silver sulfadiazine cream. J Burn Care Rehabil *16*, 241–247.

Hashimoto, T., Suzuki, Y., Tanihara, M., Kakimaru, Y., and Suzuki, K. (2004). Development of alginate wound dressings linked with hybrid peptides derived from laminin and elastin. Biomaterials *25*, 1407–1414.

Heldin, C.H. and Westermark, B. (1999). Mechanism of action and in vivo role of platelet-derived growth factor. Physiol Rev *79*, 1283–1316.

Jacinto, A., Martinez-Arias, A., and Martin, P. (2001). Mechanisms of epithelial fusion and repair. Nat Cell Biol *3*, E117–E123.

Jayakumar, R., Prabaharan, M., Nair, S.V., and Tamura, H. (2010). Novel chitin and chitosan nanofibers in biomedical applications. Biotechnol Adv *28*, 142–150.

Jiang, C.K., Magnaldo, T., Ohtsuki, M., Freedberg, I.M., Bernerd, F., and Blumenberg, M. (1993). Epidermal growth factor and transforming growth factor alpha specifically induce the activation- and hyperproliferation-associated keratins 6 and 16. Proc Natl Acad Sci U S A *90*, 6786–6790.

Jones, I., Currie, L., and Martin, R. (2002). A guide to biological skin substitutes. Br J Plast Surg *55*, 185–193.

Kim, D.H., Provenzano, P.P., Smith, C.L., and Levchenko, A. (2012). Matrix nanotopography as a regulator of cell function. J Cell Biol *197*, 351–360.

Kirker, K.R., Luo, Y., Nielson, J.H., Shelby, J., and Prestwich, G.D. (2002). Glycosaminoglycan hydrogel films as bio-interactive dressings for wound healing. Biomaterials *23*, 3661–3671.

Komori, M., Tomizawa, Y., Takada, K., and Ozaki, M. (2005). A single local application of recombinant human basic fibroblast growth factor accelerates initial angiogenesis during wound healing in rabbit ear chamber. Anesth Analg *100*, 830–834.

Kumar, V.A., Taylor, N.L., Jalan, A.A., Hwang, L.K., Wang, B.K., and Hartgerink, J.D. (2014). A nanostructured synthetic collagen mimic for hemostasis. Biomacromolecules *15*, 1484–1490.

Larjava, H., Koivisto, L., and Häkkinen, L. (2000). Keratinocyte Interactions with Fibronectin During Wound Healing In Cell Invasion, J. Heino and V.-M. Kähäri, eds. (Landes Bioscience Georgetown, TX).

Lau, K., Paus, R., Tiede, S., Day, P., and Bayat, A. (2009). Exploring the role of stem cells in cutaneous wound healing. Exp Dermatol *18*, 921–933.

Levenson, S.M., Geever, E.F., Crowley, L.V., Oates, J.F., 3rd, Berard, C.W., and Rosen, H. (1965). The healing of rat skin wounds. Ann Surg *161*, 293–308.

Linden, E., Cai, W., He, J.C., Xue, C., Li, Z., Winston, J., Vlassara, H., and Uribarri, J. (2008). Endothelial dysfunction in patients with chronic kidney disease results from advanced glycation end products (AGE)-mediated inhibition of endothelial nitric oxide synthase through RAGE activation. Clin J Am Soc Nephrol *3*, 691–698.

Lipsky, B.A., Berendt, A.R., Cornia, P.B., Pile, J.C., Peters, E.J., Armstrong, D.G., Deery, H.G., Embil, J.M., Joseph, W.S., Karchmer, A.W., et al. (2012). 2012 Infectious Diseases Society of America clinical practice guideline for the diagnosis and treatment of diabetic foot infections. Clin Infect Dis *54*, e132–e173.

Lobmann, R., Ambrosch, A., Schultz, G., Waldmann, K., Schiweck, S., and Lehnert, H. (2002). Expression of matrix-metalloproteinases and their inhibitors in the wounds of diabetic and non-diabetic patients. Diabetologia *45*, 1011–1016.

Lodish, H., Berk, A., Zipursky, S.L., Matsudaira, P., Baltimore, D., and Darnell, J. (2000). Molecular Cell Biology, 4th edn. (W. H. Freeman New York).

Lovvorn, H.N., 3rd, Cheung, D.T., Nimni, M.E., Perelman, N., Estes, J.M., and Adzick, N.S. (1999). Relative distribution and crosslinking of collagen distinguish fetal from adult sheep wound repair. J Pediatr Surg *34*, 218–223.

Lynch, S.E., Nixon, J.C., Colvin, R.B., and Antoniades, H.N. (1987). Role of platelet-derived growth factor in wound healing: synergistic effects with other growth factors. Proc Natl Acad Sci U S A *84*, 7696–7700.

Malinda, K.M., Wysocki, A.B., Koblinski, J.E., Kleinman, H.K., and Ponce, M.L. (2008). Angiogenic laminin-derived peptides stimulate wound healing. Int J Biochem Cell Biol *40*, 2771–2780.

Mammadov, R., Mammadov, B., Toksoz, S., Aydin, B., Yagci, R., Tekinay, A.B., and Guler, M.O. (2011). Heparin mimetic peptide nanofibers promote angiogenesis. Biomacromolecules *12*, 3508–3519.

Mammadov, B., Mammadov, R., Guler, M.O., and Tekinay, A.B. (2012a). Cooperative effect of heparan sulfate and laminin mimetic peptide nanofibers on the promotion of neurite outgrowth. Acta Biomater *8*, 2077–2086.

Mammadov, R., Mammadov, B., Guler, M.O., and Tekinay, A.B. (2012b). Growth factor binding on heparin mimetic peptide nanofibers. Biomacromolecules *13*, 3311–3319.

Marchese, C., Chedid, M., Dirsch, O.R., Csaky, K.G., Santanelli, F., Latini, C., LaRochelle, W.J., Torrisi, M.R., and Aaronson, S.A. (1995). Modulation of keratinocyte growth factor and its receptor in reepithelializing human skin. J Exp Med *182*, 1369–1376.

Markeson, D., Pleat, J.M., Sharpe, J.R., Harris, A.L., Seifalian, A.M., and Watt, S.M. (2015). Scarring, stem cells, scaffolds and skin repair. J Tissue Eng Regen Med *9*, 649–668.

Martin, P. (1997). Wound healing—aiming for perfect skin regeneration. Science *276*, 75–81.

Martino, M.M. and Hubbell, J.A. (2010). The 12th-14th type III repeats of fibronectin function as a highly promiscuous growth factor-binding domain. FASEB J *24*, 4711–4721.

Martino, M.M., Briquez, P.S., Guc, E., Tortelli, F., Kilarski, W.W., Metzger, S., Rice, J.J., Kuhn, G.A., Muller, R., Swartz, M.A., et al. (2014). Growth factors engineered for super-affinity to the extracellular matrix enhance tissue healing. Science *343*, 885–888.

Mast, B.A. and Schultz, G.S. (1996). Interactions of cytokines, growth factors, and proteases in acute and chronic wounds. Wound Repair Regen *4*, 411–420.

Menon, G.K. (2002). New insights into skin structure: scratching the surface. Adv Drug Deliv Rev *54*(Suppl 1), S3–S17.

Miller, S.J., Burke, E.M., Rader, M.D., Coulombe, P.A., and Lavker, R.M. (1998). Re-epithelialization of porcine skin by the sweat apparatus. J Invest Dermatol *110*, 13–19.

Min, S.K., Lee, S.C., Hong, S.D., Chung, C.P., Park, W.H., and Min, B.M. (2010). The effect of a laminin-5-derived peptide coated onto chitin microfibers on re-epithelialization in early-stage wound healing. Biomaterials *31*, 4725–4730.

Mogoşanu, G.D. and Grumezescu, A.M. (2014). Natural and synthetic polymers for wounds and burns dressing. Int J Pharm *463*, 127–136.

Mostow, E.N., Haraway, G.D., Dalsing, M., Hodde, J.P., King, D., and Group, O.V.U.S. (2005). Effectiveness of an extracellular matrix graft (OASIS wound matrix) in the treatment of chronic leg ulcers: a randomized clinical trial. J Vasc Surg *41*, 837–843.

Moura, L.I., Dias, A.M., Carvalho, E., and de Sousa, H.C. (2013). Recent advances on the development of wound dressings for diabetic foot ulcer treatment—a review. Acta Biomater *9*, 7093–7114.

O'Kane, S. and Ferguson, M.W. (1997). Transforming growth factor beta s and wound healing. Int J Biochem Cell Biol *29*, 63–78.

Pardes, J.B., Takagi, H., Martin, T.A., Ochoa, M.S., and Falanga, V. (1995). Decreased levels of alpha 1(I) procollagen mRNA in dermal fibroblasts grown on fibrin gels and in response to fibrinopeptide B. J Cell Physiol *162*, 9–14.

Phelps, E.A. and Garcia, A.J. (2009). Update on therapeutic vascularization strategies. Regen Med *4*, 65–80.

Prompers, L., Schaper, N., Apelqvist, J., Edmonds, M., Jude, E., Mauricio, D., Uccioli, L., Urbancic, V., Bakker, K., Holstein, P., et al. (2008). Prediction of outcome in individuals with diabetic foot ulcers: focus on the differences between individuals with and without peripheral arterial disease. The EURODIALE Study. Diabetologia *51*, 747–755.

Rajangam, K., Behanna, H.A., Hui, M.J., Han, X., Hulvat, J.F., Lomasney, J.W., and Stupp, S.I. (2006). Heparin binding nanostructures to promote growth of blood vessels. Nano Lett *6*, 2086–2090.

Reiber, G.E., Vileikyte, L., Boyko, E.J., del Aguila, M., Smith, D.G., Lavery, L.A., and Boulton, A.J. (1999). Causal pathways for incident lower-extremity ulcers in patients with diabetes from two settings. Diabetes Care *22*, 157–162.

Rheinwald, J.G. and Green, H. (1977). Epidermal growth factor and the multiplication of cultured human epidermal keratinocytes. Nature *265*, 421–424.

Roy, D.C., Wilke-Mounts, S.J., and Hocking, D.C. (2011). Chimeric fibronectin matrix mimetic as a functional growth- and migration-promoting adhesive substrate. Biomaterials *32*, 2077–2087.

Rustad, K.C., Wong, V.W., Sorkin, M., Glotzbach, J.P., Major, M.R., Rajadas, J., Longaker, M.T., and Gurtner, G.C. (2012). Enhancement of mesenchymal stem cell angiogenic capacity and stemness by a biomimetic hydrogel scaffold. Biomaterials *33*, 80–90.

Shah, M., Foreman, D.M., and Ferguson, M.W. (1995). Neutralisation of TGF-beta 1 and TGF-beta 2 or exogenous addition of TGF-beta 3 to cutaneous rat wounds reduces scarring. J Cell Sci *108* (Pt 3), 985–1002.

Shah, R.N., Shah, N.A., Del Rosario Lim, M.M., Hsieh, C., Nuber, G., and Stupp, S.I. (2010). Supramolecular design of self-assembling nanofibers for cartilage regeneration. Proc Natl Acad Sci U S A *107*, 3293–3298.

Silva, G.A., Czeisler, C., Niece, K.L., Beniash, E., Harrington, D.A., Kessler, J.A., and Stupp, S.I. (2004). Selective differentiation of neural progenitor cells by high-epitope density nanofibers. Science *303*, 1352–1355.

Simmons, T. and Linhardt, R. (2012). Application of Carbon Nanotubes to Wound Healing Biotechnology In Nanomaterials for Biomedicine, R. Nagarajan, ed. (American Chemical Society Washington, DC), pp. 156–174.

Singer, A.J. and Clark, R.A. (1999). Cutaneous wound healing. N Engl J Med *341*, 738–746.

Stendahl, J.C., Rao, M.S., Guler, M.O., and Stupp, S.I. (2006). Intermolecular forces in the self-assembly of peptide amphiphile nanofibers. Adv Funct Mater *16*, 499–508.

Stoscheck, C.M., Nanney, L.B., and King, L.E., Jr. (1992). Quantitative determination of EGF-R during epidermal wound healing. J Invest Dermatol *99*, 645–649.

Supp, D.M. and Boyce, S.T. (2005). Engineered skin substitutes: practices and potentials. Clin Dermatol *23*, 403–412.

Szabowski, A., Maas-Szabowski, N., Andrecht, S., Kolbus, A., Schorpp-Kistner, M., Fusenig, N.E., and Angel, P. (2000). c-Jun and JunB antagonistically control cytokine-regulated mesenchymal-epidermal interaction in skin. Cell *103*, 745–755.

Tsang, M.W., Wong, W.K., Hung, C.S., Lai, K.M., Tang, W., Cheung, E.Y., Kam, G., Leung, L., Chan, C.W., Chu, C.M., et al. (2003). Human epidermal growth factor enhances healing of diabetic foot ulcers. Diabetes Care *26*, 1856–1861.

Waldeck, H., Chung, A.S., and Kao, W.J. (2007). Interpenetrating polymer networks containing gelatin modified with PEGylated RGD and soluble KGF:

synthesis, characterization, and application in in vivo critical dermal wound. J Biomed Mater Res A *82*, 861–871.

Walker, D.J. (1999). Venous stasis wounds. Orthop Nurs *18*, 65–74, 95.

Wallis, S., Lloyd, S., Wise, I., Ireland, G., Fleming, T.P., and Garrod, D. (2000). The alpha isoform of protein kinase C is involved in signaling the response of desmosomes to wounding in cultured epithelial cells. Mol Biol Cell *11*, 1077–1092.

Wang, Y., Wang, G., Luo, X., Qiu, J., and Tang, C. (2012). Substrate stiffness regulates the proliferation, migration, and differentiation of epidermal cells. Burns *38*, 414–420.

Webber, M.J., Tongers, J., Newcomb, C.J., Marquardt, K.T., Bauersachs, J., Losordo, D.W., and Stupp, S.I. (2011). Supramolecular nanostructures that mimic VEGF as a strategy for ischemic tissue repair. Proc Natl Acad Sci U S A *108*, 13438–13443.

Werner, S. and Grose, R. (2003). Regulation of wound healing by growth factors and cytokines. Physiol Rev *83*, 835–870.

WHO (April, 2014). Fact Sheets: Burns. Available at http://www.who.int/mediacentre/factsheets/fs365/en/ (accessed on October 29, 2015).

Woo, K.M., Chen, V.J., and Ma, P.X. (2003). Nano-fibrous scaffolding architecture selectively enhances protein adsorption contributing to cell attachment. J Biomed Mater Res A *67*, 531–537.

Yildirimer, L., Thanh, N.T., and Seifalian, A.M. (2012). Skin regeneration scaffolds: a multimodal bottom-up approach. Trends Biotechnol *30*, 638–648.

Zhang, Y., Wang, B., Meng, X., Sun, G., and Gao, C. (2011). Influences of acid-treated multiwalled carbon nanotubes on fibroblasts: proliferation, adhesion, migration, and wound healing. Ann Biomed Eng *39*, 414–426.

6

NANOMATERIALS FOR BONE TISSUE REGENERATION AND ORTHOPEDIC IMPLANTS

Gulcihan Gulseren, Melis Goktas, Hakan Ceylan, Ayse B. Tekinay, and Mustafa O. Guler

Institute of Materials Science and Nanotechnology, National Nanotechnology Research Center (UNAM), Bilkent University, Ankara, Turkey

6.1 INTRODUCTION

Bone is a highly mineralized connective tissue with complex biochemical and biophysical cues including structural, compositional, and cellular properties that cooperate to direct bone function and remodeling. Hierarchical organization and specialized composition of bone extracellular matrix (ECM) control the cellular processes including proliferation, migration, and differentiation for continuous modulation and maintenance of structure (Fernandez-Yague et al., 2014). In general, bone tissue consists of inorganic components such as hydroxyapatite and a wide variety of organic components, most of which is type I collagen. Inorganic mineral phase constitutes 65% of the wet weight of the bone tissue, while the organic components contribute to the rest with water (Triffit, 1980). Organic extracellular components found in bone matrix can be classified as insoluble (e.g., collagen) and soluble (e.g., growth factors (GFs), transcription

Therapeutic Nanomaterials, First Edition. Edited by Mustafa O. Guler and Ayse B. Tekinay.

factors) factors and play a critical role in bone organization and remodeling (Athanasiou et al., 2000). Organic components, primarily collagen, provide resistance to tension, while the inorganic mineral phase contributes to resistance against compression. When the bone is demineralized, remaining organic phase provides flexibility and resistance to fracture, while removal of organic matrix makes the bone rigid and brittle. As a composite structure consisting of these organic and inorganic components, hierarchically organized bone matrix not only serves as a structural and mechanical support to cells but also provides biochemical cues that regulate cell and tissue functions.

6.2 BONE MATRIX

6.2.1 Organic Matrix and Bioactivity

6.2.1.1 Insoluble Factors The organic matrix of bone is predominantly constituted by collagens. Collagens, mostly type I along with small amounts of type V and type XII, form 90% of the organic bone matrix and the rest 10% is composed of noncollagenous matrix proteins and proteoglycans (Triffit, 1980). Generally, nonmineralized part of the bone is composed of a variety of network organizations including random network of coarse collagen bundles as well as hierarchically organized parallel fibrils and bundles. Type I collagen fibrils are bundled together to form parallel collagen fibers with a large scale in diameter ranging from 0.2 to 12 μm (Kielty and Grant, 2002). This specific organization into large diameter parallel fibrils plays the primary role in supporting bone tissue to withstand tensile stress. However, collagen not only provides mechanical strength to bone but also contains short peptide sequences that exhibit bioactive signals such as aspartic acid–glycine–glutamic acid–alanine (DGEA) peptide motif, which can activate adhesion and differentiation of osteoblastic cells and mesenchymal stem cells through interaction with cell surface receptors called integrins (Hennessy et al., 2009; Mizuno et al., 2000).

Bone also contains a wide variety of noncollagenous proteins produced by osteoblasts including osteonectin, osteocalcin, osteopontin, bone sialoprotein (BSP), and small proteoglycans, which play important roles in matrix organization, mineralization, and cellular functions. Bone cells contain a variety of transmembrane receptors such as integrins, cadherins, and selectins. Among these, integrins are the primary receptors regulating the cell–matrix interactions (Hynes, 1992). Through the interaction with ligand motifs such as arginine–glycine–aspartic acid (RGD) adhesion sequence present in the structure of many noncollagenous bone matrix proteins, integrin signaling pathways regulate cellular functions including osteoblast survival, gene expression, and matrix mineralization (Garcia and Reyes, 2005).

The most abundant noncollagenous bone matrix protein in mineralized bone is osteonectin. It has been proposed that osteonectin contains calcium and collagen binding domains that act as nucleation sites for hydroxyapatite formation and mineralization (Engel et al., 1987; Maurer et al., 1996; Young et al., 1992). Another calcium-binding matrix protein is osteocalcin, which is also called as bone Gla protein. Due to the acidic residues (γ-carboxyglutamic acid) found in its structure, this protein has a Ca^{2+}/hydroxyapatite affinity (Ducy et al., 1996). Osteocalcin also acts as a regulator of bone remodeling through suppression of bone formation by osteoblasts and promotion of osteoclast migration (ButLer, 2000). Osteopontin is another acidic glycoprotein found in bone tissue, which is involved in the regulation of mineral deposition and cell migration (de Oliveira and Nanci, 2004). It has been proposed that hydroxyapatite formation is inhibited by the suppression of crystallite growth through high-level phosphorylation and negative charge density of osteopontin preventing ionic growth via electrostatic repulsion of ions (Pampena et al., 2004). Osteopontin also contains cell-adhesive RGD amino acid sequence that provides cell attachment and regulates osteoclast motility and migration during bone resorption (ButLer, 2000; Oldberg et al., 1986). BSP constitutes 15% of the noncollagenous bone matrix proteins and contributes to bone mineralization and remodeling through facilitation of HA nucleation (Hunter et al., 1996; Wuttke et al., 2001). Also, due to the RGD peptide sequence found in its protein structure, BSP promotes osteoblast adhesion and differentiation (Oldberg et al., 1988).

6.2.1.2 Soluble Factors Bone matrix also contains soluble and mobile biochemical cues synthesized by local osteoblasts or by cells outside of the bone and then delivered to bone matrix via the bloodstream. GFs are a group of these soluble factors, which take place in the regulation of cellular functions such as cell growth, motility, proliferation, differentiation, and bone formation (Mackie, 2003; Urist, 1965). Bone matrix contains a variety of GFs including bone morphogenic proteins (BMPs), transforming growth factor-β (TGF-β) family, insulin-like growth factor I and II (IGF-I, IGF-II), fibroblast growth factors (FGFs), and vascular endothelial growth factor (VEGF) (Canalis et al., 1988; Gowen et al., 1983; Hauschka et al., 1986; Mohan and Baylink, 1991; Sporn et al., 1986; Urist et al., 1983).

BMPs are the only GFs that can stimulate differentiation of mesenchymal stem cells into both chondroblasts and osteoblasts. Therefore, they play a critical role in endochondral bone formation where cartilage forms first and then it is replaced by bone (Wozney and Rosen, 1998). Other GFs such as TGF-β, IGF, and FGFs take place in the proliferation of already differentiated bone-forming cells and matrix deposition. Also, FGFs and VEGF act in bone formation and remodeling by inducing vascularization (angiogenesis).

In addition to GFs, osteoblasts carry other soluble factors in their matrix in vesicles built from phospholipid membranes, which play an important role in nucleation of hydroxyapatite and mineralization (Bosetti et al., 2005; Eanes, 1989; Raggio et al., 1986). Matrix vesicles contain calcium, phosphorus, and alkaline phosphatase (ALP). ALP functions by hydrolyzing organic phosphate esters and producing free inorganic phosphate, which initiates biomineralization and hydroxyapatite deposition on ECM proteins (Beertsen and van den Bos, 1992; Storrie and Stupp, 2005).

6.3 INORGANIC MATRIX, MINERALIZATION, AND BONE ORGANIZATION

The inorganic matrix of bone consists of a mineral phase in the form of carbonated hydroxyapatite crystals (Reznikov et al., 2014). Mineral phase functions as an ion reservoir to keep the extracellular fluid concentrations within the proper ranges for physiological functions and also provides stiffness and strength to bone tissue to withstand the mechanical forces. Bone mineral crystals contain 99% of the total body calcium, 85% of the phosphorus, and more than 40% of the sodium and magnesium (Glimcher, 1992).

Basically, mineralization of bone occurs as a phase transformation of soluble calcium and phosphate found in the organic matrix into solid calcium phosphate crystals. Solid calcium phosphate first forms as a poor crystalline apatite and the crystallinity of bone apatite increases in time (Roberts et al., 1992). Calcification of bone first starts in specific hole regions of the collagen fibrils formed by type I collagen. Hierarchical array structure built by the triple helical molecules of type I collagen results in formation of holes within the fibrils that are further aligned to construct grooves (1.5 nm thick) in which the mineral crystals form (Traub et al., 1989). Crystals of carbonated hydroxyapatite nucleate in these grooves as a precursor phase and the growth of crystals extends to include the zone of collagen fibrils between the grooves. Eventually, mineral phase deposits all over the space within fibrils and mineralized collagen fibrils in the form of plate-shaped crystal layers are formed (Traub et al., 1992; Weiner and Traub, 1986). Mineralization usually takes place very rapidly so that 60% of the total mineral capacity is achieved within hours. After the initial phase, mineral concentration, density, and stiffness of bone increase, while the water content and the noncollagenous protein concentration decrease, but the collagen organization and concentration remain the same (Triffit, 1980).

6.3.1 Mechanical Properties and Structural Hierarchy of Bone Tissue

Bone tissue is a hierarchically organized connective tissue with specialized structural features, material properties, and cellular organization that direct bone function in an interactive manner. According to its structural organization, porosity, and mechanical properties, bone tissue can be classified as either cortical or cancellous bone. Cortical bone has less than 20% porosity and a specialized compartmentalized organization that consists of closely packed osteons and cylindrical systems called Haversian canals with a central channel enclosing a blood vessel surrounded by concentric rings (lamellae) of bone matrix (Fernandez-Yague et al., 2014). Macroscale organization of this compartmentalized system provides a characteristic mechanical anisotropy to long bones. According to the direction of applied force, mechanical stiffness of the cortical bones differs. When a force is applied along the Haversian canals, the elastic modulus of the system can be measured as approximately $E = 20\,\text{GPa}$, while it can be altered to 8 GPa along the transverse axis (Katsamenis et al., 2013). Cancellous bone (spongy bone) is more porous (>90% porosity), softer, and flexible than cortical bone. However, mechanical stiffness of cancellous bone is more isotropic depending on the density and porosity of the trabeculae and can change depending on the loading rate. Elastic modulus of cancellous bone can vary between 50 and 100 MPa (Beddoe et al., 1976; Goldstein et al., 1983). Bone cells including osteocytes, osteoblasts, osteoclasts, and osteoprogenitor cells are able to sense the mechanical properties of their environment and the mechanical stress arising from physical loading during skeletal movement, and they produce cellular biochemical responses via integrin-mediated mechanotransduction to maintain bone remodeling (Sikavitsas et al., 2001). Therefore, in order to properly simulate the mechanical physiological environment in natural bone tissue, design of a scaffold should comprise similar mechanical properties as well as the hierarchical organization of native bone since the factors such as porosity could compromise the mechanical competence of the material.

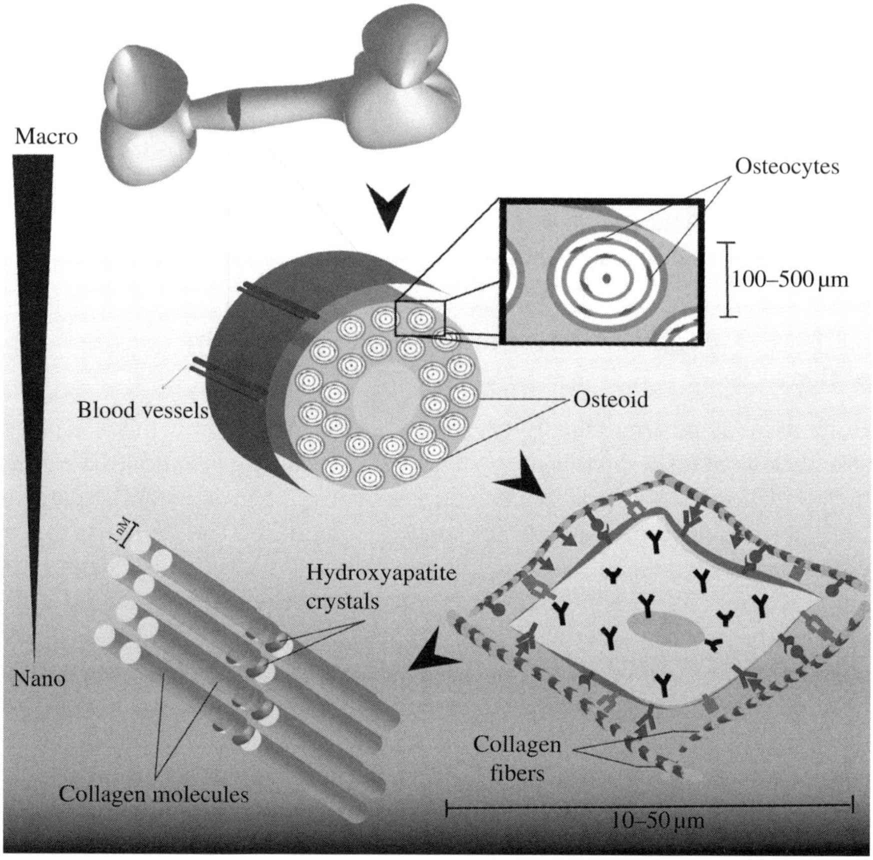

FIGURE 6.1 Hierarchical organization of bone from macro- to nanoscale.

Macro- to nanoscale hierarchical structure and porosity of the bone are also important at cell–material interface in terms of regulating mineralization, vascularization, and cellular functions such as proliferation, migration, and ECM production (Fig. 6.1). Macroscale porosity (>400 μm) and interconnectivity of pores promote osteogenesis by enabling cell migration, vascularization, nutrient transportation, and waste disposal. Micro- and nanoscale architectures also serve as a cell and mineral binding template. Organization of type I collagens into fibers with diameters of 3–10 nm and lengths of 300 μm supports the binding of nanoscale hydroxyapatite crystals (10–50 nm long) and proteins. Cell adhesion and mineral nucleation directed by microscale collagen template contribute to the resilience and strength of bone tissue (Mistry and Mikos, 2005; Nair et al., 2013).

6.4 REGULATION OF BONE MATRIX IN ADULT TISSUE

Bone is a rigid but dynamic living tissue that continuously undergoes remodeling and regeneration to maintain its structural integrity and function. Angiogenesis plays an important role in bone remodeling as well as in fracture healing. Vasculature supplies bone tissue with oxygen, nutrients, GFs, hormones, and cytokines as well as osteoblast and osteoclast precursor cells, which are important for bone remodeling processes, and also provides communication between bone and neighboring tissues (Chim et al., 2013; Kanczler and Oreffo, 2008).

Bone remodeling occurs in two important steps accomplished by osteoclasts and osteoblasts. Osteoclasts perform bone resorption by removing old and damaged bone through degradation of bone matrix. Then, mineralization and deposition of new bone matrix are carried out by osteoblasts in the resorption area. This process takes place in specialized vascularized structures called bone remodeling compartments (BRCs) (Hauge et al., 2001). BRC is a narrow sinus formed between the bone marrow and the remodeling surface, which contains osteoclasts, osteoblasts, osteocytes, bone lining cells, and the capillary blood supply. BRC serves as an anatomical structure that couples the blood supply with bone remodeling process. Osteoblast and osteoclast precursor cells as well as cytokines and systemic hormones are recruited to the remodeling site via circulating blood supply. BRC also facilitates the direct cell-to-cell contact between osteoblast and osteoclast precursors since the secretion of membrane-bound receptor activator of NF-κB ligand by osteoblasts is required for differentiation and activation of osteoclast precursors, which enter the BRC through the blood supply to start the remodeling process (Yasuda et al., 1998).

The initial stage of bone remodeling after a stimulus such as an injury is the formation of new blood vessels connecting the BRC to the existing vasculature. Early vascularization and blood flow to the injured site are essential for a proper healing and osteogenic repair since vessels provide nutrients and other factors required for the regeneration process (Fang et al., 2005). Previous studies demonstrated that angiogenesis takes place before the initiation of osteogenesis and the formation of new blood vessels is crucial for the transport of osteoblast and osteoclast precursors to remodeling sites (Sojo et al., 2005). Therefore, the intercellular communication between bone-forming cells and vessel-forming endothelial cells plays an important role in remodeling and regeneration of bone tissue. Communication between these cell types is conducted by direct or indirect cell contact and by the secretion of diffusible factors. It has been indicated that the diffusible factors produced by osteoblasts and osteoclast affect endothelial cell functions such as proliferation, survival, migration, and angiogenesis (Brandi and Collin-Osdoby, 2006). Recent studies demonstrated that osteocytes also play an important role in the initiation of remodeling process by secreting angiogenic factors including RANKL, VEGF, and BMPs (Chim et al., 2013).

6.4.1 Angiogenic Factors in Bone Remodeling

A large variety of angiogenic factors such as VEGF, FGF, BMPs, TGF-β, and platelet-derived growth factor (PDGF) are involved in bone remodeling process. Among these, VEGF is a potent mitogen and angiogenic factor for endothelial cells and it serves as a central regulator for the other diffusible factors. At the initial stage of bone remodeling, VEGF plays an important role in the formation of new blood vessels since it regulates the mobilization of endothelial progenitor cells, their differentiation into endothelial cells, and proliferation (Ferrara et al., 2003). VEGF also acts as a strong chemoattractant for osteoblasts and osteoclasts and supports their survival (Henriksen et al., 2003; Mayr-Wohlfart et al., 2002). It has been reported that inhibition of VEGF blocks angiogenic activity of FGF and BMP-2. Blocking of BMP-2 activity in turn results in inhibition of osteoblast differentiation and bone formation (Deckers et al., 2002; Peng et al., 2005). It has been reported that the inhibition of VEGF prevents *in vitro* osteoblast differentiation, while its *in vivo* inhibition results in decreased blood vessel formation and osteoclastic activity during bone remodeling, as well as reduced mineralization and trabecular bone healing (Gerber et al., 1999; Maes et al., 2002, 2004; Street et al., 2002).

FGF is another angiogenic factor produced by various cell types including fibroblasts, endothelial cells, and osteoblasts (Globus et al., 1989; Montero et al., 2000; Sato et al., 1991). FGF stimulates angiogenesis via inducing endothelial cell proliferation and migration and increases expression of GFs and integrins involved in angiogenesis (Klein et al., 1993; Seghezzi et al., 1998). It also induces osteogenesis through promoting proliferation of osteoblast precursor cells and their differentiation into mature osteoblasts (Hanada et al., 1997; Pitaru et al., 1993).

BMPs including BMP-2, BMP-4, and BMP-7 also play important roles in promotion of angiogenesis and osteogenesis. BMPs stimulate the proliferation and differentiation of MSCs and osteoprogenitor cells to promote bone regeneration (Bax et al., 1999; Fiedler et al., 2002; Lee et al., 2006; Li et al., 2005). They also induce endothelial cell proliferation, migration, and angiogenesis (David et al., 2009).

Both TGF-β and PDGF are known to be potent chemoattractants produced by degranulating platelets at the injury site. TGF-β serves as a chemoattractant for MSCs and it promotes differentiation of osteoblasts (Erlebacher et al., 1998; Tang et al., 2009), while PDGF is known to be a chemoattractant and mitogenic factor for osteoblasts at the bone remodeling sites (Mehrotra et al., 2004). TGF-β also induces VEGF expression in endothelial cells within the BRC to increase blood supply during remodeling process (He and Chen, 2005; Shao et al., 2009).

6.5 STRATEGIES FOR BONE TISSUE REGENERATION

6.5.1 Hard Grafts for Bone Regeneration

Bone implants are medical devices used to replace or fix injured bone parts for a temporary period of time or permanently. They are used extensively in dental, maxillofacial, and orthopedic surgeries (Dohan Ehrenfest et al., 2010; Smalley et al., 1988; Yeo, 2014). Bone anchoring is also an integral part of cochlear and spinal implants and artificial limbs (Gittens et al., 2014; Khan et al., 2012; Wazen et al., 2007). Global estimates suggest that more than 12 million patients receive oral implants and around half a million patients undergo hip replacement surgery per annum (Albrektsson et al., 2014; Hoption Cann et al., 2003; Mundi et al., 2014). As these devices are designed to bear and transfer mechanical loads, tight and stable bone bonding is essential for long-term functioning. In this regard, understanding and directing complex interactions occurring at the bone–implant interface has critical implications for developing effective healing strategies. When an implant comes into contact with the native bone, there are two possible outcomes. If the bone–implant interface is properly filled with a bone-like mineral, biological bone bonding, that is, osseointegration, is established. Coined first by Brånemark, osseointegration is the clinically desired state of healing where no progressive relative movement between the implant and bone tissue is observed under mechanical loading (Adell, 1985; Branemark et al., 2001; Khan et al., 2012). If this process is delayed or permanently fails, the interface is instead filled with a collagenous, nonmineralizable soft matrix, which isolates implant from the rest of the body, leading to implant failure (Le Guéhennec et al., 2007; Southam and Selwyn, 1970). In this case, a revision surgery becomes inevitable, which is usually more expensive and invasive, with increased risk factors compared to the primary intervention (Vanhegan et al., 2012). On the one hand, remarkable progress has been made over the last four decades in the understanding of short- and long-term contributors to bone bonding and functioning. Recent clinical surveys suggest that the long-term success rates are above 80% for various types of implants (Melo et al., 2006; Penarrocha-Diago et al., 2012; Simonis et al., 2010). On the other hand, there remain a number of major challenges associated with unsuccessful osseointegration, or failure in the longer period. Among the failed implants, the majority is early term failures caused by softening (Penarrocha-Diago et al., 2012). Particularly in patients with impaired osteoblastogenesis, such as osteoporosis, mineralization process takes much longer time and is more prone to failure (Alghamdi et al., 2013). This and other factors that impair the osseointegration process and long-term functioning of bone-anchored prosthetic devices are discussed in detail in the following. Current accomplishments to overcome such challenges are highlighted along with prospective technologies that would have profound impact on the recovery time and the overall life quality of patients.

After implantation of a bone-anchored prosthesis, formation of a tight, mineralized interface is vital for stable osseointegration. This process is under competitive pressure from fibrosis progression, and hence softening, so accelerated osteoblast activity is needed for proper mineralization (Daculsi et al., 1990; Lavenus et al., 2010; LeGeros, 2008; Yuan et al., 2010). Bone implants are made mainly from titanium and cobalt–chromium (molybdenum) alloys due to their suitable mechanical properties and corrosion resistance (Le Guéhennec et al., 2007; Linder, 1989; Mavrogenis et al., 2011). In terms or bioactivity, however, these materials are inert, that is, not osteoinductive, thereby requiring an additional bioinstructive support to promote osteoblast activity. As a result, the main challenge in the field is to define the toolbox of engineering implant surface so as to impart bioactivity for promoted osteoblast activity and osseointegration. However, the irreproducible complexity of dynamic interactions regulating osteoblast formation and subsequent biomineralization has limited the existing efforts to reductionist approaches where one or few bioactive components are displayed on the implant surface, endowing with a competitive advantage for the desired regenerative response. To understand mechanistic relations of surface parameters with osteoblast behavior and bone formation, various *in vitro*, that is, cell culture, and *in vivo*, that is, animal, models have been developed. Primary osteoblasts, preosteoblastic cell lines, and progenitor stem cells obtained from mice, rats, and humans have provided invaluable insight into the molecular biology of the biomineralization process (Ceylan et al., 2012, 2014; Fratzl-Zelman et al., 1998; Kalajzic et al., 2005; Kocabey et al., 2013; Marinucci et al., 2006; Olivares-Navarrete et al., 2008; Qi et al., 2003; Wang et al., 2014). Early studies toward this purpose concentrated on exploiting topography and wettability of the implant surface to modulate surface free energy at local sites and thereby altering the adsorbed matrix protein profile, that is, conformation, orientation, and composition, from the biological fluids (Arima and Iwata, 2007; Chug et al., 2013; Marinucci et al., 2006; Ruardy et al., 1995; Song and Mano, 2013; Sousa et al., 2008). Interaction of cells with the implant surface through this adsorbed layer of proteins strongly influences early adhesion, spreading, viability, proliferation, and differentiation (Chug et al., 2013; Wilson et al., 2005). For example, the presence of micrometer-scale surface roughness promotes higher osteoblast activity compared to the smooth surface (Gittens et al., 2014; Puleo and Nanci, 1999; von Wilmowsky et al., 2014). Then again, the actual method of choice for creating roughness has varying impacts on certain osteoblast behaviors. Both micro- and macrosand blasted surfaces promote cell proliferation, whereas osteoblast differentiation is induced only by macrosand blasting (Marinucci et al., 2006). Osteoblasts can also discriminate micro- and nanoscale topographic features (Olivares-Navarrete et al., 2014). Nanoscale roughness can induce differentiation of mesenchymal stem cells into mature osteoblasts and promotes osseointegration (Aboushelib et al., 2013; Bjursten et al., 2010; Mendonca et al., 2009). In addition to its direct impact on osteoblast functionality, roughening also provides greater surface area for primary bone bonding. For this reason, the majority of commercial bone implants are manufactured to have a roughness of about 1–2 μm, which leads to an approximately sixfold increase in the total surface area (Albrektsson and Wennerberg, 2004; von Wilmowsky et al., 2014). It is also noteworthy that the minimum bone ingrowth pore size on a material surface is approximately 50 μm, which implies that osseointegration is established over a much less part of the actual implant surface area (Bobyn et al., 1980). The impact of surface wettability on osteoblast behavior and osseointegration is rather vague. Several reports have linked increase in implant hydrophilicity to enhanced osteoblast functioning and osseointegration (Ceylan et al., 2012; Schwarz et al., 2009). However, depending on the actual surface chemistry, hydrophilicity might not be a striking factor for establishment of osseointegration or it can even be associated with the decrease in osteoblast adhesion *in vitro* (Unal Gulsuner et al., 2015; Vasak et al., 2014). Hydroxyapatite mineral coating is another strategy to accelerate mineralization at the bone–implant interface. Depending on the micro- and nanoscale geometry and porosity, hydroxyapatite can exhibit osteoinductivity in addition to its osteoconductive properties (Chien and Tsai, 2013; LeGeros, 2008). In this regard, precipitation of carbonated biological apatite on the implanted material would be a useful platform for promoting adhesion, survival, and osteogenic differentiation of the progenitor cells (Ceylan et al., 2014; Layrolle, 2011; LeGeros, 2008). Altogether, the interplay between different topographic features with varying surface chemistry and wettability represents a complex state of understanding, which requires more detailed investigation to develop optimal surface parameters for osteoblast behavior and stable osseointegration. Nevertheless, indirect modulation of surface bioactivity through altering the surface free energy is ultimately limited by the lack of selectivity for promoting adhesion of particular cell types, such as osteoblasts with respect to fibroblasts. Reconstitution of artificial microenvironments by means of ECM-mimetic, synthetic cues to direct cellular activities in a controlled way can provide effective guidance for osseointegration through adhesion, growth, and differentiation of osteoblasts and their progenitor cells. Targeting molecular level interactions through cell surface receptors has recently drawn broad attention. For example, a simple tetrapeptide, Asp–Gly–Glu–Ala (DGEA), derived from collagen type I $\alpha 1$ chain, can induce osteogenic differentiation of human mesenchymal stem cells and mouse preosteoblastic cells via binding to integrin receptor $\alpha 2\beta 1$. Presentation of this molecule on a solid support is critical in differentiation, adhesion, spreading, migration, and proliferation of human mesenchymal stem cells (Anderson et al., 2009b; Ceylan et al., 2014; Hennessy et al., 2009; Mizuno et al., 2000; Mizuno and Kuboki, 2001; Popov et al., 2011; Staatz et al., 1991; Yoo et al., 2011). In this approach, selecting proper cues is the crucial step to modulate desired biological response. For example, a well-established biomimetic tripeptide, Arg–Gly–Asp (RGD), of fibronectin interacts with integrin $\alpha 5\beta 1$, which is critical for adhesion and proliferation, while not having any impact on osteogenic differentiation (Hennessy et al., 2009; Olivares-Navarrete et al., 2008; Yoo et al., 2011). Furthermore, being a rather “universal” cell-adhesive molecule, RGD can also promote adhesion and growth of fibroblasts, which potentially promote fibrosis (Anselme, 2000; Gailit et al., 1997; Shu et al., 2004). In addition to the biochemical identity, surface density, spatial availability, and patterning are the other principal parameters determining the overall cellular response. For example, a threshold density of RGD ligand is necessary on a polymer substrate to induce focal adhesion of osteoblasts (Chollet et al., 2009). Nanostructuring of bioactive cues has been sought as an appealing method for increasing their display density and optimal presentation geometry to the cell surface receptors (Anderson et al., 2009a and 2009b; Englund et al., 2012; Lim et al., 2009; Niece et al., 2003; Silva et al., 2004; Yoo et al., 2011). Self-assembled peptide nanofibers have attracted special attention for this purpose because of their inherent biocompatibility and chemical versatility (Cui et al., 2010;

Hartgerink et al., 2001; Zhang, 2003). Controlled assembly of chemically well-defined building blocks further allows combinatorial display of multiple cues. Nevertheless, one major drawback of the strategy relying on the modification of implant surfaces with biomimetic cues is the adhesion chemistry that is applied in the form of a robust coating. Water molecules, dissolved ions, and polyionic biomolecules in the biological fluid as well as mechanical abrasiveness during implant placement create a challenging environment for adhesives to operate efficiently (Ceylan et al., 2013). To overcome this challenge, natural adhesives adapted to living underwater provide a plethora of inspirations toward developing biologically safe and reliable synthetic adhesives for medical applications. Mussel-inspired surface functionalization has been proposed a viable alternative for biofunctionalization of implant surfaces (Black et al., 2012; Ceylan et al., 2011, 2012; Kang et al., 2012; Lee et al., 2007). As discussed in detail in Chapter 8, development of a biomedical adhesive is of paramount importance in developing robust implant coatings. Combining the aforementioned strategies, recapitulation of a synthetic matrix coating comprised from a bioactive Lys–Arg–Ser–Arg (KRSR) sequence known to selectively bind and promote osteoblastic cell behaviors over fibroblasts, as well as from mussel-mimetic adhesive moiety, Dopa, represents one of the state-of-the-art approaches (Nelson et al., 2006).

In summary, a complex variety of parameters create an array of interactions at the bone–implant interface, the outcome of which has an integral role on the establishment of osseointegration. The aforementioned factors constitute a useful toolbox in the form of source of inspiration for guiding cellular behavior toward efficient bone bonding as well as aspiration for the prospective strategies.

6.6 SOFT GRAFTS FOR BONE REGENERATION

Native bone tissue is mainly a nanocomposite structure with physical and biological features guiding osteoinductive and osteoconductive properties. For bone tissue regeneration, peptide- or polymer-based biomaterials have offered a framework to design interactive molecules displaying bone composite properties to mimic living bone tissue. Biomimetic peptide and polymer materials can enable the appropriate matrix environment for bone regeneration, integrate desirable biological properties, and provide controlled release of multiple soluble factors or GFs during regeneration process. In the following, structure and properties of peptide- and polymer-based soft grafts for bone tissue regeneration will be reviewed with a summary of upcoming goals and challenges in the future of these versatile materials. This section basically covers types and applications of soft bone grafts, directed bone regeneration from biocompatible and bioactive biomaterials, and nanocomposite scaffolds for bone tissue regeneration.

6.6.1 Peptide-Based Bone Grafts

Degradable small molecules emerged as an alternative approach for developing implantable materials for tissue regeneration. Bone regeneration studies have been primarily focused on polymers and synthetic proteins. However, these structures have several shortcomings like toxic preparation components (monomers, cross-linking agents, etc.), shrinkage after application, and low biodegradability. Various types of supramolecular nanostructures and hydrogels attained great interest in regenerative medicine because of their significant features, such as simple production, complete biodegradability, and biocompatibility. These inert nanostructures can be applied to deformed bone tissue as implantable or injectable materials with simple procedures and minimal invasiveness. Investigation of self-assembled peptide nanostructures is a growing field with great potential to generate new, facile, and effective bone regenerative applications. Here, we describe some of the significant contributions to the field of bone regeneration with self-assembled peptide structures.

Bone formation is a well-orchestrated and complex biological process; therefore, bone regeneration strategies should meet rigid requirements to constitute proper remodeling of deformed site. During bone regeneration, osteoinduction and osteoconduction are governed by different factors (mechanical, biological, and chemical) that interplay each other, and these components should be considered carefully while constructing materials for bone regeneration. While a diverse set of materials have been utilized to build such scaffolds, inert and mechanically supportive metals and alloys have so far been used as permanent bone implants. These metallic implants and surface modification techniques lack osteoinductive properties, despite their success in osteoconductive features to accelerate the bone healing process. To improve cell attachment and to induce bone differentiation process, bioactive molecule (ECM proteins, GFs) attachments to implant surfaces are critical to obtain adequate bone healing and controlled mineralization. Proteins and GFs are large molecules bearing short peptide sequences, which can trigger downstream processes, especially cell adhesion and differentiation among many other roles. Accordingly, peptide structures with short bioactive units draw significant attention for bone remodeling studies.

In general, bioactive short peptides are cell-binding epitopes including RGDS, IKVAV, and YIGSR. The RGDS sequence has been frequently used to direct cell attachment. The RGD is found in fibronectin, osteopontin, and sialoprotein, and fibrous materials that display RGD can mimic protein function (Puleo and Bizios, 1991; Pytela et al., 1987). In early studies, adhesion of bone marrow-derived stem cells was investigated on RGDS-containing peptide amphiphile (PA) surfaces, and activity of scaffold was compared with epitope-free peptides. Stem cells were encapsulated into PAs and the coassembled system was injected *in vivo*. RGDS-containing PA gels promoted cell viability notably better than epitope-free control group (Guler et al., 2006).

The primary component of bone ECM is collagen I fibers, and bioactive short sequences derived from this protein are preferential targets for induction of bone tissue remodeling. The most studied collagen I epitope, DGEA, is found in its α1 helix, and its osteoinductive characteristics were utilized intensively for bone regeneration studies. For example, DGEA peptide-coated hydroxyapatite surfaces enhanced differentiation of mesenchymal stem cell into osteogenic fate (Harbers and Healy, 2005). However, in some cases, bioactivity is not enough for adequate regeneration, and more than one component may be required to enhance activity. For example, another study using CGGDGEAG sequence reported lack of adhesion by rat calvarial osteoblasts onto peptide surfaces (Anderson et al., 2011). To improve adhesion and osteoinductive potential of DGEA-PA, RGDS-PA, and S-PA peptides was utilized with different combinations. According to histochemical staining and PCR results, the RGDS-PA and DGEA-PA combination upregulated osteogenic differentiation (Anderson et al., 2009b). Another short peptide sequence adopted from collagen is GFOGER peptide, which binds to osteogenesis regulating integrin $\alpha2\beta1$. Differentiation of osteoprogenitor cells into osteoblasts can be triggered by using this sequence, and GFOGER-coated polycaprolactone scaffolds were shown to support remodeling of critical-sized defects in rat bone. GFOGER sequence induced significant regeneration in nonhealing femoral defects compared to peptide-free control groups (Dee et al., 1998).

In addition to cell attachment and proliferation, peptide sequences were also used as GF binding units. Binding units located onto gel scaffold can attract GFs to the desired location to sustain improvement during healing process. A well-known bone differentiation-regulating factor, bone morphogenetic protein-2 (BMP-2), which plays significant roles during osteogenesis, was targeted for this purpose. BMP receptor-binding peptides with osteopromotive domains, DWIVA and A_4G_3EDWIVA, were adequate for maintaining bone regeneration process (Gelain et al., 2006). In the ECM, glycosaminoglycans also interact with BMP-2 and sulfated GAG-mimetic peptide–BMP interaction was used to promote osteoblast maturation and mineralization of osteogenic cells. The GAG-mimicking ability of the peptide nanofibers and their interaction with BMP-2 promoted osteogenic activity and mineralization of osteoblastic cells (Webber et al., 2010).

RADA16-I is another scaffold that shows stable beta sheet formation, and these peptides can be used with different active groups to form nanostructures to enhance tissue regeneration. In one study, RADA16-I was linked to variable bioactive signal inducing short peptides like ALK (ALKRQGRTLYGF) osteogenic growth motif, DGR (DGRGDSVAYG) osteopontin-based cell adhesion sequence, and PGR (PRGDSGYRGDS) two-repeat RGD adhesion sequence to enhance osteogenic differentiation and support (Horii et al., 2007).

Mussel-inspired Dopa-mediated surface adhesion strategy was applied to direct efficient osteoblast maturation on implant surfaces. For this purpose, E_3-PA/Dopa-PA and DGEA-PA/Dopa-PA combination was used to induce osteoblast formation and subsequent bone-like mineralization. Hydroxyapatite-boosting glutamic acid and Dopa residues on fiber surfaces induced significant osteogenic activity as a consequence of significant mineral deposition, since inorganic materials like hydroxyapatite minerals can also induce osteoblast maturation during bone formation process (Tashiro et al., 1989). Therefore, mineral-inducing peptide scaffolds have important roles for bone remodeling studies.

Peptide-based soft bone grafts can also serve as structural frameworks. These materials can be designed in order to have structural components presenting a complementary effect to signaling epitope to enhance bioactivity. For example, micropatterned RGDS-modified scaffolds bearing 20–40 nm holes upregulated osteogenic marker expression compared to nonpatterned surfaces. The hole-patterned surfaces presented high-density epitopes that were able to enclose cells and stimulated differentiation (Sur et al., 2012). In another example, phosphoserine-bearing peptide amphiphile nanofibers were used to induce mineralization. This matrix presented convenient template for hydroxyapatite crystal (HA)—basic bone mineral—deposition as a result of its suitable structure for crystal packing (Anderson et al., 2009b). By the virtue of allowing nanopatterned structure design, peptides are versatile building blocks for generating templates that can induce inorganic material deposition.

6.6.2 Polymer Nanocomposites as Bone Grafts

Bone autografts and allografts are already widely used in clinical practice for bone regeneration studies. However, these grafts are not ideal scaffolds for osteogenesis, especially because of biocompatibility and handling issues. Producing demineralized bone scaffolds to regenerate native bone tissue requires tedious work, and mineralization of these grafts can be less controllable compared to synthetic scaffolds. These shortcomings of autografts and allografts boosted the interest in utilization of synthetic polymer scaffolds as another useful strategy for bone regeneration. Polymers can be constructed from various building blocks depending on the desired outcome and can be connected or decorated through covalent/noncovalent linkages to create ordered composite.

Biodegradability and biocompatibility are important parameters for bone engineering applications. Natural polymer scaffolds are basically derived from collagen, gelatin-like proteins, or alginate-, hyaluronate-, and chitin-like polysaccharides. However, despite their biocompatibility or biodegradability, these molecules have several shortcomings. These natural constructs were shown to exhibit variable physical characteristics for each batch, such as molecular weight, crystallinity, polydispersity, degradation kinetics, structure, etc., which causes different scaffold properties. On the other hand, synthetic polymeric materials (poly(glycolic acid) (PGA), poly(lactic acid) (PLA)) and their copolymers offer facile and more controllable methods for clinical bone regeneration applications.

Aliphatic polymers were the earliest examples of polymers that were used in the field of bone tissue engineering. PLA, PGA, and poly(lactic-*co*-glycolic acid) (PLGA) and their copolymers are well-known aliphatic polymer types. The most widely utilized subtypes of these polymers are D-PLA PDLA, L-PLA (PLLA) forms of PLA, and blend of D, L-PLA (PDLLA), PLA, PGA, and PLGA, especially used in bone fixation devices, sutures, drug carriers, and bioregeneration scaffolds. In addition to aliphatic polymers, high-molecular-weight aliphatic polyesters, poly(L-lactide), PLA, and PCL, are also used for similar regeneration applications. There are also amorphous polymers, poly(orthoesters), which are hydrophobic and pH-sensitive structures. Poly(orthoesters), like pH-sensitive polymers, are especially used as drug delivery systems, because of enabled pH-responsive degradation that can lead to controlled drug release for medical studies.

Mineralization is a crucial part of bone formation and CaP minerals are the building blocks of bone composite; hence, mineral-inducing matrices are important for osteogenesis process. During natural bone formation, collagen I fibrils serve as a template upon which HA deposition occurs to form crystalline bone material. Collagen matrix comprises anionic proteins that are nucleator and inhibitor of mineralization, and density of these proteins changes according to bone type or age for controlling deposition (Kocabey et al., 2013). Accordingly, anionic polymers are used to mimic mineralizable collagen matrices for bone mineral nucleation (Palmer et al., 2008).

For polymer-based mineralization studies, simulated body fluid (SBF), which mimics human plasma in terms of ion concentrations, is a widely applied solution to evaluate *ex vivo* mineral deposition. Synthetic PLGA and poly-L-lactic acid (PLLA) are early examples of scaffolds mineralized in SBF. Mineralized PLGA polymer construct implanted in a mouse calvarial defect and after 4 weeks showed bone mineral growth, whereas nonmineralized scaffolds did not induce any mineralization (Ceylan et al., 2014). Polymers can also be utilized by mixing with biomolecules; in one example, PLLA mixed with collagen I and SBF closely mimicked natural bone mineral nucleating surface (Mata et al., 2009).

Different monolayers such as dihydrogen phosphate (PO_4H_2), carboxylic acid (COOH), and methyl (CH_3) were also investigated for mineralization, and it was found that negatively charged groups are potent inducers of mineralization, where noncharged methyl monolayer did not induce mineralization (Chen et al., 2006; Cowan et al., 2004). Moreover, different monolayers showed different Ca : P ratios and different crystal morphologies, meaning that mineral phase is highly dependent charged groups on templates. Charge-functionalized surface moieties induce interaction between fibronectin and functional groups, and this cooperation attracts integrin binding by osteoblasts and probably simulates integrin binding to induce cell-dependent mineralization. In one example of anionic group-functionalized polymer, carboxymethyl attached to poly(2-hydroxyethyl methacrylate) (PHEMA) hydrogels displayed better mineralization and cell spreading compared to only PHEMA surfaces after incubation with SBF (Tanahashi and Matsuda, 1997).

Phosphorus is one of the most effective mineralization-inducing chemical units, among the other effective groups. The main role of phosphate ion is stabilization of calcium ions on the polymer surfaces. Phosphate-modified polymers mineralize faster both in *in vivo* and *in vitro* conditions (Filmon et al., 2002). Addition of these chemical units also enhances properties of polymers such as vinylphosphonic acid (VPA), which showed better swelling, protein uptake, and adhesion and proliferation of osteoblast (Tan et al., 2005). In another study, PLGA films stabilized and used for enhanced mineralization by the attachment of phosphate ion (Cowan et al., 2004).

Poly(ethylene glycol) (PEG) polymer is one of the most widely used polymers for regenerative studies. Phosphoester-containing PEG hydrogels demonstrate unique characteristics, where polymer backbone allows cleavage of phosphate group by ALP enzyme (Stancu et al., 2004). This material has shown osteoinductive and osteoconductive features by both its phosphate source and phosphate-modified surface characteristics.

However, some of phosphorus-deficient polymer materials are also capable of surface mineralization like phosphorus-containing analogs. Poly(phosphazene)-based composites interacted with hydroxyapatite due to surface hydrolysis, hydrolyzed carboxylate groups on polymer chain, and nucleated hydroxyapatite layer by interacting with Ca ions in the surrounding environment (Li et al., 2006).

TABLE 6.1 Short Peptides for Bone Regeneration

Peptide	Origin	Applications	Reference
RGDS	Found in ECM proteins, mostly in fibronectin and binds to integrin	Cell adhesion, attach cells to the bioactive surface for osteogenic activity	Webber et al. (2010)
IKVAV	Laminin	Cell adhesion, spreading, migration	Tashiro et al. (1989)
YIGSR	Laminin	Cell adhesion	Sur et al. (2012)
DGEA	Collagen type I	Osteoblast-specific binding via α2–β1 integrin	Anderson et al. (2009b)
GFOGER	Collagen IV	Collagen-mimetic sequence	Zhang et al. (2003)
KRSR	Binds to transmembrane proteoglycans	Selectively increase osteoblast adhesion with bioadhesive moiety functionalization	Anderson et al. (2009b)
GAG-PA	Heparan sulfate-mimicking peptide	Protein-based extracellular matrix components, glycosaminoglycans (GAGs) regulate bone formation	Kocabey et al. (2013)
E3-PA + Dopa-PA	Noncollagenous matrix proteins	Mineralization and osteogenic differentiation	Ceylan et al. (2014)
RADA16	Originally designed as ionic self-complementary oligopeptides		Gelain et al. (2006)
ALK	Osteogenic growth peptide	Osteogenic differentiation of osteoprogenitors	
DGR	Osteopontin		
PRG	Two-unit RGD motifs		

6.7 FUTURE PERSPECTIVES

Bone grafts can offer promising platforms for bone regeneration studies and therapeutic applications by integrating dynamics and functions of different materials to reach high responsivity to stimuli, healing potential, and environmental adaptation. The expansion of knowledge on bone structure and reconstruction at molecular level is continuing to lead to improvement of bone regeneration methods. Yet, further information is required to fill remaining gaps in this area. Particularly, short peptide-mediated regeneration requires more detailed analysis in terms of investigation of new sequences and their cellular activities during skeletal repair to elicit cellular–molecular interactions for developing novel and effective regenerative materials. Further understanding in this area, with the integration of a wide variety of material-based strategies, could be the key to develop integrated and efficient bone regeneration methods. Hence, integration and manipulation of bone regeneration strategies that approximate natural bone cascade can lead to successful treatment of many bone diseases including aging- or genetic-related illnesses and bone traumas (Table 6.1).

REFERENCES

Aboushelib, M.N., Salem, N.A., Taleb, A.L., and El Moniem, N.M. (2013). Influence of surface nano-roughness on osseointegration of zirconia implants in rabbit femur heads using selective infiltration etching technique. The Journal of Oral Implantology *39*, 583–590.

Adell, R. (1985). Tissue integrated prostheses in clinical dentistry. International Dental Journal *35*, 259–265.

Albrektsson, T., Dahlin, C., Jemt, T., Sennerby, L., Turri, A., and Wennerberg, A. (2014). Is marginal bone loss around oral implants the result of a provoked foreign body reaction? Clinical Implant Dentistry and Related Research *16*, 155–165.

Albrektsson, T. and Wennerberg, A. (2004). Oral implant surfaces: part 1—review focusing on topographic and chemical properties of different surfaces and in vivo responses to them. The International Journal of Prosthodontics *17*, 536–543.

Alghamdi, H.S., Bosco, R., van den Beucken, J.J.J.P., Walboomers, X.F., and Jansen, J.A. (2013). Osteogenicity of titanium implants coated with calcium phosphate or collagen type-I in osteoporotic rats. Biomaterials *34*, 3747–3757.

Anderson, J.M., Andukuri, A., Lim, D.J., and Jun, H.W. (2009a). Modulating the gelation properties of self-assembling peptide amphiphiles. ACS Nano *3*, 3447–3454.

Anderson, J.M., Kushwaha, M., Tambralli, A., Bellis, S.L., Camata, R.P., and Jun, H.W. (2009b). Osteogenic differentiation of human mesenchymal stem cells directed by extracellular matrix-mimicking ligands in a biomimetic self-assembled peptide amphiphile nanomatrix. Biomacromolecules *10*, 2935–2944.

Anderson, J.M., Vines, J.B., Patterson, J.L., Chen, H., Javed, A., and Jun, H.W. (2011). Osteogenic differentiation of human mesenchymal stem cells synergistically enhanced by biomimetic peptide amphiphiles combined with conditioned medium. Acta Biomaterialia *7*, 675–682.

Anselme, K. (2000). Osteoblast adhesion on biomaterials. Biomaterials *21*, 667–681.

Arima, Y. and Iwata, H. (2007). Effect of wettability and surface functional groups on protein adsorption and cell adhesion using well-defined mixed self-assembled monolayers. Biomaterials *28*, 3074–3082.

Athanasiou, K.A., Zhu, C., Lanctot, D.R., Agrawal, C.M., and Wang, X. (2000). Fundamentals of biomechanics in tissue engineering of bone. Tissue Engineering *6*, 361–381.

Bax, B.E., Wozney, J.M., and Ashhurst, D.E. (1999). Bone morphogenetic protein-2 increases the rate of callus formation after fracture of the rabbit tibia. Calcified Tissue International *65*, 83–89.

Beddoe, A.H., Darley, P.J., and Spiers, F.W. (1976). Measurements of trabecular bone structure in man. Physics in Medicine and Biology *21*, 589–607.

Beertsen, W. and van den Bos, T. (1992). Alkaline phosphatase induces the mineralization of sheets of collagen implanted subcutaneously in the rat. The Journal of Clinical Investigation *89*, 1974–1980.

Bjursten, L.M., Rasmusson, L., Oh, S., Smith, G.C., Brammer, K.S., and Jin, S. (2010). Titanium dioxide nanotubes enhance bone bonding in vivo. Journal of Biomedical Materials Research, Part A *92*, 1218–1224.

Black, K.C.L., Yi, J., Rivera, J.G., Zelasko-Leon, D.C., and Messersmith, P.B. (2012). Polydopamine-enabled surface functionalization of gold nanorods for cancer cell-targeted imaging and photothermal therapy. Nanomedicine *8*, 17–28.

Bobyn, J.D., Pilliar, R.M., Cameron, H.U., and Weatherly, G.C. (1980). The optimum pore size for the fixation of porous-surfaced metal implants by the ingrowth of bone. Clinical Orthopaedics and Related Research, 263–270.

Bosetti, M., Lloyd, A.W., Santin, M., Denyer, S.P., and Cannas, M. (2005). Effects of phosphatidylserine coatings on titanium on inflammatory cells and cell-induced mineralisation in vitro. Biomaterials *26*, 7572–7578.

Brandi, M.L. and Collin-Osdoby, P. (2006). Vascular biology and the skeleton. Journal of Bone and Mineral Research *21*, 183–192.

Branemark, R., Branemark, P.I., Rydevik, B., and Myers, R.R. (2001). Osseointegration in skeletal reconstruction and rehabilitation: a review. Journal of Rehabilitation Research and Development *38*, 175–181.

ButLer, W.T., ed. (2000). Noncollagenous proteins of bone and dentin: a brief overview (Rosemont IL:AAOS).

Canalis, E., McCarthy, T., and Centrella, M. (1988). Growth factors and the regulation of bone remodeling. The Journal of Clinical Investigation *81*, 277–281.

Ceylan, H., Kocabey, S., Tekinay, A.B., and Guler, M.O. (2012). Surface-adhesive and osteogenic self-assembled peptide nanofibers for bioinspired functionalization of titanium surfaces. Soft Matter *8*, 3929–3937.

Ceylan, H., Kocabey, S., Unal Gulsuner, H., Balcik, O.S., Guler, M.O., and Tekinay, A.B. (2014). Bone-like mineral nucleating peptide nanofibers induce differentiation of human mesenchymal stem cells into mature osteoblasts. Biomacromolecules *15*, 2407–2418.

Ceylan, H., Tekinay, A.B., and Guler, M.O. (2011). Selective adhesion and growth of vascular endothelial cells on bioactive peptide nanofiber functionalized stainless steel surface. Biomaterials *32*, 8797–8805.

Ceylan, H., Tekinay, A.B., and Guler, M.O. (2013). Mussel-inspired adhesive interfaces for biomedical applications. In Biological and biomimetic adhesives: challenges and opportunities, R. Santos, N. Aldred, S. Gorb, and P. Flammang, eds. (Cambridge: The Royal Society of Chemistry), pp. 103–116.

Chen, Y., Mak, A.F., Wang, M., and Li, J. (2006). Composite coating of bonelike apatite particles and collagen fibers on poly-L-lactic acid formed through an accelerated biomimetic coprecipitation process. Journal of Biomedical Materials Research Part B: Applied Biomaterials *77*, 315–322.

Chien, C.-Y. and Tsai, W.-B. (2013). Poly(dopamine)-assisted immobilization of Arg-Gly-Asp peptides, hydroxyapatite, and bone morphogenic protein-2 on titanium to improve the osteogenesis of bone marrow stem cells. ACS Applied Materials & Interfaces *5*, 6975–6983.

Chim, S.M., Tickner, J., Chow, S.T., Kuek, V., Guo, B., Zhang, G., Rosen, V., Erber, W., and Xu, J. (2013). Angiogenic factors in bone local environment. Cytokine & Growth Factor Reviews *24*, 297–310.

Chollet, C., Chanseau, C., Remy, M., Guignandon, A., Bareille, R., Labrugere, C., Bordenave, L., and Durrieu, M.C. (2009). The effect of RGD density on osteoblast and endothelial cell behavior on RGD-grafted polyethylene terephthalate surfaces. Biomaterials *30*, 711–720.

Chug, A., Shukla, S., Mahesh, L., and Jadwani, S. (2013). Osseointegration—molecular events at the bone–implant interface: a review. Journal of Oral and Maxillofacial Surgery, Medicine, and Pathology *25*, 1–4.

Cowan, C.M., Shi, Y.Y., Aalami, O.O., Chou, Y.F., Mari, C., Thomas, R., Quarto, N., Contag, C.H., Wu, B., and Longaker, M.T. (2004). Adipose-derived adult stromal cells heal critical-size mouse calvarial defects. Nature Biotechnology *22*, 560–567.

Cui, H., Webber, M.J., and Stupp, S.I. (2010). Self-assembly of peptide amphiphiles: from molecules to nanostructures to biomaterials. Biopolymers *94*, 1–18.

Daculsi, G., Passuti, N., Martin, S., Deudon, C., Legeros, R.Z., and Raher, S. (1990). Macroporous calcium phosphate ceramic for long bone surgery in humans and dogs. Clinical and histological study. Journal of Biomedical Materials Research *24*, 379–396.

David, L., Feige, J.J., and Bailly, S. (2009). Emerging role of bone morphogenetic proteins in angiogenesis. Cytokine & Growth Factor Reviews *20*, 203–212.

de Oliveira, P.T. and Nanci, A. (2004). Nanotexturing of titanium-based surfaces upregulates expression of bone sialoprotein and osteopontin by cultured osteogenic cells. Biomaterials *25*, 403–413.

Deckers, M.M., van Bezooijen, R.L., van der Horst, G., Hoogendam, J., van Der Bent, C., Papapoulos, S.E., and Lowik, C.W. (2002). Bone morphogenetic proteins stimulate angiogenesis through osteoblast-derived vascular endothelial growth factor A. Endocrinology *143*, 1545–1553.

Dee, K.C., Andersen, T.T., and Bizios, R. (1998). Design and function of novel osteoblast-adhesive peptides for chemical modification of biomaterials. Journal of Biomedical Materials Research *40*, 371–377.

Dohan Ehrenfest, D.M., Coelho, P.G., Kang, B.-S., Sul, Y.-T., and Albrektsson, T. (2010). Classification of osseointegrated implant surfaces: materials, chemistry and topography. Trends in Biotechnology *28*, 198–206.

Ducy, P., Geoffroy, V., and Karsenty, G. (1996). Study of osteoblast-specific expression of one mouse osteocalcin gene: characterization of the factor binding to OSE2. Connective Tissue Research *35*, 7–14.

Eanes, E.D. (1989). Biophysical aspects of lipid interaction with mineral: liposome model studies. The Anatomical Record *224*, 220–225.

Engel, J., Taylor, W., Paulsson, M., Sage, H., and Hogan, B. (1987). Calcium binding domains and calcium-induced conformational transition of SPARC/BM-40/osteonectin, an extracellular glycoprotein expressed in mineralized and nonmineralized tissues. Biochemistry *26*, 6958–6965.

Englund, E.A., Wang, D., Fujigaki, H., Sakai, H., Micklitsch, C.M., Ghirlando, R., Martin-Manso, G., Pendrak, M.L., Roberts, D.D., Durell, S.R., et al. (2012). Programmable multivalent display of receptor ligands using peptide nucleic acid nanoscaffolds. Nature Communications *3*, 614.

Erlebacher, A., Filvaroff, E.H., Ye, J.Q., and Derynck, R. (1998). Osteoblastic responses to TGF-beta during bone remodeling. Molecular Biology of the Cell *9*, 1903–1918.

Fang, T.D., Salim, A., Xia, W., Nacamuli, R.P., Guccione, S., Song, H.M., Carano, R.A., Filvaroff, E.H., Bednarski, M.D., Giaccia, A.J., et al. (2005). Angiogenesis is required for successful bone induction during distraction osteogenesis. Journal of Bone and Mineral Research *20*, 1114–1124.

Fernandez-Yague, M.A., Abbah, S.A., McNamara, L., Zeugolis, D.I., Pandit, A., and Biggs, M.J. (2014). Biomimetic approaches in bone tissue engineering: integrating biological and physicomechanical strategies. Advanced Drug Delivery Reviews *84*, 1–29.

Ferrara, N., Gerber, H.P., and LeCouter, J. (2003). The biology of VEGF and its receptors. Nature Medicine *9*, 669–676.

Fiedler, J., Roderer, G., Gunther, K.P., and Brenner, R.E. (2002). BMP-2, BMP-4, and PDGF-bb stimulate chemotactic migration of primary human mesenchymal progenitor cells. Journal of Cellular Biochemistry *87*, 305–312.

Filmon, R., Grizon, F., Basle, M.F., and Chappaard, D. (2002). Effects of negatively charged groups (carboxymethyl) on the calcification of poly(2-hydroxyethyl methacrylate). Biomaterials *23*, 3053–3059.

Fratzl-Zelman, N., Fratzl, P., Horandner, H., Grabner, B., Varga, F., Ellinger, A., and Klaushofer, K. (1998). Matrix mineralization in MC3T3-E1 cell cultures initiated by beta-glycerophosphate pulse. Bone *23*, 511–520.

Gailit, J., Clarke, C., Newman, D., Tonnesen, M.G., Mosesson, M.W., and Clark, R.A. (1997). Human fibroblasts bind directly to fibrinogen at RGD sites through integrin alpha(v)beta3. Experimental Cell Research *232*, 118–126.

Garcia, A.J. and Reyes, C.D. (2005). Bio-adhesive surfaces to promote osteoblast differentiation and bone formation. Journal of Dental Research *84*, 407–413.

Gelain, F., Bottai, D., Vescovi, A., and Zhang, S. (2006). Designer self-assembling peptide nanofiber scaffolds for adult mouse neural stem cell 3-dimensional cultures. PLoS One *1*, e119.

Gerber, H.P., Vu, T.H., Ryan, A.M., Kowalski, J., Werb, Z., and Ferrara, N. (1999). VEGF couples hypertrophic cartilage remodeling, ossification and angiogenesis during endochondral bone formation. Nature Medicine *5*, 623–628.

Gittens, R.A., Olivares-Navarrete, R., Schwartz, Z., and Boyan, B.D. (2014). Implant osseointegration and the role of microroughness and nanostructures: lessons for spine implants. Acta Biomaterialia *10*, 3363–3371.

Glimcher, M.J. (1992). The nature of the mineral component of bone and the mechanisms of calcification (New York: Raven Press).

Globus, R.K., Plouet, J., and Gospodarowicz, D. (1989). Cultured bovine bone cells synthesize basic fibroblast growth factor and store it in their extracellular matrix. Endocrinology *124*, 1539–1547.

Goldstein, S.A., Wilson, D.L., Sonstegard, D.A., and Matthews, L.S. (1983). The mechanical properties of human tibial trabecular bone as a function of metaphyseal location. Journal of Biomechanics *16*, 965–969.

Gowen, M., Wood, D.D., Ihrie, E.J., Mcguire, M.K.B., and Russell, R.G.G. (1983). An interleukin-1 like factor stimulates bone-resorption in vitro. Nature *306*, 378–380.

Guler, M.O., Hsu, L., Soukasene, S., Harrington, D.A., Hulvat, J.F., and Stupp, S.I. (2006). Presentation of RGDS epitopes on self-assembled nanofibers of branched peptide amphiphiles. Biomacromolecules *7*, 1855–1863.

Hanada, K., Dennis, J.E., and Caplan, A.I. (1997). Stimulatory effects of basic fibroblast growth factor and bone morphogenetic protein-2 on osteogenic differentiation of rat bone marrow-derived mesenchymal stem cells. Journal of Bone and Mineral Research *12*, 1606–1614.

Harbers, G.M. and Healy, K.E. (2005). The effect of ligand type and density on osteoblast adhesion, proliferation, and matrix mineralization. Journal of Biomedical Materials Research, Part A *75*, 855–869.

Hartgerink, J.D., Beniash, E., and Stupp, S.I. (2001). Self-assembly and mineralization of peptide-amphiphile nanofibers. Science *294*, 1684–1688.

Hauge, E.M., Qvesel, D., Eriksen, E.F., Mosekilde, L., and Melsen, F. (2001). Cancellous bone remodeling occurs in specialized compartments lined by cells expressing osteoblastic markers. Journal of Bone and Mineral Research *16*, 1575–1582.

Hauschka, P.V., Mavrakos, A.E., Iafrati, M.D., Doleman, S.E., and Klagsbrun, M. (1986). Growth-factors in bone-matrix—isolation of multiple types by affinity-chromatography on heparin-Sepharose. Journal of Biological Chemistry *261*, 2665–2674.

He, C. and Chen, X. (2005). Transcription regulation of the vegf gene by the BMP/Smad pathway in the angioblast of zebrafish embryos. Biochemical and Biophysical Research Communications *329*, 324–330.

Hennessy, K.M., Pollot, B.E., Clem, W.C., Phipps, M.C., Sawyer, A.A., Culpepper, B.K., and Bellis, S.L. (2009). The effect of collagen I mimetic peptides on mesenchymal stem cell adhesion and differentiation, and on bone formation at hydroxyapatite surfaces. Biomaterials *30*, 1898–1909.

Henriksen, K., Karsdal, M., Delaisse, J.M., and Engsig, M.T. (2003). RANKL and vascular endothelial growth factor (VEGF) induce osteoclast chemotaxis through an ERK1/2-dependent mechanism. Journal of Biological Chemistry *278*, 48745–48753.

Hoption Cann, S., van Netten, J., and van Netten, C. (2003). Dr. William Coley and tumour regression: a place in history or in the future. Postgraduate Medical Journal *79*, 672–680.

Horii, A., Wang, X., Gelain, F., and Zhang, S. (2007). Biological designer self-assembling peptide nanofiber scaffolds significantly enhance osteoblast proliferation, differentiation and 3-D migration. PLoS One *2*, e190.

Hunter, G.K., Hauschka, P.V., Poole, A.R., Rosenberg, L.C., and Goldberg, H.A. (1996). Nucleation and inhibition of hydroxyapatite formation by mineralized tissue proteins. The Biochemical Journal *317* (Pt 1), 59–64.

Hynes, R.O. (1992). Integrins: versatility, modulation, and signaling in cell adhesion. Cell *69*, 11–25.

Kalajzic, I., Staal, A., Yang, W.-P., Wu, Y., Johnson, S.E., Feyen, J.H.M., Krueger, W., Maye, P., Yu, F., Zhao, Y., et al. (2005). Expression profile of osteoblast lineage at defined stages of differentiation. Journal of Biological Chemistry *280*, 24618–24626.

Kanczler, J.M. and Oreffo, R.O. (2008). Osteogenesis and angiogenesis: the potential for engineering bone. European Cells & Materials *15*, 100–114.

Kang, S.M., Hwang, N.S., Yeom, J., Park, S.Y., Messersmith, P.B., Choi, I.S., Langer, R., Anderson, D.G., and Lee, H. (2012). One-step multipurpose surface functionalization by adhesive catecholamine. Advanced Functional Materials *22*, 2949–2955.

Katsamenis, O.L., Chong, H.M.H., Andriotis, O.G., and Thurner, P.J. (2013). Load-bearing in cortical bone microstructure: selective stiffening and heterogeneous strain distribution at the lamellar level. Journal of the Mechanical Behavior of Biomedical Materials *17*, 152–165.

Khan, S.N., Ramachandran, M., Senthil Kumar, S., Krishnan, V., and Sundaram, R. (2012). Osseointegration and more—a review of literature. Indian Journal of Dentistry *3*, 72–76.

Kielty, C.M. and Grant, M. E. (2002). The collagen family: structure, assembly, and organization in the extracellular matrix. In Connective tissue and its heritable disorders: molecular, genetic, and medical aspects, 2nd edn. (eds P. M. Royce and B. Steinmann, John Wiley & Sons, Inc., Hoboken).

Klein, S., Giancotti, F.G., Presta, M., Albelda, S.M., Buck, C.A., and Rifkin, D.B. (1993). Basic fibroblast growth factor modulates integrin expression in microvascular endothelial cells. Molecular Biology of the Cell *4*, 973–982.

Kocabey, S., Ceylan, H., Tekinay, A.B., and Guler, M.O. (2013). Glycosaminoglycan mimetic peptide nanofibers promote mineralization by osteogenic cells. Acta Biomaterialia *9*, 9075–9085.

Lavenus, S., Louarn, G., and Layrolle, P. (2010). Nanotechnology and dental implants. International Journal of Biomaterials *2010*, 9.

Layrolle, P. (2011). 1.112—Calcium phosphate coatings. In Comprehensive biomaterials, D. Paul, ed. (Oxford: Elsevier), pp. 223–229.

Le Guéhennec, L., Soueidan, A., Layrolle, P., and Amouriq, Y. (2007). Surface treatments of titanium dental implants for rapid osseointegration. Dental Materials *23*, 844–854.

Lee, D.H., Park, B.J., Lee, M.S., Lee, J.W., Kim, J.K., Yang, H.C., and Park, J.C. (2006). Chemotactic migration of human mesenchymal stem cells and MC3T3-E1 osteoblast-like cells induced by COS-7 cell line expressing rhBMP-7. Tissue Engineering *12*, 1577–1586.

Lee, H., Dellatore, S.M., Miller, W.M., and Messersmith, P.B. (2007). Mussel-inspired surface chemistry for multifunctional coatings. Science *318*, 426–430.

LeGeros, R.Z. (2008). Calcium phosphate-based osteoinductive materials. Chemical Reviews *108*, 4742–4753.

Li, G., Cui, Y., McIlmurray, L., Allen, W.E., and Wang, H. (2005). rhBMP-2, rhVEGF(165), rhPTN and thrombin-related peptide, TP508 induce chemotaxis of human osteoblasts and microvascular endothelial cells. Journal of Orthopaedic Research *23*, 680–685.

Li, Q., Wang, J., Shahani, S., Sun, D.D., Sharma, B., Elisseeff, J.H., and Leong, K.W. (2006). Biodegradable and photocrosslinkable polyphosphoester hydrogel. Biomaterials *27*, 1027–1034.

Lim, Y.-b., Moon, K.-S., and Lee, M. (2009). Recent advances in functional supramolecular nanostructures assembled from bioactive building blocks. Chemical Society Reviews *38*, 925–934.

Linder, L. (1989). Osseointegration of metallic implants. I. Light microscopy in the rabbit. Acta Orthopaedica Scandinavica *60*, 129–134.

Mackie, E.J. (2003). Osteoblasts: novel roles in orchestration of skeletal architecture. The International Journal of Biochemistry & Cell Biology *35*, 1301–1305.

Maes, C., Carmeliet, P., Moermans, K., Stockmans, I., Smets, N., Collen, D., Bouillon, R., and Carmeliet, G. (2002). Impaired angiogenesis and endochondral bone formation in mice lacking the vascular endothelial growth factor isoforms VEGF164 and VEGF188. Mechanisms of Development *111*, 61–73.

Maes, C., Stockmans, I., Moermans, K., Van Looveren, R., Smets, N., Carmeliet, P., Bouillon, R., and Carmeliet, G. (2004). Soluble VEGF isoforms are essential for establishing epiphyseal vascularization and regulating chondrocyte development and survival. The Journal of Clinical Investigation *113*, 188–199.

Marinucci, L., Balloni, S., Becchetti, E., Belcastro, S., Guerra, M., Calvitti, M., Lilli, C., Calvi, E.M., and Locci, P. (2006). Effect of titanium surface roughness on human osteoblast proliferation and gene expression in vitro. The International Journal of Oral & Maxillofacial Implants *21*, 719–725.

Mata, A., Hsu, L., Capito, R., Aparicio, C., Henrikson, K., and Stupp, S.I. (2009). Micropatterning of bioactive self-assembling gels. Soft Matter *5*, 1228–1236.

Maurer, P., Hohenester, E., and Engel, J. (1996). Extracellular calcium-binding proteins. Current Opinion in Cell Biology *8*, 609–617.

Mavrogenis, A.F., Papagelopoulos, P.J., and Babis, G.C. (2011). Osseointegration of cobalt-chrome alloy implants. Journal of Long-Term Effects of Medical Implants *21*, 349–358.

Mayr-Wohlfart, U., Waltenberger, J., Hausser, H., Kessler, S., Gunther, K.P., Dehio, C., Puhl, W., and Brenner, R.E. (2002). Vascular endothelial growth factor stimulates chemotactic migration of primary human osteoblasts. Bone *30*, 472–477.

Mehrotra, M., Krane, S.M., Walters, K., and Pilbeam, C. (2004). Differential regulation of platelet-derived growth factor stimulated migration and proliferation in osteoblastic cells. Journal of Cellular Biochemistry *93*, 741–752.

Melo, M.D., Shafie, H., and Obeid, G. (2006). Implant survival rates for oral and maxillofacial surgery residents: a retrospective clinical review with analysis of resident level of training on implant survival. Journal of Oral and Maxillofacial Surgery *64*, 1185–1189.

Mendonca, G., Mendonca, D.B., Simoes, L.G., Araujo, A.L., Leite, E.R., Duarte, W.R., Aragao, F.J., and Cooper, L.F. (2009). The effects of implant surface nanoscale features on osteoblast-specific gene expression. Biomaterials *30*, 4053–4062.

Mistry, A.S. and Mikos, A.G. (2005). Tissue engineering strategies for bone regeneration. Advances in Biochemical Engineering/Biotechnology *94*, 1–22.

Mizuno, M., Fujisawa, R., and Kuboki, Y. (2000). Type I collagen-induced osteoblastic differentiation of bone-marrow cells mediated by collagen-alpha2beta1 integrin interaction. Journal of Cellular Physiology *184*, 207–213.

Mizuno, M. and Kuboki, Y. (2001). Osteoblast-related gene expression of bone marrow cells during the osteoblastic differentiation induced by type I collagen. Journal of Biochemistry *129*, 133–138.

Mohan, S. and Baylink, D.J. (1991). Bone growth factors. Clinical Orthopaedics and Related Research *263*, 30–48.

Montero, A., Okada, Y., Tomita, M., Ito, M., Tsurukami, H., Nakamura, T., Doetschman, T., Coffin, J.D., and Hurley, M.M. (2000). Disruption of the fibroblast growth factor-2 gene results in decreased bone mass and bone formation. The Journal of Clinical Investigation *105*, 1085–1093.

Mundi, S., Pindiprolu, B., Simunovic, N., and Bhandari, M. (2014). Similar mortality rates in hip fracture patients over the past 31 years. Acta Orthopaedica *85*, 54–59.

Nair, A.K., Gautieri, A., Chang, S.W., and Buehler, M.J. (2013). Molecular mechanics of mineralized collagen fibrils in bone. Nature Communications *4*, 1724.

Nelson, M., Balasundaram, G., and Webster, T.J. (2006). Increased osteoblast adhesion on nanoparticulate crystalline hydroxyapatite functionalized with KRSR. International Journal of Nanomedicine *1*, 339–349.

Niece, K.L., Hartgerink, J.D., Donners, J.J.J.M., and Stupp, S.I. (2003). Self-assembly combining two bioactive peptide-amphiphile molecules into nanofibers by electrostatic attraction. Journal of the American Chemical Society *125*, 7146–7147.

Oldberg, A., Franzen, A., and Heinegard, D. (1986). Cloning and sequence analysis of rat bone sialoprotein (osteopontin) cDNA reveals an Arg-Gly-Asp cell-binding sequence. Proceedings of the National Academy of Sciences of the United States of America *83*, 8819–8823.

Oldberg, A., Franzen, A., Heinegard, D., Pierschbacher, M., and Ruoslahti, E. (1988). Identification of a bone sialoprotein receptor in osteosarcoma cells. The Journal of Biological Chemistry *263*, 19433–19436.

Olivares-Navarrete, R., Hyzy, S.L., Berg, M.E., Schneider, J.M., Hotchkiss, K., Schwartz, Z., and Boyan, B.D. (2014). Osteoblast lineage cells can discriminate microscale topographic features on titanium-aluminum-vanadium surfaces. Annals of Biomedical Engineering *42*, 2551–2561.

Olivares-Navarrete, R., Raz, P., Zhao, G., Chen, J., Wieland, M., Cochran, D.L., Chaudhri, R.A., Ornoy, A., Boyan, B.D., and Schwartz, Z. (2008). Integrin $\alpha 2\beta 1$ plays a critical role in osteoblast response to micron-scale surface structure and surface energy of titanium substrates. Proceedings of the National Academy of Sciences *105*, 15767–15772.

Palmer, L.C., Newcomb, C.J., Kaltz, S.R., Spoerke, E.D., and Stupp, S.I. (2008). Biomimetic systems for hydroxyapatite mineralization inspired by bone and enamel. Chemistry Review *108*, 4754–4783.

Pampena, D.A., Robertson, K.A., Litvinova, O., Lajoie, G., Goldberg, H.A., and Hunter, G.K. (2004). Inhibition of hydroxyapatite formation by osteopontin phosphopeptides. The Biochemical Journal *378*, 1083–1087.

Penarrocha-Diago, M., Demarchi, C.L., Maestre-Ferrin, L., Carrillo, C., Penarrocha-Oltra, D., and Penarrocha-Diago, M.A. (2012). A retrospective comparison of 1,022 implants: immediate versus nonimmediate. The International Journal of Oral & Maxillofacial Implants *27*, 421–427.

Peng, H., Usas, A., Olshanski, A., Ho, A.M., Gearhart, B., Cooper, G.M., and Huard, J. (2005). VEGF improves, whereas sFlt1 inhibits, BMP2-induced bone formation and bone healing through modulation of angiogenesis. Journal of Bone and Mineral Research *20*, 2017–2027.

Pitaru, S., Kotev-Emeth, S., Noff, D., Kaffuler, S., and Savion, N. (1993). Effect of basic fibroblast growth factor on the growth and differentiation of adult stromal bone marrow cells: enhanced development of mineralized bone-like tissue in culture. Journal of Bone and Mineral Research *8*, 919–929.

Popov, C., Radic, T., Haasters, F., Prall, W.C., Aszodi, A., Gullberg, D., Schieker, M., and Docheva, D. (2011). Integrins [alpha]2[beta]1 and [alpha]11[beta]1 regulate the survival of mesenchymal stem cells on collagen I. Cell Death and Disease *2*, e186.

Puleo, D.A. and Bizios, R. (1991). RGDS tetrapeptide binds to osteoblasts and inhibits fibronectin-mediated adhesion. Bone *12*, 271–276.

Puleo, D.A. and Nanci, A. (1999). Understanding and controlling the bone–implant interface. Biomaterials *20*, 2311–2321.

Pytela, R., Pierschbacher, M.D., Argraves, S., Suzuki, S., and Ruoslahti, E. (1987). Arginine-glycine-aspartic acid adhesion receptors. Methods in Enzymology *144*, 475–489.

Qi, H., Aguiar, D.J., Williams, S.M., La Pean, A., Pan, W., and Verfaillie, C.M. (2003). Identification of genes responsible for osteoblast differentiation from human mesodermal progenitor cells. Proceedings of the National Academy of Sciences *100*, 3305–3310.

Raggio, C.L., Boyan, B.D., and Boskey, A.L. (1986). In vivo hydroxyapatite formation induced by lipids. Journal of Bone and Mineral Research *1*, 409–415.

Reznikov, N., Shahar, R., and Weiner, S. (2014). Bone hierarchical structure in three dimensions. Acta Biomaterialia *10*, 3815–3826.

Roberts, J.E., Bonar, L.C., Griffin, R.G., and Glimcher, M.J. (1992). Characterization of very young mineral phases of bone by solid state 31phosphorus magic angle sample spinning nuclear magnetic resonance and X-ray diffraction. Calcified Tissue International *50*, 42–48.

Ruardy, T.G., Schakenraad, J.M., van der Mei, H.C., and Busscher, H.J. (1995). Adhesion and spreading of human skin fibroblasts on physicochemically characterized gradient surfaces. Journal of Biomedical Materials Research *29*, 1415–1423.

Sato, Y., Shimada, T., and Takaki, R. (1991). Autocrinological role of basic fibroblast growth factor on tube formation of vascular endothelial cells in vitro. Biochemical and Biophysical Research Communications *180*, 1098–1102.

Schwarz, F., Wieland, M., Schwartz, Z., Zhao, G., Rupp, F., Geis-Gerstorfer, J., Schedle, A., Broggini, N., Bornstein, M.M., Buser, D., et al. (2009). Potential of chemically modified hydrophilic surface characteristics to support tissue integration of titanium dental implants. Journal of Biomedical Materials Research Part B: Applied Biomaterials *88*, 544–557.

Seghezzi, G., Patel, S., Ren, C.J., Gualandris, A., Pintucci, G., Robbins, E.S., Shapiro, R.L., Galloway, A.C., Rifkin, D.B., and Mignatti, P. (1998). Fibroblast growth factor-2 (FGF-2) induces vascular endothelial growth factor (VEGF) expression in the endothelial cells of forming capillaries: an autocrine mechanism contributing to angiogenesis. The Journal of Cell Biology *141*, 1659–1673.

Shao, E.S., Lin, L., Yao, Y., and Bostrom, K.I. (2009). Expression of vascular endothelial growth factor is coordinately regulated by the activin-like kinase receptors 1 and 5 in endothelial cells. Blood *114*, 2197–2206.

Shu, X.Z., Ghosh, K., Liu, Y., Palumbo, F.S., Luo, Y., Clark, R.A., and Prestwich, G.D. (2004). Attachment and spreading of fibroblasts on an RGD peptide-modified injectable hyaluronan hydrogel. Journal of Biomedical Materials Research, Part A *68*, 365–375.

Sikavitsas, V.I., Temenoff, J.S., and Mikos, A.G. (2001). Biomaterials and bone mechanotransduction. Biomaterials *22*, 2581–2593.

Silva, G.A., Czeisler, C., Niece, K.L., Beniash, E., Harrington, D.A., Kessler, J.A., and Stupp, S.I. (2004). Selective differentiation of neural progenitor cells by high-epitope density nanofibers. Science *303*, 1352–1355.

Simonis, P., Dufour, T., and Tenenbaum, H. (2010). Long-term implant survival and success: a 10–16-year follow-up of non-submerged dental implants. Clinical Oral Implants Research *21*, 772–777.

Smalley, W.M., Shapiro, P.A., Hohl, T.H., Kokich, V.G., and Branemark, P.I. (1988). Osseointegrated titanium implants for maxillofacial protraction in monkeys. American Journal of Orthodontics and Dentofacial Orthopedics *94*, 285–295.

Sojo, K., Sawaki, Y., Hattori, H., Mizutani, H., and Ueda, M. (2005). Immunohistochemical study of vascular endothelial growth factor (VEGF) and bone morphogenetic protein-2,-4 (BMP-2,-4) on lengthened rat femurs. Journal of Cranio-Maxillofacial Surgery *33*, 238–245.

Song, W. and Mano, J.F. (2013). Interactions between cells or proteins and surfaces exhibiting extreme wettabilities. Soft Matter *9*, 2985–2999.

Sousa, S.R., Lamghari, M., Sampaio, P., Moradas-Ferreira, P., and Barbosa, M.A. (2008). Osteoblast adhesion and morphology on TiO_2 depends on the competitive preadsorption of albumin and fibronectin. Journal of Biomedical Materials Research, Part A *84*, 281–290.

Southam, J.C. and Selwyn, P. (1970). Structural changes around screws used in the treatment of fractured human mandibles. British Journal of Oral Surgery *8*, 211–221.

Sporn, M.B., Roberts, A.B., Wakefield, L.M., and Assoian, R.K. (1986). Transforming growth factor-beta: biological function and chemical structure. Science *233*, 532–534.

Staatz, W.D., Fok, K.F., Zutter, M.M., Adams, S.P., Rodriguez, B.A., and Santoro, S.A. (1991). Identification of a tetrapeptide recognition sequence for the alpha 2 beta 1 integrin in collagen. Journal of Biological Chemistry *266*, 7363–7367.

Stancu, I.C., Filmon, R., Grizon, F., Zaharia, C., Cincu, C., Basle, M.F., and Chappard, D. (2004). The in vivo calcification capacity of a copolymer, based on methacryloyloxyethyl phosphate, does not favor osteoconduction. Journal of Biomedical Materials Research, Part A *69*, 584–589.

Storrie, H. and Stupp, S.I. (2005). Cellular response to zinc-containing organoapatite: an in vitro study of proliferation, alkaline phosphatase activity and biomineralization. Biomaterials *26*, 5492–5499.

Street, J., Bao, M., deGuzman, L., Bunting, S., Peale, F.V., Jr., Ferrara, N., Steinmetz, H., Hoeffel, J., Cleland, J.L., Daugherty, A., et al. (2002). Vascular endothelial growth factor stimulates bone repair by promoting angiogenesis and bone turnover. Proceedings of the National Academy of Sciences of the United States of America *99*, 9656–9661.

Sur, S., Pashuck, E.T., Guler, M.O., Ito, M., Stupp, S.I., and Launey, T. (2012). A hybrid nanofiber matrix to control the survival and maturation of brain neurons. Biomaterials *33*, 545–555.

Tan, J., Gemeinhart, R.A., Ma, M., and Saltzman, W.M. (2005). Improved cell adhesion and proliferation on synthetic phosphonic acid-containing hydrogels. Biomaterials *26*, 3663–3671.

Tanahashi, M. and Matsuda, T. (1997). Surface functional group dependence on apatite formation on self-assembled monolayers in a simulated body fluid. Journal of Biomedical Materials Research *34*, 305–315.

Tang, Y., Wu, X., Lei, W., Pang, L., Wan, C., Shi, Z., Zhao, L., Nagy, T.R., Peng, X., Hu, J., et al. (2009). TGF-β1-induced migration of bone mesenchymal stem cells couples bone resorption with formation. Nature Medicine *15*, 757–765.

Tashiro, K., Sephel, G.C., Weeks, B., Sasaki, M., Martin, G.R., Kleinman, H.K., and Yamada, Y. (1989). A synthetic peptide containing the IKVAV sequence from the A chain of laminin mediates cell attachment, migration, and neurite outgrowth. The Journal of Biological Chemistry *264*, 16174–16182.

Traub, W., Arad, T., and Weiner, S. (1989). Three-dimensional ordered distribution of crystals in turkey tendon collagen fibers. Proceedings of the National Academy of Sciences of the United States of America *86*, 9822–9826.

Traub, W., Arad, T., and Weiner, S. (1992). Origin of mineral crystal growth in collagen fibrils. Matrix *12*, 251–255.

Triffit, J.T. (1980). The organic matrix of bone tissue (Philadelphia, PA: J. B. Lippincott).

Unal Gulsuner, H., Atalay Gengec, N., Kilinc, M., Erbil, H.Y., and Tekinay, A.B. (2015). Osteoselection supported by phase separated polymer blend films. Journal of Biomedical Materials Research, Part A *103*, 154–161.

Urist, M.R. (1965). Bone: formation by autoinduction. Science *150*, 893–899.

Urist, M.R., DeLange, R.J., and Finerman, G.A. (1983). Bone cell differentiation and growth factors. Science *220*, 680–686.

Vanhegan, I.S., Malik, A.K., Jayakumar, P., Ul Islam, S., and Haddad, F.S. (2012). A financial analysis of revision hip arthroplasty: the economic burden in relation to the national tariff. The Journal of Bone and Joint Surgery *94*, 619–623.

Vasak, C., Busenlechner, D., Schwarze, U.Y., Leitner, H.F., Munoz Guzon, F., Hefti, T., Schlottig, F., and Gruber, R. (2014). Early bone apposition to hydrophilic and hydrophobic titanium implant surfaces: a histologic and histomorphometric study in minipigs. Clinical Oral Implants Research *25*, 1378–1385.

von Wilmowsky, C., Moest, T., Nkenke, E., Stelzle, F., and Schlegel, K.A. (2014). Implants in bone: part I. A current overview about tissue response, surface modifications and future perspectives. Oral and Maxillofacial Surgery *18*, 243–257.

Wang, X., Tolba, E., Schroder, H.C., Neufurth, M., Feng, Q., Diehl-Seifert, B., and Muller, W.E. (2014). Effect of bioglass on growth and biomineralization of SaOS-2 cells in hydrogel after 3D cell bioprinting. PLoS One *9*, e112497.

Wazen, J.J., Gupta, R., Ghossaini, S., Spitzer, J., Farrugia, M., and Tjellstrom, A. (2007). Osseointegration timing for Baha system loading. The Laryngoscope *117*, 794–796.

Webber, M.J., Tongers, J., Renault, M.A., Roncalli, J.G., Losordo, D.W., and Stupp, S.I. (2010). Development of bioactive peptide amphiphiles for therapeutic cell delivery. Acta Biomaterialia *6*, 3–11.

Weiner, S. and Traub, W. (1986). Organization of hydroxyapatite crystals within collagen fibrils. FEBS Letters *206*, 262–266.

Wilson, C.J., Clegg, R.E., Leavesley, D.I., and Pearcy, M.J. (2005). Mediation of biomaterial-cell interactions by adsorbed proteins: a review. Tissue Engineering *11*, 1–18.

Wozney, J.M. and Rosen, V. (1998). Bone morphogenetic protein and bone morphogenetic protein gene family in bone formation and repair. Clinical Orthopaedics and Related Research *346*, 26–37.

Wuttke, M., Muller, S., Nitsche, D.P., Paulsson, M., Hanisch, F.G., and Maurer, P. (2001). Structural characterization of human recombinant and bone-derived bone sialoprotein. Functional implications for cell attachment and hydroxyapatite binding. The Journal of Biological Chemistry *276*, 36839–36848.

Yasuda, H., Shima, N., Nakagawa, N., Yamaguchi, K., Kinosaki, M., Mochizuki, S., Tomoyasu, A., Yano, K., Goto, M., Murakami, A., et al. (1998). Osteoclast differentiation factor is a ligand for osteoprotegerin/osteoclastogenesis-inhibitory factor and is identical to TRANCE/RANKL. Proceedings of the National Academy of Sciences of the United States of America *95*, 3597–3602.

Yeo, I.S. (2014). Reality of dental implant surface modification: a short literature review. The Open Biomedical Engineering Journal *8*, 114–119.

Yoo, S.Y., Kobayashi, M., Lee, P.P., and Lee, S.-W. (2011). Early osteogenic differentiation of mouse preosteoblasts induced by collagen-derived DGEA-peptide on nanofibrous phage tissue matrices. Biomacromolecules *12*, 987–996.

Young, M.F., Kerr, J.M., Ibaraki, K., Heegaard, A.M., and Robey, P.G. (1992). Structure, expression, and regulation of the major noncollagenous matrix proteins of bone. Clinical Orthopaedics and Related Research *281*, 275–294.

Yuan, H., Fernandes, H., Habibovic, P., de Boer, J., Barradas, A.M.C., de Ruiter, A., Walsh, W.R., van Blitterswijk, C.A., and de Bruijn, J.D. (2010). Osteoinductive ceramics as a synthetic alternative to autologous bone grafting. Proceedings of the National Academy of Sciences United States of America *107*, 13614–13619.

Zhang, S. (2003). Fabrication of novel biomaterials through molecular self-assembly. Nature Biotechnology *21*, 1171–1178.

Zhang, W-M, Käpylä, J., Puranen, J.S., Knight, C.G., Tiger, C.F., Pentikäinen O.T., Johnson, M.S., Farndale, R.W., Heino, J. and Gullberg. D. (2003) alpha-11 beta-1 Integrin recognizes the GFOGER sequence in interstitial collagens. J. Biol. Chem. *278*, 7270–7277.

7

NANOMATERIALS FOR THE REPAIR AND REGENERATION OF DENTAL TISSUES

Gulistan Tansik, Alper Devrim Ozkan, Mustafa O. Guler, and Ayse B. Tekinay

Institute of Materials Science and Nanotechnology, National Nanotechnology Research Center (UNAM), Bilkent University, Ankara, Turkey

7.1 INTRODUCTION

Among all tissues of the human body, teeth are perhaps the most frequently exposed to the greatest variety of environmental stresses. Owing to the diversity of food items available in the present world, teeth must not only endure the wear created by the crushing and grinding of food but also resist considerable variances in temperature and acidity, as well as invasions by pathogenic bacteria that may compromise the integrity of the tooth structure through demineralization of the overlaying enamel. Constant exposure to these stresses, combined with the fact that adult human teeth cannot produce enamel and may only regenerate lost dentin (Nguyen et al., 2013), trigger the formation of regions conductive to the growth of caries. Once formed, dental cavities cannot be regenerated and may only be excised and replaced with a suitable filling material. More

Therapeutic Nanomaterials, First Edition. Edited by Mustafa O. Guler and Ayse B. Tekinay.

major tooth injuries, caused by trauma or periodontal disease, may likewise necessitate the artificial replacement of one or more teeth. As such, dental care is a major issue in the modern world and much effort has been spent in fluoridation and dental health campaigns to reduce the prevalence of tooth decay (Mullen, 2005).

Due to their unique function in masticating food, teeth exhibit a structure quite unlike other mineralized tissues. The outermost layer of the tooth consists of a heavily mineralized, wear-resistant substance called *enamel*, which is produced through the activity of specialized matrix-secreting cells called *ameloblasts* (Deutsch et al., 1995). In humans, ameloblasts are present only during the embryonic formation of teeth and undergo apoptosis following this process; as such, mature human teeth are incapable of regenerating enamel (although rodents retain a population of ameloblasts in their incisor teeth (Warshawsky and Smith, 1974), allowing their continuous growth throughout the life of the animal). Although no new enamel deposition occurs in the adult human tooth, demineralized enamel is nonetheless capable of accumulating minerals back into its structure, restoring its structural integrity over time (Lippert et al., 2004). Despite its hardness, however, enamel is brittle and requires structural support, which is provided through the underlying layer of *dentin*.

Unlike enamel, dentin is a softer and more elastic substance and is produced continuously in the human tooth, both to replace lost enamel and as a response mechanism against caries-forming bacteria. *Odontoblasts*, the cells responsible for dentin formation, reside within the inner periphery of the dentin layer and communicate with the pulp through *dental tubules*, which allow the transport of oxygen and nutrients to cells that otherwise have no access to blood flow (Pashley, 1989). The *dental pulp* itself is directly underneath the odontoblast layer and plays a crucial role in the maintenance of the dental cell population, as it is equipped with blood vessels and facilitates nutrient exchange. In addition, the pulp is innervated and thus allows the tooth to sense temperature changes and physical impacts. Lastly, the *cementum* is populated by *cementoblasts* and anchors the tooth to its socket through a fibrous *periodontal ligament* (Han, 2009). Cementoblasts are notable in that they (like bone and unlike the majority of the cells in the pulp) typically embed themselves completely within the matrix they produce; they are called *cementocytes* in this state (Nanci and Bosshardt, 2006).

The regenerative capacity (or lack thereof) of dental tissues renders it difficult to reverse the effects of caries, as enamel does not regenerate and dentin is soft and highly susceptible to decay. Nonetheless, efforts have been made to artificially induce the biomineralization of teeth to reverse the process of tooth decay, to restore the function of enamel, or even to grow entire replacement teeth under *in vitro* conditions. These efforts often attempt to recapitulate the events that occur during embryonic tooth formation and incorporate a variety of stem and progenitor cells of both dental and alien origin, which are often implanted within a bioactive matrix to trigger their differentiation into the desired cell type. Likewise, the use of bioactive materials for the effective integration of artificial implants into the alveolar bone is an active area of research. Although such attempts are complicated by the layered structure of the tooth, considerable successes have been reported in the field of artificial tooth regeneration, and this chapter will detail the recent advances concerning the use of scaffolds and nanomaterials for this purpose.

7.2 FORMATION OF DENTAL AND OSSEOUS TISSUES

Biomineralization is the process responsible for the formation of hard tissues and involves the cell-mediated deposition of inorganic materials onto a specialized extracellular matrix. It occurs almost exclusively in teeth and the skeletal system and is directed by the enzymes and molecular scaffolds secreted by the native cells of these tissues (Goldberg et al., 2011). Collagen is often a major player in the formation of such scaffolds: The organic matrices of bone, dentin, and cementum are composed of collagens, other fibrous proteins, and lesser amounts of nonfibrous, noncollagenous material, while enamel is exceptional in that it largely lacks collagen. No matter its exact composition, the soft, proteinaceous scaffold must also be reinforced through a mineral component to create the rigid, durable composite material that is characteristic of bones and teeth (Boskey, 2007). Hydroxyapatite (HA) is the principal inorganic component of both dental and osseous tissues, although its precise structure differs between (or even within) tissue types. HA found in dentin and bone features a number of vacancies and substitutions in its matrix, leading to a Ca/P molar ratio that is distinct from the error-free crystal structure, while the mineral component of enamel is closer to stoichiometric HA and consequently displays a Ca/P ratio closer to the "ideal" value of 1.67 (Boskey, 2006, 2007).

Osseous tissues can recover from injury through the bone remodeling process, which involves the removal of damaged tissue and redeposition of a fresh protein matrix for subsequent mineralization (Sims and Gooi, 2008). While dentin can also undergo this process in response to pathogenic bacteria or the erosion of the enamel layer, its regenerative capacity is lower than that of bone. Enamel, in contrast to bone and dentin, is entirely incapable of regeneration in the strict sense, although demineralized enamel can reaccumulate its inorganic matrix in a suitable environment (Featherstone et al., 1990). The defect-free composition and nonregenerative properties of enamel are a result of its unique and structurally demanding function: Enamel is the hardest structure in the human body (Beniash et al., 2009) and must continuously endure the stresses associated with the chewing of food, and its placement over the dentin layer precludes any location suitable for the maintenance of a living cell population (rodents regenerate enamel only in a specialized region at the base of the incisors (Ohshima et al., 2005)). The functional specialization of enamel is also evident in the structure of its constituent HA crystals: Bone and dentin typically feature rod-shaped crystals of approximately 20–50 nm in length and approximately 12–20 nm in width, while enamel crystals are 10-fold larger in all dimensions (Glimcher, 2006; Kirkham et al., 1998).

The embryonic formation of the tooth in general, and enamel in particular, is a well-regulated process. Tooth development begins with the formation of a dental lamina, which grows inward to the mesenchyme and eventually creates a tooth bud. Odontoblasts are produced from the mesenchymal cells that associate with the bud, while ameloblasts develop from the epithelial cells of the dental lamina (Peters and Balling, 1999). The secretion of enamel is closely linked to that of dentin: Although there is evidence that initial secretion of the two layers may be independent (Diekwisch et al., 1995), the coordinated activity of odontoblasts and ameloblasts is nonetheless necessary for tooth development and results in the formation of the dentinoenamel junction, where enamel extends within and fills the dentin structure. Odontoblast extensions that remain in this layer are embedded into the enamel, forming structures that are called enamel spindles and suspected to be sensory in nature (Simmer and Hu, 2001). The mature enamel contains very little organic material and is composed of well-ordered crystals.

7.3 DENTAL IMPLANTS

The natural process of tooth formation is complex, and the structure itself is capable of regenerating only to a limited degree. As such, defects of the tooth are typically treated by removing the damaged region in its entirety and replacing it with a variety of artificial materials. While simpler forms of damage, such as minor caries, are easily treated, more serious injuries warrant the removal of the tooth and its subsequent replacement with a suitable *dental implant*. These implants are typically made of inert, alloplastic materials and may be embedded in the maxilla or the mandible; they are commonly used for the replacement of orofacial structures lost due to trauma, neoplasia, congenital defects, and other diseases (Pye et al., 2009). Dental implants can be classified based on their shape, material (metallic, ceramic, or polymeric), location (endosseous, transosseous, or subperiosseous), or the technique used for their placement (single-stage or two-stage).

A dental implant consists of a *crown*, which replicates the function of the tooth; an *abutment* region, which connects the crown to the implant proper; and the *implant* itself, which affixes the crown and abutment to the maxilla or mandible. The implant may be covered entirely by the jawbone or feature an additional length of material for the ease of attaching the abutment and the crown; this property determines the type of surgery used in implant placement. Implants that are positioned within the jawbone require two surgeries for tooth replacement, as they are typically left without crown or abutment to allow the jawbone to heal around the site of surgery. The overlaying gum tissue is stitched over the implant during this time period and necessitates a second, minor surgery for the subsequent attachment of the abutment and the crown. Single-stage implants, in contrast, use a longer, one-piece implant that protrudes through the gingiva, eliminating the need to stitch the gum tissue (although these implants are also left to heal prior to the attachment of the abutment and the crown).

Implants may be placed on, within, or through the jawbone; this also determines the suitable type of surgery for their attachment. *Subperiosteal implants* are typically positioned within the gum tissue and on the jawbone, and they are typically attached using single-stage procedures (Wingrove, 2013). They are advantageous in that they allow the secure attachment of dentures in individuals who do not have sufficient bone height. *Endosteal implants*, in contrast, are usually affixed in two-stage surgeries; they are shaped like a screw or cylinder and made out of metal, ceramic, or ceramic-covered metal. Designed to replace the roots of teeth, they are implanted into the jawbone, and they may be screw-shaped, threaded, cylindrical, smooth, or bladed depending on the tooth operated and the defect involved (Wingrove, 2013). Lastly, *transosteal* or *stable implants* are metallic and inserted through the jawbone, and they are useful when both teeth and the mandible are atrophied (Wingrove, 2013). Considerable variation exists in the exact sizes and morphologies of these implants, and some are custom-made depending on the injury in question.

Implant materials are likewise highly variable. Aluminum, silver, gold, porcelain, and platinum were among the first industrial materials used for replacing teeth (Barfeie et al., 2015); however, most of these substrates raise substantial immune responses and trigger the formation of fibrous tissue. As such, they are rarely used in modern dentistry (Donath et al., 1992). Dental implant materials can be categorized according to their chemical properties or the biological responses they produce. From a purely chemical point of view, implants are typically either metallic, ceramic, or polymeric (Legeros and Craig, 1993). However, since implants remain embedded within live tissues for considerable periods of time, their biocompatibility is another important aspect for their practical use, and materials with similar chemical properties may trigger vastly different biological responses. As such, implants are also classified by their biocompatibility as *biotolerant* (the material is not necessarily rejected by host tissue, but is nonetheless surrounded by a fibrous capsule), *bioinert* (the material allows the close apposition of bone on its surface), and *bioactive* (the material triggers the formation of new bone on its surface and creates chemical bonds along its interface with the host tissue) (Pilliar, 1990).

Metallic, polymeric, and ceramic materials used in the construction of implants are listed as follows.

7.3.1 Metallic Implants

Titanium alloys, including Ti-6 aluminum-4 vanadium (Ti-6Al-4V), were among the first modern materials used for dental implants (Triplett et al., 2003). Upon contact with air, metallic titanium forms a surface oxide layer that can reach a thickness of 2–10 nm within a short period of time. This stable oxide layer renders titanium biocompatible (Ducheyne, 1987; Lautenschlager and Monaghan, 1993) and provides it with a high corrosion resistance (Donley and Gillette, 1991; Parr et al., 1985). The modulus of elasticity of titanium and its alloys are comparable to that of bone, which allows titanium implants to serve as implants in sites that regularly bear strong loads (Kasemo and Lausmaa, 1985; Meffert et al., 1992). Zirconium, gold, and Ti–aluminum–vanadium alloys are other metallic materials that have been used for osseointegration; however, some of these alloys are known to insufficiently support bone-to-implant connections (Triplett et al., 2003).

7.3.2 Ceramic Implants

Despite their low strength, ceramics are highly biocompatible and integrate well into host tissues, which make them popular candidates for the manufacture of dental implants (Triplett et al., 2003). In addition to their use in stand-alone implants, ceramics can also be used as bioactive coatings to support the osseointegration of other materials. Plasma-sprayed HA is a common surface coating for bone implants, and it has been demonstrated that it and other bioactive ceramic coatings can enhance the chemical bonding of the implant with bone (Lacefield, 1998). Tricalcium phosphate and aluminum oxide ceramics are also currently utilized as plasma-sprayed coatings, usually over a metallic core (Triplett et al., 2003).

7.3.3 Polymeric Implants

Polymers are softer and more flexible than the other classes of biomaterials but can nonetheless be used in dental and bone implants (Triplett et al., 2003). A variety of polymers, including polyurethane, polymethyl methacrylate, polyamide fibers, and polytetrafluoroethylene, have been used to manufacture dental implants (Lemons, 1990). The low mechanical strength of polymers makes them prone to mechanical fractures under high loading forces. In addition, polymers have been reported to trigger adverse immunological reactions and display subpar adhesion capacities to living tissues (Chapman and Kirsch, 1989; Kawahara, 1983). As such, polymeric materials have very few applications in implant dentistry and are only used for the production of shock-absorbing components to be placed between the implant and its suprastructure (Triplett et al., 2003).

7.4 OSSEOINTEGRATION OF DENTAL IMPLANTS

An estimated one million endosseous dental implants are placed annually worldwide (Brunski, 1999; Jokstad et al., 2003). Not all implants are successful, however, as the implant must integrate into the jawbone in order to function. The Brånemark system was introduced for dental implants in 1971 and defines osseointegration as the ability of the bone-to-implant contact to function under load (Brånemark et al., 1983; Hobo et al., 1989). Osseointegration occurs in two steps: primary and secondary (Natali et al., 2009). Primary osseointegration is the mechanical attachment of an implant to the surrounding bone following its insertion, while secondary osseointegration (biological stability) involves bone regeneration and remodeling around the implant (Greenstein et al., 2008; Natali et al., 2009).

Primary stability is a critical determinant of the long-term success of dental implants (Rabel et al., 2007). The success of osseointegration is also affected by the material used in the implant, the machining conditions, the surface finish, the type of bone that receives the implant, the surgical technique, the design of the prosthesis, and patient care (Elias, 2011). Surface properties of implants are extremely important for controlling the biological response that the implant will trigger and can be modified to improve the performance of the implant. These properties involve the attraction, repulsion, adsorption, and absorption capacity of the implant toward cells and proteins, as well as its roughness, wettability, electrical charge, chemical composition, surface energy, residual stresses, and morphology (Elias, 2011).

Various surface modifications have been used to enhance the osseointegration of implants. These modifications typically aim to provide metal implants with surface properties capable of facilitating the adsorption of proteins, adhesion and differentiation of cells, and integration into living tissues. The success of titanium implants, for example, has been shown to depend heavily on their surface topography (Le Guéhennec et al., 2007), including macroscopic, microscopic, and nanometric characteristics. These effects may be caused by the surface preferences and mechanosensory behavior of the cells responsible for facilitating osseointegration. Schwartz et al. have reported that osteoblast proliferation is increased on rough surfaces (Schwartz et al., 1996), while Albrektsson and Wennerberg likewise showed that the differentiation and adhesion of osteoblasts are enhanced on rough surfaces, although fibroblast adhesion was weaker (Schwartz et al., 1996).

7.5 USES OF NANOTECHNOLOGY IN THE DEVELOPMENT OF DENTAL IMPLANTS

Greater control over the topography and chemistry of implant surfaces would assist greatly in understanding the nature of biological interactions that occur on material surfaces and developing novel implants that display enhanced tissue-integrative properties. Such materials can be produced with the assistance of nanotechnology, since more textured surface topographies increase the surface energy at the nanoscale, which in turn enhances the wettability of the surface to blood and the adhesion of cells to the surface. Nanotopography can promote cell differentiation, migration, and proliferation and therefore enhance the wound healing and osseointegration process following implant placement (Dalby et al., 2008; Ehrenfest et al., 2010). Various methods exist for the fabrication of materials with nanometer-scale roughnesses; grit blasting, ionization, and acid etching are among the more common. Dental implants have also begun to use similar methods to increase surface roughness and promote protein adsorption and cell adhesion. In addition, biomimetic calcium phosphate coatings and growth factor-releasing scaffolds are also under development for bone and tooth regeneration (Le Guéhennec et al., 2007).

7.5.1 Enhancement of the Osseointegration Process

The surface of Ti dental implants can be coated with bone-stimulating agents such as growth factors (transforming growth factor-β, bone morphogenetic proteins [BMPs], platelet-derived growth factors, and insulin-like growth factor [IGF]-1 and 2) and antiresorptive drugs (bisphosphonates) in order to locally enhance the bone healing process (Le Guéhennec et al., 2007; Tomsia et al., 2011). Schliephake et al., for example, have reported that a titanium implant coated with type I collagen and BMP-2 displayed greater peri-implant bone formation within the grooves of an endosseous screw, compared to an implant coated with collagen alone. Implant surfaces can also be loaded with molecules that modulate the bone remodeling process to further enhance their osseointegration.

The incorporation of bone antiresorptive drugs, such as bisphosphonates, into implants might be very relevant in clinical cases lacking bone support, for example, resorbed alveolar ridges. It has recently been shown that a bisphosphonate-containing titanium implant could locally increase bone density in the peri-implant region (Josse et al., 2004). The effect of antiresorptive drugs seems to be limited to the vicinity of the implant, and *in vivo* studies suggest that dental implants functionalized with bisphosphonates have little to no side effects despite only displaying a slight increase in osteointegrative capacity (Meraw and Reeve, 1999; Meraw et al., 1999). Plasma-sprayed HA-coated dental implants immersed in pamidronate or zoledronate, however, could trigger a significant increase in bone contact area (Kajiwara et al., 2005; Peter et al., 2005; Yoshinari et al., 2001). Bisphosphonates display a high chemical affinity for calcium phosphate surfaces, and their incorporation onto dental implants can be achieved easily by using a biomimetic coating method at room temperature. However, the dose of the drug will nonetheless have to be optimized on a case-by-case basis, as bisphosphonate-mediated increases in peri-implant bone densities are concentration dependent (Peter et al., 2005).

Growth factors and biomolecules can also be immobilized onto implants to enhance tissue growth and integration. TiO_2 nanotubes produced by anodization have been proposed as drug-eluting coatings for implantable devices (Popat et al., 2007). The surfaces of these tubes can be functionalized to attach biomolecules, such as bovine serum albumin. A Ti-based implant, for example, has been functionalized with BMP to enhance its bioactivity and bone formation capacity (Puleo et al., 2002). The advantage of immobilizing BMP is that it allows the controlled administration of the hormone and avoids the problems associated with overdosing. Implants have been coated with nanocrystalline diamonds to increase the surface area and facilitate the immobilization of BMP (Kloss et al., 2008). The enhanced differentiation and proliferation of cells can be achieved without changing the overall texture of the implant using these diamonds (Specht et al., 2004).

Studies have also demonstrated that biphasic calcium phosphate grit-blasted surfaces can provide a more rapid osseointegration in comparison to smooth surfaces. Osseointegration can also be promoted by applying a calcium phosphate coat onto the implant through plasma spraying or biomimetic and electrophoretic deposition (Lavenus et al., 2010). Calcium phosphate residues on implant surface release calcium and phosphate ions to their immediate environment, potentially assisting in the precipitation of biological apatite nanocrystals and the adsorption of various proteins onto these structures. This protein matrix may in turn promote cell adhesion, osteoblast differentiation, and the synthesis of mineralized collagen. Osteoclast cells are also activated in response to calcium phosphate coatings, allowing the formation of bone tissue and the establishment of a direct bone-to-implant contact without an intervening layer of connective tissue (Lavenus et al., 2010). Another interesting approach involves the use of molecular self-assembled monolayers, which are formed by the spontaneous assembly of a single layer of molecules on a surface. These molecules expose only their end-chain groups to the environment, and these chains can be designed with osteoinductive or cell-adhesive properties, such as by the use of RGD peptides.

7.5.2 Pulp and Dentin Tissue Regeneration

Traumatic dental injuries are often irreversible and may require the excision of even the healthy portions of the tooth prior to their filling or replacement. The need for these surgeries would be reduced greatly if the natural regeneration of dental pulp cells can be enhanced. Regeneration of pulp tissue ordinarily proceeds at a slow pace, as the dental pulp has minimal collateral blood supply and the immune system cannot adequately defend against bacterial entry into the pulp (Huang, 2009). In addition, odontoblasts are postmitotic cells and exhibit only a limited ability to proliferate (Arana-Chavez and Massa, 2004). However, modern tissue engineering methods and especially the discovery of dental stem cells have allowed the development of techniques for the regeneration of pulp and dentin (Huang, 2009). The interplay between nanotechnology and stem cell biology allows the selective differentiation of stem cells into specific lineages through artificial scaffolds that incorporate the active sequences of factors involved in the lineage commitment process. The administration of dynamic biological agents composed of stem cells, bioactive scaffolds, and/or nanoparticles to patients is an effective means of increasing the regenerative ability of damaged dental tissues, although many of these methods are still under development and do not currently see clinical use (Mitsiadis et al., 2012).

Scaffolds are three-dimensional structures that provide an initial framework for the growth and recruitment of cells (Muschler et al., 2004). They are commonly used in regenerative medicine and typically produced to mimic the gross morphology of the missing section of tissue. A number of factors should be taken into consideration for the design of tooth (and other tissue) scaffolds, including vascularization, cell–matrix interactions, growth factor incorporation, matrix degradation, mineralization capacity, and the risk of contamination or undesirable immune responses (Galler et al., 2011a). As with implants, a wide variety of scaffold materials can be used for the tissue engineering of teeth. These include long-lasting porous HA ceramics, inherently transpiring molecules (e.g., collagen and chitosan) and biodegradable polymers such as polyglycolic acid (PGA), polylactic acid (PLA), polyglycolic acid–poly-L-lactic acid (PGA–PLLA), and poly(lactic-*co*-glycolic acid) (PLGA) (Zhang et al., 2013). In addition to the effects of its raw material and implanted stem cells, a scaffold may also be functionalized through secondary chemical modifications and the slow release of regeneration-enhancing biological factors.

Certain types of isolated pulp cells have been shown to differentiate into odontoblast-like cells and generate a dentin-like mineral structure under *in vitro* conditions (About et al., 2000; Tsukamoto et al., 1992). These are collectively called dental stem cells and include dental pulp stem cells (DPSCs), stem cells from human exfoliated deciduous teeth (SHED), and stem cells from apical papilla (SCAP). These cells are derived either from mature pulp tissue or its embryonic precursors and may potentially serve as sources for the regeneration of pulp and dentin, especially when integrated into a scaffold suitable for their growth and differentiation (tissue banks have been founded to preserve the stem cells in deciduous teeth for that purpose) (Huang, 2009). DPSCs and SCAPs form a pulp–dentin complex when transplanted into immunocompromised mice (Gronthos et al., 2000; Sonoyama et al., 2006), while SHED produce a mineralized tissue matrix without a distinct pulp–dentin complex (Miura et al., 2003). The activity of these cells can also be enhanced through the use of growth factors, which are integral to the formation and repair of dentin and pulp tissues. In addition to the factors involved in the embryonic development of teeth, the dentin matrix naturally contains odontoblast-secreted growth factors, primarily of the TGF-β superfamily (Smith et al., 2008), that play important roles in signaling events leading to the formation of tertiary dentin in mature teeth (Tziafas, 1995).

Although scaffolds used in regenerative medicine need not display nanoscale features, these features generally assist in the function of the scaffold by increasing its surface area and therefore enhancing its interaction with cells and the surrounding tissue. Wang et al., for example, tested the effects of a novel nanoscale bioactive glass (n-BG) on the proliferation, apoptosis, chemotaxis, mineralization, and differentiative capacity of human dental pulp cells (hDPCs) and found that a combination of these effects allows n-BG to induce dentin formation more effectively than a microscale equivalent (microscale bioactive glass (m-BG)), which was attributed to the greater number of binding sites, faster dissolution rate, and other biochemical properties associated with the larger surface area of n-BG (Wang et al., 2014). Nanostructured materials therefore allow more effective presentation of the active groups present on their surfaces.

Self-assembled peptide nanofiber gels are one of the more common forms of smart materials used in regenerative medicine. Rational design of the peptide sequence enables the precise control of material stiffness and allows the material itself to participate in biomineralization or induce the differentiation of surrounding cells. Cell adhesion motifs, enzyme-cleavable sites, and the incorporation of growth factors into the gel structure further enhance the ability of peptide amphiphile scaffolds to elicit specific cellular responses. Inductive scaffolds can also be seeded with stem cells to increase the rate at which the pulp–dentin complex is regenerated (Galler et al., 2011a). Peptide gels can also be designed to exhibit antimicrobial properties; for example, lysine-rich surfaces may facilitate the electrostatic interaction of the peptide with negatively charged bacterial surfaces, resulting in the disruption of the bacterial membrane (Salick et al., 2007). Cell-free approaches using peptide and other types of scaffolds have also been developed; these scaffolds recruit stem cells and growth and differentiation factors from the dentin, the pulp, or the periapical region (Galler et al., 2014).

A combination of the aforementioned approaches can also be utilized to increase the repair efficiency of a scaffold. Galler et al., for example, used a cell-adhesive, enzyme-cleavable hydrogel composed of self-assembling peptide nanofibers, DPSCs, and three growth factors (basic fibroblast growth factor, transforming growth factor β1, and vascular endothelial growth) in tooth regeneration and further transplanted the gel within a dentin cylinder to enhance its integration into the native tissue. Their system was able to promote cell proliferation, differentiation, and angiogenesis and further supported the formation of a vascularized soft connective tissue with a structure similar to that found in the dental pulp. In addition to increasing the proliferation and differentiation of the seeded DPSCs, growth factors by themselves were also able to attract host cells into the peptide-based bioactive scaffolds (Galler et al., 2011b).

7.5.3 Whole Tooth Regeneration

Although implanted scaffolds are useful in cases where tooth structure stays partially intact, they cannot be used to replace teeth that have been lost in their entirety. The formation of complete replacement teeth would therefore be of great utility in regenerative dentistry. Such teeth could either be grown under *in vitro* conditions and subsequently implanted into empty sockets or produced *in vivo* directly on the maxilla or mandible, accomplishing in both cases the near-complete restoration of the original function of the tooth. As the development of teeth is a complex and well-regulated process, research in this area is still in its infancy and clinical applications are lacking, but the successful generation of tooth structures has been reported in both *in vitro* scaffolds and *in vivo* rodent and swine models. These efforts generally involve a combination of regenerative matrices, dental or mesenchymal stem cells, and growth factors that together mimic the environment in which the embryonic tooth is developed.

Tooth bud cells from rats have been used in the artificial generation of tooth crowns in a scaffold matrix implanted in rat omenta (Duailibi et al., 2004), while the fact that these stem cells could be cultured under *in vitro* conditions prior to implantation was highlighted as an indicator that the *in vitro* expansion of tooth bud cells is feasible (Duailibi et al., 2006). Likewise, SCAPs and periodontal ligament cells were able to produce the root structures necessary for anchoring an artificial crown (Sonoyama et al., 2006). Tooth bud implants in both pigs and rats appear to follow a developmental process similar to the embryonic teeth, as the times required for the production of functional teeth in implanted tooth buds are similar to the duration of tooth formation in the embryonic animals (Nakahara and Idei, 2007). Nondental stem cells were also shown to trigger the development of partial or complete tooth structures. Although embryonic and neural stem cells were unable to produce teeth, bone marrow cells could generate functional tooth crowns with layers of enamel, dentin, and pulp (Nakahara and Idei, 2007; Ohazama et al., 2004). Adipose-derived stem cells have also been suggested as alternate cell sources for the regeneration of teeth (Jing et al., 2008).

7.6 CONCLUSIONS AND FUTURE PERSPECTIVES

The use of nanotechnology for the functionalization of dental implants has become widespread in the recent decade. Nevertheless, the development of safer and more effective coatings is still an active area of research, and advances in nanotechnology will no doubt uncover a greater diversity of material types and surface architectures for use in the modification of implants. In addition, biological signals are now being tested for their potential effect in modulating the osseointegration of implants; if successful, these materials may allow the implant surface to truly behave as native tissue for the attachment of cells. While these developments have greatly improved dental implants, efforts involving stem cells and regenerative scaffolds are tackling the problem from a different angle: By regrowing teeth from scratch, the entire rationale in using a foreign material as an implant would be eliminated, and although these methods are still in their infancy, their advancement may produce a new generation of tooth implants. Overall, nanotechnology has assisted in the production of more effective implants under more reliable methods, and while most of their applications are experimental, nanostructured materials are nonetheless promising candidates for use as implants in the following decades.

REFERENCES

About, I., Bottero, M.-J., de Denato, P., Camps, J., Franquin, J.-C., and Mitsiadis, T.A. (2000). Human dentin production in vitro. Exp Cell Res *258*, 33–41.

Arana-Chavez, V.E. and Massa, L.F. (2004). Odontoblasts: the cells forming and maintaining dentine. Int J Biochem Cell Biol *36*, 1367–1373.

Barfeie, A., Wilson, J., and Rees, J. (2015). Implant surface characteristics and their effect on osseointegration. Br Dent J *218*, E9.

Beniash, E., Metzler, R.A., Lam, R.S., and Gilbert, P.U. (2009). Transient amorphous calcium phosphate in forming enamel. J Struct Biol *166*, 133–143.

Boskey, A.L. (2006). Assessment of bone mineral and matrix using backscatter electron imaging and FTIR imaging. Curr Osteoporos Rep *4*, 71–75.

Boskey, A.L. (2007). Mineralization of bones and teeth. Elements *3*, 385–391.

Brånemark, P., Adell, R., Albrektsson, T., Lekholm, U., Lundkvist, S., and Rockler, B. (1983). Osseointegrated titanium fixtures in the treatment of edentulousness. Biomaterials *4*, 25–28.

Brunski, J.B. (1999). In vivo bone response to biomechanical loading at the bone/dental-implant interface. Adv Dent Res *13*, 99–119.

Chapman, R. and Kirsch, A. (1989). Variations in occlusal forces with a resilient internal implant shock absorber. Int J Oral Maxillofac Implants *5*, 369–374.

Dalby, M.J., Andar, A., Nag, A., Affrossman, S., Tare, R., McFarlane, S., and Oreffo, R.O. (2008). Genomic expression of mesenchymal stem cells to altered nanoscale topographies. J R Soc Interface *5*, 1055–1065.

Deutsch, D., CatalanoSherman, J., Dafni, L., David, S., and Palmon, A. (1995). Enamel matrix proteins and ameloblast biology. Connect Tissue Res *32*, 97–107.

Diekwisch, T.G., Berman, B.J., Gentner, S., and Slavkin, H.C. (1995). Initial enamel crystals are not spatially associated with mineralized dentine. Cell Tissue Res *279*, 149–167.

Donath, K., Laaß, M., and Günzl, H.-J. (1992). The histopathology of different foreign-body reactions in oral soft tissue and bone tissue. Virchows Arch A Pathol Anat Histopathol *420*, 131–137.

Donley, T.G. and Gillette, W.B. (1991). Titanium endosseous implant-soft tissue interface: a literature review. J Periodontol *62*, 153–160.

Duailibi, M., Duailibi, S., Young, C., Bartlett, J., Vacanti, J., and Yelick, P. (2004). Bioengineered teeth from cultured rat tooth bud cells. J Dent Res *83*, 523–528.

Duailibi, S., Duailibi, M., Vacanti, J., and Yelick, P. (2006). Prospects for tooth regeneration. Periodontol 2000 *41*, 177–187.

Ducheyne, P. (1987). Titanium and calcium phosphate ceramic dental implants, surfaces, coatings and interfaces. J Oral Implantol *14*, 325–340.

Ehrenfest, D.M.D., Coelho, P.G., Kang, B.-S., Sul, Y.-T., and Albrektsson, T. (2010). Classification of osseointegrated implant surfaces: materials, chemistry and topography. Trends Biotechnol *28*, 198–206.

Elias, C.N. (2011). Factors affecting the success of dental implants, implant dentistry—a rapidly evolving practice (I. Turkyilmaz, ed. INTECH). Available at http://www.intechopen.com/books/implant-dentistry-a-rapidly-evolving-practice/factors-affecting-the-success-of-dental-implants (accessed on October 29, 2015).

Featherstone, J.D., Glena, R., Shariati, M., and Shields, C.P. (1990). Dependence of in vitro demineralization of apatite and remineralization of dental enamel on fluoride concentration. J Dent Res *69* Spec No, 620–625.

Galler, K.M., D'Souza, R.N., Hartgerink, J.D., and Schmalz, G. (2011a). Scaffolds for dental pulp tissue engineering. Adv Dent Res *23*, 333–339.

Galler, K.M., Hartgerink, J.D., Cavender, A.C., Schmalz, G., and D'Souza, R.N. (2011b). A customized self-assembling peptide hydrogel for dental pulp tissue engineering. Tissue Eng Part A *18*, 176–184.

Galler, K.M., Eidt, A., and Schmalz, G. (2014). Cell-free approaches for dental pulp tissue engineering. J Endod *40*, S41–S45.

Glimcher, M.J. (2006). Bone: nature of the calcium phosphate crystals and cellular, structural, and physical chemical mechanisms in their formation. Rev. Mineral Geochem *64*, 223–282.

Goldberg, M., Kulkarni, A.B., Young, M., and Boskey, A. (2011). Dentin: structure, composition and mineralization: the role of dentin ECM in dentin formation and mineralization. Front Biosci (Elite Ed) *3*, 711.

Greenstein, G., Cavallaro, J., Romanos, G., and Tarnow, D. (2008). Clinical recommendations for avoiding and managing surgical complications associated with implant dentistry: a review. J Periodontol *79*, 1317–1329.

Gronthos, S., Mankani, M., Brahim, J., Robey, P.G., and Shi, S. (2000). Postnatal human dental pulp stem cells (DPSCs) in vitro and in vivo. Proc Natl Acad Sci USA *97*, 13625–13630.

Han, H.-S. (2009). Design of new root-form endosseous dental implant and evaluation of fatigue strength using finite element analysis. Theses and Dissertations (Iowa City: University of Iowa), 294.

Hobo, S., Ichida, E., and Garcia, L.T. (1989). Osseointegration and occlusal rehabilitation (Tokyo/Chicago, IL: Quintessence Pub Co).

Huang, G.T. (2009). Pulp and dentin tissue engineering and regeneration: current progress. Regen Med *4*, 697–707.

Jing, W., Wu, L., Lin, Y., Liu, L., Tang, W., and Tian, W. (2008). Odontogenic differentiation of adipose-derived stem cells for tooth regeneration: necessity, possibility, and strategy. Med Hypotheses *70*, 540–542.

Jokstad, A., Braegger, U., Brunski, J.B., Carr, A.B., Naert, I., and Wennerberg, A. (2003). Quality of dental implants. Int Dent J *53*, 409–443.

Josse, S., Faucheux, C., Soueidan, A., Grimandi, G., Massiot, D., Alonso, B., Janvier, P., Laïb, S., Gauthier, O., and Daculsi, G. (2004). Chemically modified calcium phosphates as novel materials for bisphosphonate delivery. Adv Mater *16*, 1423–1427.

Kajiwara, H., Yamaza, T., Yoshinari, M., Goto, T., Iyama, S., Atsuta, I., Kido, M.A., and Tanaka, T. (2005). The bisphosphonate pamidronate on the surface of titanium stimulates bone formation around tibial implants in rats. Biomaterials *26*, 581–587.

Kasemo, B. and Lausmaa, J. (1985). Metal selection and surface characteristics. In P-I. Brånemark, G.A. Zarb, and T. Albrektsson, eds. Tissue-integrated prostheses: osseointegration in clinical dentistry (Chicago, IL: Quintessence), 99–116.

Kawahara, H. (1983). Cellular responses to implant materials: biological, physical and chemical factors. Int Dent J *33*, 350–375.

Kirkham, J., Brookes, S., Shore, R., Bonass, W., Smith, D., Wallwork, M., and Robinson, C. (1998). Atomic force microscopy studies of crystal surface topology during enamel development. Connect Tissue Res *38*, 91–100.

Kloss, F.R., Gassner, R., Preiner, J., Ebner, A., Larsson, K., Hächl, O., Tuli, T., Rasse, M., Moser, D., and Laimer, K. (2008). The role of oxygen termination of nanocrystalline diamond on immobilisation of BMP-2 and subsequent bone formation. Biomaterials *29*, 2433–2442.

Lacefield, W.R. (1998). Current status of ceramic coatings for dental implants. Implant Dent *7*, 315–322.

Lautenschlager, E.P. and Monaghan, P. (1993). Titanium and titanium alloys as dental materials. Int Dent J *43*, 245–253.

Lavenus, S., Louarn, G., and Layrolle, P. (2010). Nanotechnology and dental implants. Int J Biomater *2010*. doi:10.1155/2010/915327

Le Guéhennec, L., Soueidan, A., Layrolle, P., and Amouriq, Y. (2007). Surface treatments of titanium dental implants for rapid osseointegration. Dent Mater *23*, 844–854.

Legeros, R.Z. and Craig, R.G. (1993). Strategies to affect bone remodeling: osteointegration. J Bone Miner Res *8*, S583–S596.

Lemons, J.E. (1990). Dental implant biomaterials. J Am Dent Assoc *121*, 716–719.

Lippert, F., Parker, D.M., and Jandt, K.D. (2004). In vitro demineralization/remineralization cycles at human tooth enamel surfaces investigated by AFM and nanoindentation. J Colloid Interface Sci *280*, 442–448.

Meffert, R.M., Langer, B., and Fritz, M.E. (1992). Dental implants: a review. J Periodontol *63*, 859–870.

Meraw, S.J. and Reeve, C.M. (1999). Qualitative analysis of peripheral peri-implant bone and influence of alendronate sodium on early bone regeneration. J Periodontol *70*, 1228–1233.

Meraw, S.J., Reeve, C.M., and Wollan, P.C. (1999). Use of alendronate in peri-implant defect regeneration. J Periodontol *70*, 151–158.

Mitsiadis, T.A., Woloszyk, A., and Jiménez-Rojo, L. (2012). Nanodentistry: combining nanostructured materials and stem cells for dental tissue regeneration. Nanomedicine *7*, 1743–1753.

Miura, M., Gronthos, S., Zhao, M., Lu, B., Fisher, L.W., Robey, P.G., and Shi, S. (2003). SHED: stem cells from human exfoliated deciduous teeth. Proc Natl Acad Sci USA *100*, 5807–5812.

Mullen, J. (2005). History of water fluoridation. Br Dent J *199*, 1–4.

Muschler, G.F., Nakamoto, C., and Griffith, L.G. (2004). Engineering principles of clinical cell-based tissue engineering. J Bone Joint Surg Am *86*, 1541–1558.

Nakahara, T. and Idei, Y. (2007). Tooth regeneration: Implications for the use of bioengineered organs in first-wave organ replacement. Hum Cell *20*, 63–70.

Nanci, A. and Bosshardt, D.D. (2006). Structure of periodontal tissues in health and disease. Periodontol 2000 *40*, 11–28.

Natali, A.N., Carniel, E.L., and Pavan, P.G. (2009). Investigation of viscoelasto-plastic response of bone tissue in oral implants press fit process. J Biomed Mater Res B Appl Biomater *91B*, 868–875.

Nguyen, T.T., Mui, B., Mehrabzadeh, M., Chea, Y., Chaudhry, Z., Chaudhry, K., and Tran, S.D. (2013). Regeneration of tissues of the oral complex: current clinical trends and research advances. J Can Dent Assoc *79*, d1.

Ohazama, A., Modino, S., Miletich, I., and Sharpe, P. (2004). Stem-cell-based tissue engineering of murine teeth. J Dent Res *83*, 518–522.

Ohshima, H., Nakasone, N., Hashimoto, E., Sakai, H., Nakakura-Ohshima, K., and Harada, H. (2005). The eternal tooth germ is formed at the apical end of continuously growing teeth. Arch Oral Biol *50*, 153–157.

Parr, G.R., Gardner, L.K., and Toth, R.W. (1985). Titanium: the mystery metal of implant dentistry. Dental materials aspects. J Prosthet Dent *54*, 410–414.

Pashley, D.H. (1989). Dentin: a dynamic substrate—a review. Scanning Microsc *3*, 161–174; discussion 174–166.

Peter, B., Pioletti, D.P., Laib, S., Bujoli, B., Pilet, P., Janvier, P., Guicheux, J., Zambelli, P.-Y., Bouler, J.-M., and Gauthier, O. (2005). Calcium phosphate drug delivery system: influence of local zoledronate release on bone implant osteointegration. Bone *36*, 52–60.

Peters, H. and Balling, R. (1999). Teeth. Where and how to make them. Trends Genet *15*, 59–65.

Pilliar, R. (1990). Dental implants: materials and design. J Can Dent Assoc *56*, 857–861.

Popat, K.C., Eltgroth, M., LaTempa, T.J., Grimes, C.A., and Desai, T.A. (2007). Titania nanotubes: a novel platform for drug–eluting coatings for medical implants? Small *3*, 1878–1881.

Puleo, D., Kissling, R., and Sheu, M.-S. (2002). A technique to immobilize bioactive proteins, including bone morphogenetic protein-4 (BMP-4), on titanium alloy. Biomaterials *23*, 2079–2087.

Pye, A., Lockhart, D., Dawson, M., Murray, C., and Smith, A. (2009). A review of dental implants and infection. J Hosp Infect *72*, 104–110.

Rabel, A., Köhler, S., and Schmidt-Westhausen, A. (2007). Clinical study on the primary stability of two dental implant systems with resonance frequency analysis. Clin Oral Investig *11*, 257–265.

Salick, D.A., Kretsinger, J.K., Pochan, D.J., and Schneider, J.P. (2007). Inherent antibacterial activity of a peptide-based β-hairpin hydrogel. J Am Chem Soc *129*, 14793–14799.

Schwartz, Z., Martin, J., Dean, D., Simpson, J., Cochran, D., and Boyan, B. (1996). Effect of titanium surface roughness on chondrocyte proliferation, matrix production, and differentiation depends on the state of cell maturation. J Biomed Mater Res *30*, 145–155.

Simmer, J.P. and Hu, J.C. (2001). Dental enamel formation and its impact on clinical dentistry. J Dent Educ *65*, 896–905.

Sims, N. and Gooi, J. (2008). Bone remodeling: multiple cellular interactions required for coupling of bone formation and resorption. Semin Cell Dev Biol *19*, 444–451.

Smith, A., Lumley, P., Tomson, P., and Cooper, P. (2008). Dental regeneration and materials—a partnership. Clin Oral Investig *12*, 103–108.

Sonoyama, W., Liu, Y., Fang, D., Yamaza, T., Seo, B., Zhang, C., Liu, H., Gronthos, S., Wang, C., Shi, S., et al. (2006). Mesenchymal stem cell-mediated functional tooth regeneration in swine. PLoS One *1*, e79.

Specht, C.G., Williams, O.A., Jackman, R.B., and Schoepfer, R. (2004). Ordered growth of neurons on diamond. Biomaterials *25*, 4073–4078.

Tomsia, A.P., Launey, M.E., Lee, J.S., Mankani, M.H., Wegst, U.G., and Saiz, E. (2011). Nanotechnology approaches for better dental implants. Int J Oral Maxillofac Implants *26*, 25.

Triplett, R.G., Frohberg, U., Sykaras, N., and Woody, R.D. (2003). Implant materials, design, and surface topographies: their influence on osseointegration of dental implants. J Long Term Eff Med Implants *13*, 485–501.

Tsukamoto, Y., Fukutani, S., Shin-Ike, T., Kubota, T., Sato, S., Suzuki, Y., and Mori, M. (1992). Mineralized nodule formation by cultures of human dental pulp-derived fibroblasts. Arch Oral Biol *37*, 1045–1055.

Tziafas, D. (1995). Basic mechanisms of cytodifferentiation and dentinogenesis during dental pulp repair. Int J Dev Biol *39*, 281–290.

Wang, S., Gao, X., Gong, W., Zhang, Z., Chen, X., and Dong, Y. (2014). Odontogenic differentiation and dentin formation of dental pulp cells under nanobioactive glass induction. Acta Biomater *10*, 2792–2803.

Warshawsky, H. and Smith, C.E. (1974). Morphological classification of rat incisor ameloblasts. Anat Rec *179*, 423–446.

Wingrove, S.S. (2013). Peri-implant therapy for the dental hygienist: clinical guide to maintenance and disease complications (Ames, IA: John Wiley & Sons).

Yoshinari, M., Oda, Y., Ueki, H., and Yokose, S. (2001). Immobilization of bisphosphonates on surface modified titanium. Biomaterials *22*, 709–715.

Zhang, L., Morsi, Y., Wang, Y., Li, Y., and Ramakrishna, S. (2013). Review scaffold design and stem cells for tooth regeneration. Jpn Dent Sci Rev *49*, 14–26.

8

NANOMATERIALS AS TISSUE ADHESIVES

I. CEREN YASA, HAKAN CEYLAN, AYSE B. TEKINAY, AND MUSTAFA O. GULER

Institute of Materials Science and Nanotechnology, National Nanotechnology Research Center (UNAM), Bilkent University, Ankara, Turkey

8.1 INTRODUCTION

Tissue adhesives are used to create functional bonding interfaces between injured tissue parts and between tissues and implanted biomaterials. Wound closure and biointegration of prosthetics or synthetic tissue constructs are the major applications within this context (Annabi et al., 2014; Lauto et al., 2008; Meddahi-Pellé et al., 2014; Sharma et al., 2013). Hemostats, which are used to stop bleeding, and sealants, which are used to create an interfacial barrier against air or fluid, can also be considered along the line of tissue adhesive concept as they essentially require *in vivo* bonding and biointerface formation. Skin adhesives have also become an emerging concept for sensory applications, particularly in real-time healthcare monitoring (Kwak et al., 2011).

Millions of patients across the developed and developing countries suffer from wounds or surgical incisions that need proper closure followed by accelerated healing and tissue regeneration (Annabi et al., 2014; Bre et al., 2013; Sanders and Nagatomi, 2014). These wounds and incisions mainly

---

Therapeutic Nanomaterials, First Edition. Edited by Mustafa O. Guler and Ayse B. Tekinay.

include skin and the underlying tissues, such as connective tissues, muscles, and tendons, which have limited regeneration capacity. Currently, there are several medically approved options for bonding tissue parts. The global share of adhesive market shows that surgical sutures and other mechanical bonding methods dominate the market despite their poor efficiencies, particularly in compliant and inaccessible tissues, particularly laparoscopic procedures, in addition to their invasive application and considerable toxicity and hypersensitivity risks (Annabi et al., 2014; Bre et al., 2013). Moreover, in some cases, suturing can impair the tissue healing (Bre et al., 2013). Sutures also fall short to stop the leakage of body fluids. Further, application of surgical sutures requires extensive expertise in the level of surgeons. Alternative bonding methods relying on mechanical interlocking and staples also present risks of damaging the surrounding tissues and the vasculature (Yang et al., 2013). On the other hand, ease of application of a glue substance, in principle, should significantly broaden the use of such types of materials in various types of medical and surgical applications. Tissue adhesives can be applied over the wound or incision site, act as hemostats or sealants, create an antimicrobial barrier, and can even induce proper tissue regeneration. In light of these perspectives, recent studies estimate that medical adhesives will increase their portion in the market by 2017 (Report S190, 2012). Therefore, the efforts in this field have focused on the development of new medical adhesives and improving the existing ones in accordance with defined design principles. A typical tissue adhesive for wound healing should fulfill certain criteria to find wide applicability in medical and surgical practices (Bouten et al., 2014; Páez et al., 2004):

1. Biocompatibility and biodegradability, including both the adhesive and its degradation products
2. Ease of administration
3. Timely functioning, optimized curing kinetics
4. Sufficient mechanical strength for the desired application, an optimal balance of cohesiveness and adhesiveness of the material underwater
5. Low degree of swelling
6. Long shelf life and stability
7. Cost-effectiveness

Depending on the surgical procedure, the characteristics of the adhesives might vary. In bleeding, for example, tissue parts approximated via an adhesive interface is also expected to function as a hemostatic agent. On the other hand, improvement in one parameter usually comes with a trade-off.

For example, when the adhesion strength is increased, toxicity becomes intolerably high (Lauto et al., 2008). Consequently, optimizing individual parameters while preserving the overall material performance in a dedicated medical or surgical procedure is the gold standard.

Another major application of tissue adhesives is the mechanically stable and biocompatible integration of foreign biomaterials, such as metal prosthetics or synthetic soft tissue constructs, into the sites of implantation (Gittens et al., 2014; Sharma et al., 2013). In this regard, the biochemical nature of the bonding surface of a foreign biomaterial is of utmost significance for the long-term effectiveness of the surgical intervention. Once the implant is placed, a foreign body reaction is raised by the immune system that works to isolate and remove the material from the body. An ideal adhesive, therefore, should create an interface that bond the biomaterial to the tissue while it is recognized as self in the body and functional tissue will grow on it. In order to overcome this challenge, strong underwater adhesives should also be able to modulate cellular activities through biologically active signals (Avila et al., 2009; Liu et al., 2006; Richards et al., 2012). A state-of-the-art research in this field concentrates on developing hybrid systems that consist of a base adhesive recapitulating the cellular microenvironments by mimicking the architecture and biology of the extracellular matrix of the tissue of interest (Ceylan et al., 2011, 2014; Kang et al., 2012; Sharma et al., 2013).

Adhesion *in vivo* presents considerable challenges arising mainly from the aqueous and dynamic chemistry of the living environment. Highly polar water molecules and polyionic species interact both with the surface and the adhesive molecules, thereby interfering with the adhesion process. Based on the source of the raw material and inspiration of the design, technologically available tissue adhesives can be categorized as natural, synthetic, and bioinspired. Most of the efforts for designing and manufacturing of tissue adhesives have been on curable functional polymers and derivatives of natural adhesives. Synthetic adhesives have strong mechanical properties, yet low degradation rate and cytotoxic by-products upon degradation restrict their clinical application (Agarwal et al., 2007). Although natural adhesives including fibrin glues and collagen-based products are effective in adhesion, their availability is limited due to autologous isolation from tissue. Moreover, they often have weak tensile and adhesive strength (Kull et al., 2009), which is critical for long-term adhesion. In this regard, nanoenabled materials, such as nanoparticles, and bioinspired synthetic designs promise opportunities to produce new types of tissue adhesives with superior characteristics. In this chapter, we provide a critical review of the conventional tissue adhesive materials and their shortcomings. We then concentrate on the design approaches of emerging tissue adhesive technologies, particularly those using nanotechnology, with a primary focus on their clinical applicability.

8.2 TISSUE ADHESIVES BASED ON SYNTHETIC POLYMERS

Various classes of synthetic polymers have been developed and applied as tissue adhesives. Their defined chemistry and tailorable material properties, such as adhesion strength, curing kinetics, and mechanical properties, drove the motivation in this approach. The two major classes of synthetic tissue adhesives discussed in this chapter are acrylate-based adhesives and polyurethanes. Typically, synthetic polymer-based tissue adhesives are characterized by high mechanical properties, fast curing time, poor biocompatibility, and moderate tissue adhesion strength.

Traditional and commonly used synthetic adhesives are acrylic adhesives and urethane-based ones. Acrylic adhesives are based on polymerization of acrylate and methacrylate monomers (Fig. 8.1a) by a hardener or free radical initiators to cure with time, heat, or ionizing radiation. The main properties

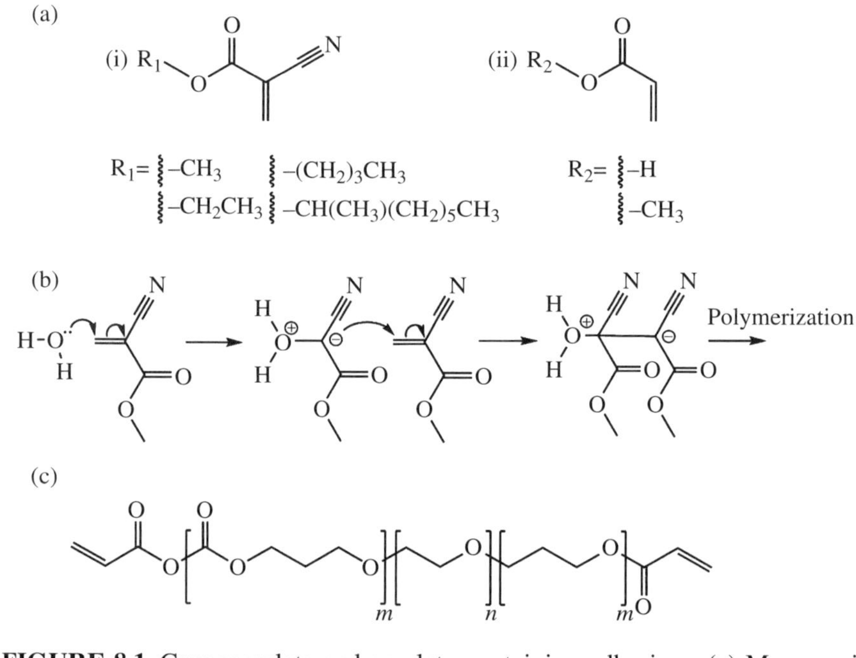

FIGURE 8.1 Cyanoacrylate and acrylate-containing adhesives. (a) Monomeric units of cyanoacrylates (i) and acrylates (ii) used as tissue adhesives. (b) Anionic polymerization (curing) of methyl cyanoacrylate in the presence of water. (c) Chemical structure of $PTMC_m$–PEG_m–$PTMC_m$ diacrylate adhesive.

of acrylics, which make them ideal for developing adhesives, are their ability to make bond with many substrates, controllable rate of cure, appropriate strength, simplicity in use, and low degree of toxicity (Lees, 1979). Cyanoacrylates and acrylic polyethylene glycol (PEG) derivatives are commonly used acrylic materials to develop adhesives.

Cyanoacrylate-based adhesives have been used to assist in wound management, and they are also popular in dentistry for bonding orthodontic brackets and repairing dentures. What makes them one of the primary choices of synthetic adhesive materials is their adhesive properties including strong adhesion, rapid curing time, instantaneous adhesion to tissue, and ease of use (Leggat et al., 2004; Trott, 1997).

Cyanoacrylate monomers including methyl cyanoacrylate, ethyl cyanoacrylate, *n*-butyl cyanoacrylate, and 2-octyl cyanoacrylate form polycyanoacrylates by polymerization reaction (Singer et al., 2008). Double bond in the acrylate is important for the polymerization of this glue. In principle, susceptibility of electron-withdrawing nitrile group to nucleophilic attack by weak bases causes rapid polymerization upon contact with water or blood (Fig. 8.1b). After initiation by water, the monomers polymerize via anionic polymerization (Bouten et al., 2014; Seewald et al., 2002), and adhesion of tissue is accomplished via formation of covalent bonds between the cyanoacrylate and functional groups in the tissue proteins, such as primary amines of the lysine side chains (Bouten et al., 2014). However, these polymers can degrade and result in accumulation of the degradation products in tissue, which causes histotoxic hypoxia. Therefore, the use of the short side chain and methyl and ethyl cyanoacrylates is mostly abandoned due to such side effects (Trott, 1997).

Chemical adhesives, including cyanoacrylate, adhere strongly to tissues by a reactive exothermic covalent cross-linking reaction. These adhesives may suffer from desired time for adhesion event and may release toxic chemicals such as formaldehyde, which can cause an inflammatory response (Vote and Elder, 2000) within *ca.* 8 weeks of degradation period. Despite these drawbacks, due to fast polymerization and strong adhesion, cyanoacrylate tissue adhesives have been approved in Canada and Europe, and several products are commercially available, such as Dermabond, Histoacryl, IFAbond, and Omnex (Bhatia, 2010) since the 1980s.

These cyanoacrylate adhesives have been used in various clinical applications including gastric variceal bleeding, esophageal variceal bleeding, peptic ulcer bleeding, and closure of fistula (Maluf et al., 2001; Ryan et al., 2004; Ryou and Thompson, 2006). Also closure of smooth and fresh skin wounds and endoscopic incisions are another indicated applications.

PEG polymers and its acrylic derivatives are also used for tissue adhesion. PEG-based tissue adhesives consist of chemically functionalized linear or branched PEG molecules, which can be cross-linked either by chemical cross-linking or photo-cross-linking, depending on available chemical groups. To be cross-linked through irradiation and form an adhesive hydrogel, PEGs should be modified with photoreactive elements such as acrylate groups (Mehdizadeh and Yang, 2013). Due to this modifiable chemistry of PEGs, the adhesion mechanism in PEG-based systems presents high variability; therefore, it is dependent on the formulation (Shazly et al., 2008). FocalSeal-L (Genzyme Biosurgery, Inc., Cambridge, Massachusetts) is the FDA-approved product in this category, which is a photopolymerizable adhesive based on self-assembly principle. Basically, as a primary solution, PEG–PLA polymer is used due to its low viscosity and applied to tissue, after which PTMC–PEG polymer (Fig. 8.1c), the sealant solution, is added and polymerized using irradiation in the presence of a photoinitiator (Bouten et al., 2014; Reece et al., 2001). Cross-links between the acrylate groups of the polymers result in the formation of hydrogel. However, application of the compound is difficult and nearly impossible in hemorrhage situations due to photoactivation requirement. In order to get rid of same activation sources, similar products are being developed. One such PEG polymer, CoSeal (Cohesion Technologies, Inc., Palo Alto, CA), consists of two PEG polymers that rapidly cross-link with proteins in tissue and mechanically adhere to synthetic graft materials. CoSeal is being used in Europe and it provides both cardiac adhesion prevention and anastomotic sealing. It is also indicated for reinforcement of suture and staple lines that can effectively cover and adhere to areas, which could be the source of an air leak around a lung resection (Napoleone et al., 2009). The main disadvantages of this system are its high degree of swelling after application and its weak adhesion to the surrounding tissue (Bhatia, 2010).

Another FDA-approved PEG sealant is Duraseal, which consists of PEG ester and trilysine amine solutions. Mixing of these two components results in the formation of a nontoxic hydrogel with 3D network due to cross-linking of PEG molecules by trilysine. It is used for closure of cerebrospinal fluid (CSF) leakage following neurosurgery (Spotnitz and Burks, 2008). PEG polymers are also used in combination with a chondroitin sulfate and resulting biodegradable CS-PEG hydrogels provide covalent adhesion to tissue (Strehin et al., 2010). For PEG-based systems, biodegradation takes 4–12 weeks (Lauto et al., 2008).

Another type of synthetic tissue adhesive material is urethane. Urethane-based adhesives have also been considered for use as soft tissue adhesives or sealants, because of their thermal stability at physiological temperature and absence of hemolytic behavior (Ferreira et al., 2007). Possible biocompatibility and biodegradability as well as the ability to form urea linkages through reaction

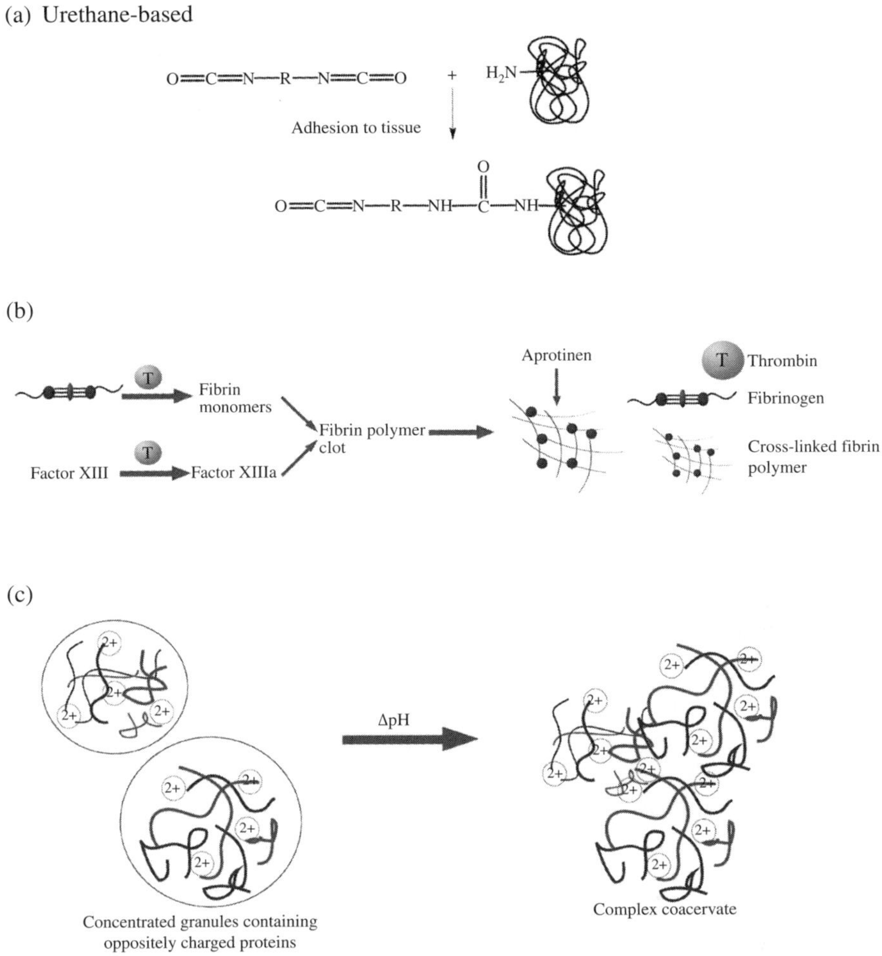

FIGURE 8.2 Mechanism of (a) adhesion of urethane-based adhesives through reacting with amine groups of proteins and water, (b) cross-linked fibrin curing, and (c) schematic illustration of complex coacervation.

with the amino groups present in the biological molecules are the main properties of the urethane-based adhesives that make them appealing candidates.

In order to synthesize urethane-based tissue adhesives, aromatic and aliphatic polyisocyanates with different polyether/polyester diols have been used in recent decades. Isocyanate groups have affinity to nucleophilic attack by hydroxyl or amine groups. As such, reaction of isocyanate-terminated prepolymer with amine groups in tissue proteins and formation of urea bond result in tissue adhesion, while cross-linking occurs through reacting with water in the tissue, summarized in Figure 8.2a (Mehdizadeh and Yang, 2013).

Polyurethanes are formed through polyaddition reaction between diisocyanates and diol groups. By virtue of their adhesion mechanism, urethane-based sealants were tested in renal surgery, endocrinology, and pancreatic occlusion and no toxic effects were observed (Bre et al., 2013). They are also being used for bone fixation, hemostasis, and sealing of vascular grafts in several surgical procedures (Heiss et al., 2006).

TissuGlu® surgical adhesive is one of the commercially available urethane-based products used for abdominal tissue bonding. It consists of a hyperbranched polymer with isocyanate end groups containing about 50 wt% of lysine (Gilbert et al., 2008). In the presence of moisture in the tissue, cross-linking occurs with itself and with the tissue surface, which forms a strong bond between layers. This adhesive can be degraded through hydrolysis and enzymatic reactions in the lysine linkages. The degradation products are usually polyols (e.g., glycerol), lysine, ethanol, and carbon dioxide, which are water soluble and easily cleared from the body (Gilbert et al., 2008).

8.3 NATURALLY DERIVED TISSUE ADHESIVES

Although many biological organisms produce their own adhesive materials that hold the above criteria, synthetic materials that have been developed so far do not fulfill all the requirements. Tissue adhesives developed based on naturally available sources benefit from inherent biocompatibility. In addition, structural and biochemical signals mimicking the native extracellular matrix could accelerate tissue regeneration (Annabi et al., 2014). On the other hand, high production cost, limited material availability, and potential immunogenicity are the major factors that hinder their wide range of use in medical and surgical operations.

Fibrin-based glues are most widely used natural tissue adhesives in clinical applications, and their use as local hemostatic agents was first reported in as early as 1910s (Leggat et al., 2004). Fibrin glues act similar to the last stage of blood clotting, where fibrinogen is converted to fibrin clot through a complex coagulation cascade. It creates a deliverable clot for use as both a hemostatic and a sealant. The fibrin sealants consist of two components: thrombin in combination with calcium chloride solution and fibrinogen (together with factor XIII and some other blood plasma proteins) (Brennan, 1991). Thrombin converts fibrinogen to fibrin monomers and activates factor XIII to factor XIIIa in the presence of calcium chloride. Then factor XIIIa cross-links fibrin molecules by the formation of amide bonds and stabilizes the polymeric network.

In a broad range of clinical applications, fibrin glues have been utilized for mainly control of bleeding and leaking after surgeries (Wheat and Wolf, 2009). TISSEEL is an approved fibrin family sealant produced to be used in patients undergoing surgery when conventional surgical methods are impractical for hemostasis. In some case studies, it was reported to be used for closure of esophageal perforations (Mutignani et al., 2006) and gastrocutaneous fistulae after gastrostomy tube removal (Rabago et al., 2002). Fibrin glue is also used to fix skin grafts in plastic and reconstructive surgery. Crosseal™ is the human protein, bovine component-free fibrin sealant, which reduces ecchymoses and hematoma formation in rhytidectomy patients efficiently (Lee et al., 2009b). Other commercial fibrin glues include Beriplast, Biocol, and Tissucol, which are intended to be used for hemorrhage, wound healing, and tissue gluing (Currie et al., 2001). Biodegradation of fibrin glues currently used in clinical procedures takes 2 weeks.

Despite of having many advantages, autologous isolation from tissue limits the availability of fibrin and makes fibrin glues relatively expensive. There are also some safety concerns regarding the use of thrombin from bovine sources, including allergic reactions and transmission of infection. Also, while clinically utilized fibrin glue can effectively bond to wet tissue without a significant inflammatory response, it exhibits low strength of adhesion due to poor cohesive properties and can be challenging to apply as the polymerization time is difficult to control during placement (Lauto et al., 2008).

Another choice of natural tissue glues is collagen-based adhesives, which represent a relatively new class. Mechanism of adhesion mainly relies on ability of collagen to adsorb coagulation products in blood and trap them in its fibers and interstices. This results in adhesion to the wound by inducing platelet adhesion and activating coagulation factors (Farndale et al., 2004).

Two agents have been approved for use in the United States: FloSeal (Sulzer Spine-tech, Anaheim, CA) and Proceed (Fusion Medical Technologies, Mountain View, CA), which includes combination of bovine thrombin and bovine collagen (Ryou and Thompson, 2006). FloSeal is produced for hemostasis in vascular surgery, whereas Proceed is for prevention and treatment of CSF leaks.

Another novel collagen product is CoStasis (Cohesion Technologies, Inc., Palo Alto, CA), which uses a combination of autologous human plasma obtained from patients' blood and a mixture of bovine collagen and thrombin. It is used as a spray for surgical hemostasis and also for endoscopic control of gastrointestinal bleeding in which it provides significant improvement

(Milkes et al., 2002). However, similar to all natural adhesives, due to bovine and porcine origin, there is a risk of allergic reaction and transmission of infection. Another drawback is the swelling with tissue similar to PEG hydrogel systems.

Another family of commercially available natural adhesives and sealants for clinical use is gelatin–glutaraldehyde systems, which undergo cross-linking reaction upon exposure to proper cross-linking agents. Gelatin is a biocompatible and bioabsorbable material and forms gels and films, which are flexible, strong, and transparent. However, in order to be used within physiological systems and due to their low stability in aqueous environment, gelatin networks need to be cross-linked. It was shown that high adhesive and tensile strength are achieved when gelatin is photochemically cross-linked (Elvin et al., 2010).

One of the adhesives of this type is gelatin–resorcin–formaldehyde/glutaraldehyde (GRF or GRFG). Through polycondensation reaction, aldehyde cross-links gelatin chains and forms bonds with amine groups of proteins. Currently, there is one such compound approved in the United States called BioGlue (CryoLife, Inc., Kennesaw, GA), which is a combination of bovine albumin and glutaraldehyde glue, used in nephron sparing surgery and aortic dissection (Hidas et al., 2006; Küçükaksu et al., 2000).

ProGel® pleural air leak sealant is another type of adhesive that circumvents the use of glutaraldehyde, as it contains human albumin that is cross-linked with NHS end group-functionalized PEG. This sealant is approved to seal air leaks on lung tissue after surgery (Kobayashi et al., 2001). However, as for GRFG glues, similar safety concerns about using aldehyde-containing products have limited their wide utilization (Spotnitz and Burks, 2008). Macrophage phagocytosis degrades the adhesives, which starts 56 days after sealing air leaks in lung surgery, yet the degradation is slow, and only 20% of the glue is degraded after 6 months (Nomori et al., 1999). Another disadvantage is the requirement of preheating to 45°C before the application (Albes et al., 1993).

8.4 BIOINSPIRED STRATEGIES

Biological organisms produce highly sophisticated dry and underwater adhesives in remarkable diversity. Driven by the stringent natural selection process, physical or chemical demands occurring in a particular environment are met by elegant designs that are assembled from various components in a hierarchical organization across nano- to centimeter length scales. Furthermore, biological adhesives are usually noninvasive and highly economical in amount. These properties of biological adhesives inspire novel synthetic technologies and more efficient engineering principles, as the performance of them outperforms their man-made counterparts. In this regard, engineering new types of adhesives has been shaped around hybridization of natural design principles with advances in the synthetic technologies. This approach is principally advantageous for developing flexible and chemically well-defined adhesives with optimized material performance based on the demand in the medical and surgical applications. Besides, the use of a synthetic backbone with well-defined chemistry offers minimum batch-to-batch variation compared to natural adhesive extracts.

Underwater adhesion of mussels has become a valuable source of inspiration for developing new types of adhesive materials (Lee et al., 2011; Waite et al., 2005). Mussels are nonmotile organisms that cling themselves onto underwater rocks, wood, and ship hulls, even in the highly challenging intertidal zones of the seashores (Bre et al., 2013). The adhesion capacity of mussels comprehends virtually all types of low- and high-energy organic and inorganic surfaces, including superhydrophobic surfaces (Crisp et al., 1985; Kang et al., 2010). Currently, there is no synthetic glue that can bond to such a broad variety of surface types. In order to achieve this, mussels produce a special polyphenolic adhesive containing hierarchically organized proteins with varying content of 3,4-dihydroxy-L-phenylalanine (Dopa) residues (Cha et al., 2008; Papov et al., 1995; Zhao and Waite, 2006). Spatiotemporal evolution of this residue within the wet adhesive precursor plays a predominant role in mussel adhesion (Wilker, 2010). Catechol side chain of Dopa can form strong coordination complexes with metals and metal oxides, which drives the adhesion on such surfaces (Lee et al., 2007a). Hydroxyl groups on catechol can mediate hydrogen bonds with polar organic surfaces. Hydrophobic interaction is also a major contributor to the bonding to different surfaces. Furthermore, Dopa residues can undergo reversible and irreversible covalent couplings within the adhesive bulk, which significantly contribute to the cohesive and self-healing characters of the mussel glue (Ceylan et al., 2013b; Harrington et al., 2010; Monahan and Wilker, 2004; Xu et al., 2012). For an in-depth discussion of mussel adhesion, interested readers are directed to several excellent reviews (Ceylan et al., 2013a; Lee et al., 2011). In summary, versatility of the substrates enabled by Dopa-mediated adhesion chemistry and its biocompatible and water-resistant properties have motivated many researchers for developing mussel-inspired synthetic adhesives for biomedical applications. In addition, Dopa has a very simple chemical structure that can be easily grafted onto synthetic systems (Ceylan et al., 2011; Xu et al., 2012). Consequently, recapitulation of various functionalities in a Dopa-functionalized base adhesive can enable multifunctional complex designs with superior material performance. For example, Dopa-mediated adhesion was successfully exploited for functionalization of cardiovascular stent and bone implant surfaces with extracellular matrix-mimetic microenvironment actively interacting with the cells and directing their behavior (Ceylan et al., 2011, 2012, 2014). Another mussel-inspired, Dopa-mimetic unit molecule, dopamine, undergoes self-assembly and autopolymerization reaction in water that results in polydopamine (Hong et al., 2012). Any type of surface can be coated with polydopamine by simply dipping it into dopamine solution at pH approximately 8.5 (Kang et al., 2012; Lee et al., 2007a). Polydopamine coating also allows for secondary modifications through coupling to nucleophiles (Lee et al., 2009a). Dopa-mediated coating can mediate bone-like hydroxyapatite crystallization under physiologically simulated ionic conditions, thereby increasing the biocompatibility and osseointegration of bone implants (Ceylan et al., 2014; Ryu et al., 2010). Catechol-functionalized chitosan/pluronic thermoresponsive and injectable hydrogels were utilized as hemostatic agents (Ryu et al., 2011). Dopa-grafted four-armed PEG hydrogel was applied to transplant pancreatic islets in a murine model where no suture was necessary for the placement of the islets (Brubaker et al., 2010).

Sandcastle worm (*Phragmatopoma californica*) is another inspirational organism for developing underwater adhesives. It is a polychaete worm that secretes cement to build tube reef by gluing sand particles to each other. The glue is composed of a mixture of polyacidic and polybasic proteins and Mg^{2+} and Ca^{2+} ions (Zhao et al., 2005). Foamy structure and waterborne underwater adhesive capability of the cement is explained by complex coacervation of oppositely charged proteins and divalent cations, which exhibit electrostatic interactions at physiological pH (Stewart et al., 2004). Sandcastle worm glue analogs have been synthesized by different approaches, one of which is synthesis by incorporating similar molar ratios of phosphate, primary amine, and catechol side chains to native proteins, PC1 and PC3. Mixing of these protein analogs results in complex coacervation, which is able to bond wet cortical bone specimens (Shao et al., 2009). In another example, Mann et al. used coacervate adhesive to seal fetal membrane patches in *in vitro* model of fetoscopic defect (Mann et al., 2012). In this study, nontoxic, complex coacervate was formed by mixing poly(acrylamide-*co*-aminopropyl methacrylamide)-PEG-dA with poly(MOEP-*co*-DMA)-PEG-dA, which contains a certain ratio of Ca^{2+} to phosphate side chains in conjunction with

1 wt% nanosilica fillers that increased the efficacy. This system effectively sealed the defect in an aqueous environment. Sealing ability and nontoxic nature of the developed adhesives inspired from complex coacervation seem to make it a possible strategy promising for clinical applications, yet further studies in animal models are required to evaluate the efficacy and durability of bioinspired adhesives. Unique underwater adaptations of other aquatic organisms also offer potential opportunities and novel inspirations for the purpose of developing advanced adhesives for biomedical and surgical applications.

Gecko-inspired adhesives represent the most important class of bioinspired dry adhesives. Gecko's foot has a complex and hierarchical organization of keratinous hairs, called setae, and its projections, which make up the adhesive arrays and provide adhesive properties to vertical surfaces with varying roughness (Ruibal and Ernst, 1965). Climbing up on vertical surfaces requiring strong attachment and rapid detachment is provided by adhesion/friction forces between surface and gecko's foot. Van der Waals forces also play a driving role in the adhesion, which is not strongly affected by the actual surface chemistry (Autumn et al., 2000, 2002). On the other hand, gecko adhesion is reduced dramatically underwater (Sun et al., 2005).

Gecko foot nanostructure is an inspiration source for development of dry adhesives for skin patch applications and wound dressings. Design and fabrication of such adhesives that stick to skin with enhanced durability and repeatable use over long periods of time by taking different properties of wounded and aged skin into consideration are vital for skin adhesive applications. It is therefore ideal for skin and wound patches to maintain moisture, provide antimicrobial effect, help healing process, and not cause any trauma upon removal (Rippon et al., 2007; Kwak et al., 2011). Despite the remarkable success of Gecko-inspired adhesives for dry conditions, it needs further modifications for wet adhesions. A prominent strategy introduced a hybrid system inspired from both mussel and gecko adhesions. Using poly(dopamine methacrylamide-comethoxyethyl acrylate), a reversible dry/wet adhesive platform was developed through combining mussel-mimetic Dopa adhesion with gecko-mimetic polydimethylsiloxane pillars (Lee et al., 2007b). Polymer nanopillars were produced and coated with a thin layer of a synthetic polymer that provides wet adhesion by mimicking proteins found in mussel holdfasts (Mahdavi et al., 2008). Due to having nanostructure and mussel-mimetic polymer, it showed reversible and effective adhesion in both wet and dry conditions. For developing effective tissue adhesives inspired from gecko footpad, further studies are necessary to provide wet adhesion, which is one of the most important requirements for strong tissue adhesion. Developing nanostructured surfaces combined with thin film coatings for underwater adhesion is a promising strategy for further studies and translation into biomedical applications.

8.5 NANOENABLED ADHESIVES

Adhesion of hydrogels through nanoparticle-embedded interfaces has emerged as a recent concept (Fig. 8.3) (Meddahi-Pellé et al., 2014; Rose et al., 2014). Network chains are adsorbed on nanoparticles at the interface of the bonding gels, or tissue parts, such as liver and heart by gently pressing onto one another. Therefore, nanoparticles act as connectors between the pieces and assure adhesion. Under mechanical stress, polymer chains reorganize and dissipate energy that withstands the rupture. Application of the adhesive is also quite straightforward: An arbitrary drop of nanoparticle dispersion is simply casted onto one substrate (either hydrogel or tissue part) and then the other substrate is pressed against the adsorbed substrate for less than a minute. Animal experiments suggest no wound leakage, infection, or inflammatory reactions with nanoparticle-mediated wound closure, which is comparable with suturing. On the other hand, the wound edges fail to bond correctly. Liver, heart, and skin bonding was demonstrated within this context (Meddahi-Pellé et al., 2014). Self-assembled mussel-mimetic nanofibrous networks that emulate the extracellular matrix via conjugated ligands represent another state-of-the-art class of nanoenabled adhesives that have been mostly applied as regenerative implant coatings, such as cardiovascular stents and bone implants (Ceylan et al., 2011, 2012, 2013a, 2014).

8.6 CONCLUSION AND FUTURE PROSPECTS

There are only few tissue adhesives that exhibit multiple functionalities in addition to bonding. In particular, guiding proper tissue regeneration at the wound/incision site would greatly aid in accelerated patient recovery as well as lesser scar tissue formation. Another significant improvement will be antibacterial adhesives, as the infection is becoming a major concern in surgical intervention (Li et al., 2011). Biomimetic materials are expected to continue to be a central theme by

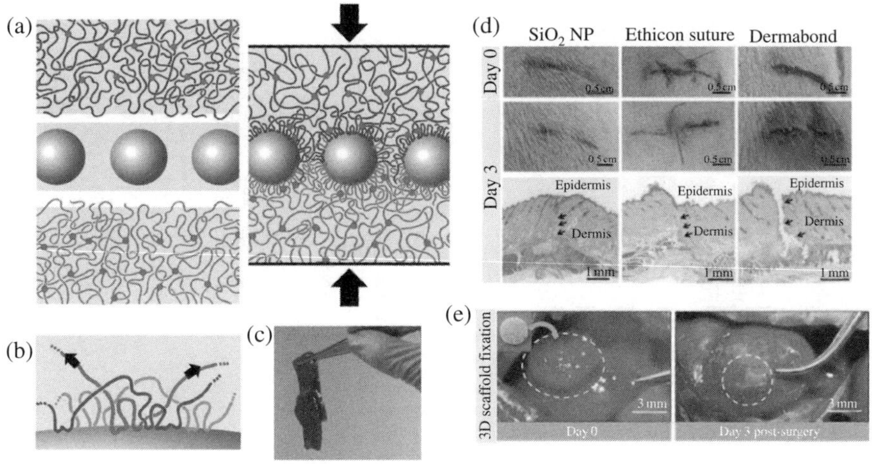

FIGURE 8.3 Nanoparticles can function as soft tissue adhesives. (a) Schematic illustration of nanoparticle-mediated bonding of polymer networks. Network chains are adsorbed on nanoparticles, via squeezing (see the arrows), at the interface of the bonding gels. Source: Rose et al. (2014). Reproduced with permission from John Wiley & Sons, Inc. and Nature Publishing Group. (b) Particle adsorption creates anchorage points to the gel networks. Under stress, one of the chains that detaches from a particle surface, can cause energy dissipation, thereby allowing for large deformations. (c) Bonding two calf liver parts using TM50 silica solution casting in between them and then gently pressing for 30 s. (d) *In vivo* application of SiO_2 nanoparticles to bond the dorsal skin in a rat model. An arbitrary drop of nanoparticle dispersion is simply casted onto the wound edges and then the two ends are gently pressed for about 1 min. Histological examination on day 3 shows no wound leakages, infection, or inflammatory reactions with nanoparticle-mediated wound closure, which is comparable with suturing. On the other hand, the wound edges fail to bond correctly. Source: Meddahi-Pellé et al. (2014). Reproduced with permission from John Wiley & Sons, Inc. and Nature Publishing Group. (e) Implantation of a 3D scaffold to a beating rat heart using Ludox TM50 silica nanoparticles (Meddahi-Pellé et al., 2014). The scaffold remains attached to the heart at the end of 3 days.

converging disciplines to reconstitute adaptive characteristics of biological systems in synthetic adhesives with nanoscale adaptations. Although natural adhesives exhibit exceptional performance under highly abrasive conditions, the high cost of obtaining adhesives from their original natural sources and high batch-to-batch variation encourages alternative biomimetic solutions in synthetic constructs (Table 8.1).

TABLE 8.1 Currently Available Tissue Adhesives

Type of Adhesive	Application Areas	Drawbacks	Tested Products	References
Cyanoacrylates	Skin, cosmetic surgery	Cytotoxicity	Dermabond, Histoacryl	Bhatia (2010), Ferreira et al. (2008), Trott (1997)
	Cornea	Limited to topical uses	IFAbond	
	Gastric therapy		Glubran2	
PEG	Cardiovascular	Cross-linker toxicity	FocalSeal-L	Napoleone et al. (2009), Reece et al. (2001), Spotnitz and Burks (2008)
CS-PEG	Liver	Swelling risk	CoSeal	
	Cartilage		Duraseal	
	Dura mater			
Urethane	Bone fixation	Curing time is long	TissuGlu	Bre et al. (2013), Ferreira et al. (2008), Gilbert et al. (2008), Heiss et al. (2006)
	Hemostasis			
Fibrin	Hemorrhage control, wound closure tissue anastomoses	Risk of disease transmission	TISSEEL	Currie et al. (2001), Spotnitz (2010), Wheat and Wolf (2009)
		Low mechanical strength	Crosseal	
		Poor tissue adhesion Relatively expensive	Beriplast	
			Biocol	
			Tissucol	
Collagen	Hemostasis	Swell with tissue compression	FloSeal	Committee, C.M.-c.C.W (2001), Milkes et al. (2002), Ryou and Thompson (2006)
	Cardiovascular, hepatic, orthopedic surgery		CoStasis	
Gelatin	Hemostasis	Cross-linker toxicity	ProGel	Hidas et al. (2006), Nomori et al. (1999), Oz et al. (2003)
	Variety of surgical procedures	Low bioabsorption	BioGlue®	
		Slow degradation	Floseal	

REFERENCES

Agarwal, S., Beckman, E.J., Buckley, M., and Zhang, J. (2007). Medical adhesive and methods of tissue adhesion, US 7264823 B2 (Google Patents).

Albes, J.M., Krettek, C., Hausen, B., Rohde, R., Haverich, A., and Borst, H.-G. (1993). Biophysical properties of the gelatin-resorcin-formaldehyde/glutaraldehyde adhesive. The Annals of Thoracic Surgery *56*, 910–915.

Annabi, N., Tamayol, A., Shin, S.R., Ghaemmaghami, A.M., Peppas, N.A., and Khademhosseini, A. (2014). Surgical materials: current challenges and nano-enabled solutions. Nano Today *9*, 574–589.

Autumn, K., Liang, Y.A., Hsieh, S.T., Zesch, W., Chan, W.P., Kenny, T.W., Fearing, R., and Full, R.J. (2000). Adhesive force of a single gecko foot-hair. Nature *405*, 681–685.

Autumn, K., Sitti, M., Liang, Y.A., Peattie, A.M., Hansen, W.R., Sponberg, S., Kenny, T.W., Fearing, R., Israelachvili, J.N., and Full, R.J. (2002). Evidence for van der Waals adhesion in gecko setae. Proceedings of the National Academy of Sciences of the United States of America *99*, 12252–12256.

Avila, G., Misch, K., Galindo-Moreno, P., and Wang, H.L. (2009). Implant surface treatment using biomimetic agents. Implant Dentistry *18*, 17–26.

Bhatia, S.K. (2010). Biomaterials for clinical applications (New York: Springer).

Bouten, P.J.M., Zonjee, M., Bender, J., Yauw, S.T.K., van Goor, H., van Hest, J.C.M., and Hoogenboom, R. (2014). The chemistry of tissue adhesive materials. Progress in Polymer Science *39*, 1375–1405.

Bre, L.P., Zheng, Y., Pego, A.P., and Wang, W. (2013). Taking tissue adhesives to the future: from traditional synthetic to new biomimetic approaches. Biomaterials Science *1*, 239–253.

Brennan, M. (1991). Fibrin glue. Blood Reviews *5*, 240–244.

Brubaker, C.E., Kissler, H., Wang, L.-J., Kaufman, D.B., and Messersmith, P.B. (2010). Biological performance of mussel-inspired adhesive in extrahepatic islet transplantation. Biomaterials *31*, 420–427.

Ceylan, H., Tekinay, A.B., and Guler, M.O. (2011). Selective adhesion and growth of vascular endothelial cells on bioactive peptide nanofiber functionalized stainless steel surface. Biomaterials *32*, 8797–8805.

Ceylan, H., Kocabey, S., Tekinay, A.B., and Guler, M.O. (2012). Surface-adhesive and osteogenic self-assembled peptide nanofibers for bioinspired functionalization of titanium surfaces. Soft Matter *8*, 3929–3937.

Ceylan, H., Tekinay, A.B., and Guler, M.O. (2013a). Mussel-inspired Adhesive Interfaces for Biomedical Applications. In Biological and biomimetic adhesives: challenges and opportunities (P. Flammang, R. Santos, N. Aldred, S. Gorb, eds., Cambridge: The Royal Society of Chemistry), pp. 103–116.

Ceylan, H., Urel, M., Erkal, T.S., Tekinay, A.B., Dana, A., and Guler, M.O. (2013b). Mussel inspired dynamic cross-linking of self-healing peptide nanofiber network. Advanced Functional Materials *23*, 2081–2090.

Ceylan, H., Kocabey, S., Unal Gulsuner, H., Balcik, O.S., Guler, M.O., and Tekinay, A.B. (2014). Bone-like mineral nucleating peptide nanofibers induce differentiation of human mesenchymal stem cells into mature osteoblasts. Biomacromolecules *15*, 2407–2418.

Cha, H.J., Hwang, D.S., and Lim, S. (2008). Development of bioadhesives from marine mussels. Biotechnology Journal *3*, 631–638.

Committee, C.M.-c.C.W. (2001). A novel collagen-based composite offers effective hemostasis for multiple surgical indications: results of a randomized controlled trial. Surgery *129*, 445–450.

Crisp, D.J., Walker, G., Young, G.A., and Yule, A.B. (1985). Adhesion and substrate choice in mussels and barnacles. Journal of Colloid and Interface Science *104*, 40–50.

Currie, L.J., Sharpe, J.R., and Martin, R. (2001). The use of fibrin glue in skin grafts and tissue-engineered skin replacements. Plastic and Reconstructive Surgery *108*, 1713–1726.

Elvin, C.M., Vuocolo, T., Brownlee, A.G., Sando, L., Huson, M.G., Liyou, N.E., Stockwell, P.R., Lyons, R.E., Kim, M., Edwards, G.A., et al. (2010). A highly elastic tissue sealant based on photopolymerised gelatin. Biomaterials *31*, 8323–8331.

Farndale, R., Sixma, J., Barnes, M., and De Groot, P. (2004). The role of collagen in thrombosis and hemostasis. Journal of Thrombosis and Haemostasis *2*, 561–573.

Ferreira, P., Pereira, R., Coelho, J.F.J., Silva, A.F.M., and Gil, M.H. (2007). Modification of the biopolymer castor oil with free isocyanate groups to be applied as bioadhesive. International Journal of Biological Macromolecules *40*, 144–152.

Ferreira, P., Coelho, J.F.J., and Gil, M.H. (2008). Development of a new photocrosslinkable biodegradable bioadhesive. International Journal of Pharmaceutics *352*, 172–181.

Gilbert, T.W., Badylak, S.F., Gusenoff, J., Beckman, E.J., Clower, D.M., Daly, P., and Rubin, J.P. (2008). Lysine-derived urethane surgical adhesive prevents seroma formation in a canine abdominoplasty model. Plastic and Reconstructive Surgery *122*, 95–102.

Gittens, R.A., Olivares-Navarrete, R., Schwartz, Z., and Boyan, B.D. (2014). Implant osseointegration and the role of microroughness and nanostructures: lessons for spine implants. Acta Biomaterialia *10*, 3363–3371.

Harrington, M.J., Masic, A., Holten-Andersen, N., Waite, J.H., and Fratzl, P. (2010). Iron-clad fibers: a metal-based biological strategy for hard flexible coatings. Science *328*, 216–220.

Heiss, C., Kraus, R., Schluckebier, D., Stiller, A.-C., Wenisch, S., and Schnettler, R. (2006). Bone adhesives in trauma and orthopedic surgery. European Journal of Trauma *32*, 141–148.

Hidas, G., Kastin, A., Mullerad, M., Shental, J., Moskovitz, B., and Nativ, O. (2006). Sutureless nephron-sparing surgery: use of albumin glutaraldehyde tissue adhesive (BioGlue). Urology *67*, 697–700.

Hong, S., Na, Y.S., Choi, S., Song, I.T., Kim, W.Y., and Lee, H. (2012). Non-covalent self-assembly and covalent polymerization co-contribute to polydopamine formation. Advanced Functional Materials *22*, 4711–4717.

Kang, S.M., You, I., Cho, W.K., Shon, H.K., Lee, T.G., Choi, I.S., Karp, J.M., and Lee, H. (2010). One-step modification of superhydrophobic surfaces by a mussel-inspired polymer coating. Angewandte Chemie International Edition *49*, 9401–9404.

Kang, S.M., Hwang, N.S., Yeom, J., Park, S.Y., Messersmith, P.B., Choi, I.S., Langer, R., Anderson, D.G., and Lee, H. (2012). One-step multipurpose surface functionalization by adhesive catecholamine. Advanced Functional Materials *22*, 2949–2955.

Kobayashi, H., Sekine, T., Nakamura, T., and Shimizu, Y. (2001). In vivo evaluation of a new sealant material on a rat lung air leak model. Journal of Biomedical Materials Research *58*, 658–665.

Küçükaksu, D.S., Akgül, A., Çağli, K., and Taşdemir, O. (2000). Beneficial effect of BioGlue® surgical adhesive in repair of iatrogenic aortic dissection. Texas Heart Institute Journal *27*, 307.

Kull, S., Martinelli, I., Briganti, E., Losi, P., Spiller, D., Tonlorenzi, S., and Soldani, G. (2009). Glubran2 surgical glue: in vitro evaluation of adhesive and mechanical properties. Journal of Surgical Research *157*, e15–e21.

Kwak, M.K., Jeong, H.-E., and Suh, K.Y. (2011). Rational design and enhanced biocompatibility of a dry adhesive medical skin patch. Advanced Materials *23*, 3949–3953.

Lauto, A., Mawad, D., and Foster, L.J.R. (2008). Adhesive biomaterials for tissue reconstruction. Journal of Chemical Technology & Biotechnology *83*, 464–472.

Lee, H., Dellatore, S.M., Miller, W.M., and Messersmith, P.B. (2007a). Mussel-inspired surface chemistry for multifunctional coatings. Science *318*, 426–430.

Lee, H., Lee, B.P., and Messersmith, P.B. (2007b). A reversible wet/dry adhesive inspired by mussels and geckos. Nature *448*, 338–341.

Lee, H., Rho, J., and Messersmith, P.B. (2009a). Facile conjugation of biomolecules onto surfaces via mussel adhesive protein inspired coatings. Advanced Materials *21*, 431–434.

Lee, S., Pham, A.M., Pryor, S.G., Tollefson, T., and Sykes, J.M. (2009b). Efficacy of Crosseal fibrin sealant (human) in rhytidectomy. Archives of Facial Plastic Surgery *11*, 29–33.

Lee, B.P., Messersmith, P.B., Israelachvili, J.N., and Waite, J.H. (2011). Mussel-inspired adhesives and coatings. Annual Review of Materials Research *41*, 99–132.

Lees, W.A. (1979). The science of acrylic adhesives. British Polymer Journal *11*, 64–71.

Leggat, P.A., Kedjarune, U., and Smith, D.R. (2004). Toxicity of cyanoacrylate adhesives and their occupational impacts for dental staff. Industrial Health *42*, 207–211.

Li, W.-R., Xie, X.-B., Shi, Q.-S., Duan, S.-S., Ouyang, Y.-S., and Chen, Y.-B. (2011). Antibacterial effect of silver nanoparticles on *Staphylococcus aureus*. Biometals *24*, 135–141.

Liu, Y., Li, J.P., Hunziker, E.B., and de Groot, K. (2006). Incorporation of growth factors into medical devices via biomimetic coatings. Philosophical Transactions. Series A, Mathematical, Physical, and Engineering Sciences *364*, 233–248.

Mahdavi, A., Ferreira, L., Sundback, C., Nichol, J.W., Chan, E.P., Carter, D.J.D., Bettinger, C.J., Patanavanich, S., Chignozha, L., Ben-Joseph, E., et al. (2008). A biodegradable and biocompatible gecko-inspired tissue adhesive. Proceedings of the National Academy of Sciences of the United States of America *105*, 2307–2312.

Maluf, F., Sakai, P., Ishioka, S., and Matuguma, S.E. (2001). Endoscopic sclerosis versus cyanoacrylate endoscopic injection for the first episode of variceal bleeding: a prospective, controlled, and randomized study in Child-Pugh class C patients. Endoscopy *33*, 421–427.

Mann, L.K., Papanna, R., Moise Jr, K.J., Byrd, R.H., Popek, E.J., Kaur, S., Tseng, S.C.G., and Stewart, R.J. (2012). Fetal membrane patch and biomimetic adhesive coacervates as a sealant for fetoscopic defects. Acta Biomaterialia *8*, 2160–2165.

Meddahi-Pellé, A., Legrand, A., Marcellan, A., Louedec, L., Letourneur, D., and Leibler, L. (2014). Organ repair, hemostasis, and in vivo bonding of medical devices by aqueous solutions of nanoparticles. Angewandte Chemie International Edition *53*, 6369–6373.

Mehdizadeh, M. and Yang, J. (2013). Design strategies and applications of tissue bioadhesives. Macromolecular Bioscience *13*, 271–288.

Milkes, D.E., Friedland, S., Lin, O.S., Reid, T.R., and Soetikno, R.M. (2002). A novel method to control severe upper GI bleeding from metastatic cancer with a hemostatic sealant: the CoStasis surgical hemostat. Gastrointestinal Endoscopy *55*, 735–740.

Monahan, J. and Wilker, J.J. (2004). Cross-linking the protein precursor of marine mussel adhesives: bulk measurements and reagents for curing. Langmuir *20*, 3724–3729.

Mutignani, M., Iacopini, F., Dokas, S., Larghi, A., Familiari, P., Tringali, A., and Costamagna, G. (2006). Successful endoscopic closure of a lateral duodenal perforation at ERCP with fibrin glue. Gastrointestinal Endoscopy *63*, 725–727.

Napoleone, C.P., Valori, A., Crupi, G., Ocello, S., Santoro, F., Vouhé, P., Weerasena, N., and Gargiulo, G. (2009). An observational study of CoSeal® for the prevention of adhesions in pediatric cardiac surgery. Interactive Cardiovascular and Thoracic Surgery *9*, 978–982.

Nomori, H., Horio, H., Morinaga, S., and Suemasu, K. (1999). Gelatin-resorcinol–formaldehyde-glutaraldehyde glue for sealing pulmonary air leaks during thoracoscopic operation. The Annals of Thoracic Surgery *67*, 212–216.

Oz, M.C., Rondinone, J.F., and Shargill, N.S. (2003). Floseal matrix. Journal of Cardiac Surgery *18*, 486–493.

Páez, J.M.G., Herrero, E.J., Rocha, A., Maestro, M., Castillo-Olivares, J.L., Millan, I., Sanmartin, A.C., and Cordon, A. (2004). Comparative study of the mechanical behaviour of a cyanoacrylate and a bioadhesive. Journal of Materials Science: Materials in Medicine *15*, 109–115.

Papov, V.V., Diamond, T.V., Biemann, K., and Waite, J.H. (1995). Hydroxyarginine-containing polyphenolic proteins in the adhesive plaques of the marine mussel *Mytilus edulis*. Journal of Biological Chemistry *270*, 20183–20192.

Rabago, L., Ventosa, N., Castro, J., Marco, J., Herrera, N., and Gea, F. (2002). Endoscopic treatment of postoperative fistulas resistant to conservative management using biological fibrin glue. Endoscopy *34*, 632–638.

Reece, T.B., Maxey, T.S., and Kron, I.L. (2001). A prospectus on tissue adhesives. The American Journal of Surgery *182*, S40–S44.

Report S190. (2012). Worldwide surgical sealants, glues, wound closure and anti-adhesion markets, 2010–2017. (Foothill Ranch, CA: MedMarket Diligence, LLC).

Richards, R.G., Moriarty, T.F., Miclau, T., McClellan, R.T., and Grainger, D.W. (2012). Advances in biomaterials and surface technologies. Journal of Orthopaedic Trauma *26*, 703–707.

Rippon, M., White, R., and Davies, P. (2007). Skin adhesives and their role in wound dressings. WOUNDS UK *3*, 76.

Rose, S., Prevoteau, A., Elziere, P., Hourdet, D., Marcellan, A., and Leibler, L. (2014). Nanoparticle solutions as adhesives for gels and biological tissues. Nature *505*, 382–385.

Ruibal, R. and Ernst, V. (1965). The structure of the digital setae of lizards. Journal of Morphology *117*, 271–293.

Ryan, B.M., Stockbrugger, R.W., and Ryan, J.M. (2004). A pathophysiologic, gastroenterologic, and radiologic approach to the management of gastric varices. Gastroenterology *126*, 1175–1189.

Ryou, M. and Thompson, C.C. (2006). Tissue adhesives: a review. Techniques in Gastrointestinal Endoscopy *8*, 33–37.

Ryu, J., Ku, S.H., Lee, H., and Park, C.B. (2010). Mussel-inspired polydopamine coating as a universal route to hydroxyapatite crystallization. Advanced Functional Materials *20*, 2132–2139.

Ryu, J.H., Lee, Y., Kong, W.H., Kim, T.G., Park, T.G., and Lee, H. (2011). Catechol-functionalized chitosan/pluronic hydrogels for tissue adhesives and hemostatic materials. Biomacromolecules *12*, 2653–2659.

Sanders, L. and Nagatomi, J. (2014). Clinical applications of surgical adhesives and sealants. Critical Reviews in Biomedical Engineering *42*, 271–292.

Seewald, S., Sriram, P.V.J., Naga, M., Fennerty, M.B., Boyer, J., Oberti, F., and Soehendra, N. (2002). Cyanoacrylate glue in gastric variceal bleeding. Endoscopy *34*, 926–932.

Shao, H., Bachus, K.N., and Stewart, R.J. (2009). A water-borne adhesive modeled after the sandcastle glue of *P. californica*. Macromolecular Bioscience *9*, 464–471.

Sharma, B., Fermanian, S., Gibson, M., Unterman, S., Herzka, D.A., Cascio, B., Coburn, J., Hui, A.Y., Marcus, N., Gold, G.E., et al. (2013). Human cartilage repair with a photoreactive adhesive-hydrogel composite. Science Translational Medicine *5*, 167ra166.

Shazly, T.M., Artzi, N., Boehning, F., and Edelman, E.R. (2008). Viscoelastic adhesive mechanics of aldehyde-mediated soft tissue sealants. Biomaterials *29*, 4584–4591.

Singer, A.J., Quinn, J.V., and Hollander, J.E. (2008). The cyanoacrylate topical skin adhesives. The American Journal of Emergency Medicine *26*, 490–496.

Spotnitz, W. (2010). Fibrin sealant: past, present, and future: a brief review. World Journal of Surgery *34*, 632–634.

Spotnitz, W.D. and Burks, S. (2008). Hemostats, sealants, and adhesives: components of the surgical toolbox. Transfusion *48*, 1502–1516.

Stewart, R.J., Weaver, J.C., Morse, D.E., and Waite, J.H. (2004). The tube cement of *Phragmatopoma californica*: a solid foam. The Journal of Experimental Biology *207*, 4727–4734.

Strehin, I., Nahas, Z., Arora, K., Nguyen, T., and Elisseeff, J. (2010). A versatile pH sensitive chondroitin sulfate–PEG tissue adhesive and hydrogel. Biomaterials *31*, 2788–2797.

Sun, W., Neuzil, P., Kustandi, T.S., Oh, S., and Samper, V.D. (2005). The nature of the gecko lizard adhesive force. Biophysical Journal *89*, L14–L17.

Trott, A.T. (1997). Cyanoacrylate tissue adhesives: an advance in wound care. JAMA *277*, 1559–1560.

Vote, B.J. and Elder, M.J. (2000). Cyanoacrylate glue for corneal perforations: a description of a surgical technique and a review of the literature. Clinical & Experimental Ophthalmology *28*, 437–442.

Waite, J.H., Andersen, N.H., Jewhurst, S., and Sun, C. (2005). Mussel adhesion: finding the tricks worth mimicking. The Journal of Adhesion *81*, 297–317.

Wheat, J.C. and Wolf Jr, J.S. (2009). Advances in bioadhesives, tissue sealants, and hemostatic agents. Urologic Clinics of North America *36*, 265–275.

Wilker, J.J. (2010). The iron-fortified adhesive system of marine mussels. Angewandte Chemie International Edition *49*, 8076–8078.

Xu, H., Nishida, J., Ma, W., Wu, H., Kobayashi, M., Otsuka, H., and Takahara, A. (2012). Competition between oxidation and coordination in cross-linking of polystyrene copolymer containing catechol groups. ACS Macro Letters *1*, 457–460.

Yang, S.Y., O'Cearbhaill, E.D., Sisk, G.C., Park, K.M., Cho, W.K., Villiger, M., Bouma, B.E., Pomahac, B., and Karp, J.M. (2013). A bio-inspired swellable microneedle adhesive for mechanical interlocking with tissue. Nature Communications *4*, 1702.

Zhao, H. and Waite, J.H. (2006). Linking adhesive and structural proteins in the attachment plaque of *Mytilus californianus*. Journal of Biological Chemistry *281*, 26150–26158.

Zhao, H., Sun, C.J., Stewart, R.J., and Waite, J.H. (2005). Cement proteins of the tube-building polychaete *Phragmatopoma californica*. Journal of Biological Chemistry *280*, 42938–42944.

9

ADVANCES IN NANOPARTICLE-BASED MEDICAL DIAGNOSTIC AND THERAPEUTIC TECHNIQUES

Melis Sardan, Alper Devrim Ozkan, Aygul Zengin, Ayse B. Tekinay, and Mustafa O. Guler

Institute of Materials Science and Nanotechnology, National Nanotechnology Research Center (UNAM), Bilkent University, Ankara, Turkey

9.1 INTRODUCTION

Advances in modern medicine have eliminated several major causes of human mortality and considerably extended life expectancies around the world; however, this increase in the global age average has also boosted the incidences of age-associated disorders. These conditions, such as cancer, neurodegenerative disorders, and cardiovascular disease, severely decrease the quality of life for the affected but are highly polymorphic and often difficult to treat. Their variable nature often necessitates the customization of treatment methods for each individual patient, which in turn requires close monitoring of the disease in question. In addition, a good prognosis is typically possible only when an early diagnosis has been made, which further emphasizes the need for rapid and accurate diagnostic methods. As medical diagnosis involves a combination of biological (e.g., throat cultures, blood and urine tests), chemical (e.g., immunosorbent and other colorimetric

Therapeutic Nanomaterials, First Edition. Edited by Mustafa O. Guler and Ayse B. Tekinay.

assays), and physical (e.g., imaging techniques) components, disease detection methods can be advanced through any of these channels, and the multidisciplinary nature and obvious importance of disease diagnostics attracted much attention to this topic over the last few decades.

Imaging techniques are of paramount importance for both *in vivo* and *ex vivo* diagnostic efforts and, alongside biosensors and biochemical assays, form the backbone of nearly all medical analysis. Histopathological analyses are usually performed by light microscopy and chemical stains that visualize different cell types and metabolic states in an excised tissue section; however, such an approach is by necessity invasive and therefore is not amenable to the diagnosis of every disease. In addition, histological analysis often requires a preliminary diagnosis, as biopsy samples are typically taken from regions suspected to be diseased (e.g., based on the patient's symptoms or in areas with gross appearances that suggest disease). As such, diseases such as early stage cancers, which are largely asymptomatic and indistinguishable from the surrounding tissue, are difficult to detect by biopsy or visual investigation (Betz et al., 2002; Strong et al., 1968). Medical imaging techniques, in contrast, are capable of yielding information without any surgical intervention and may therefore detect early stages of a disease through regular screenings. These techniques vary in the electromagnetic region they use but generally work by scanning or rotating an electromagnetic field over the target area and measuring the response given either by the affected tissue or a specific tracer administered prior to imaging (Table 9.1).

Although medical imaging techniques offer invaluable benefits as noninvasive diagnostic methods, they also suffer from a set of limitations. They are limited in resolution compared to histological analysis and therefore cannot detect diseased tissues below a certain threshold volume (this volume changes depending on the technique and the wavelength or field strength used). By extension, they are also limited in their sensitivity and early detection capacity, as tumors may be below this detection threshold during their early development (Ekberg et al., 1988). They are further limited in their ability to identify the physiological characteristics of the disease they detect; while magnetic resonance imaging (MRI) and positron emission tomography–computed tomography (PET/CT) are capable of distinguishing between healthy and damaged tissues, immunoassays are nonetheless an easier means of evaluating the physiological state of the organ of interest. In addition, as medical imaging often requires the administration of contrast agents (CAs) or exposure to potentially harmful electromagnetic radiation, safety concerns have been raised against the prospect of their overuse (Dawson, 1985).

TABLE 9.1 Medical imaging techniques and the contrast agents used for their enhancement

Technique	Resolution	Tracers and Contrast Agents	References
MRI	*c*. 100 μm (ideally), millimeter scale (in practice)	Gadolinium chelates; manganese chelates; nanoparticles of gadolinium, manganese, and iron oxides (with or without cobalt, nickel, or manganese doping); metal alloys	Eastwood et al. (2014), Pan et al. (2011), Yablonskiy and Sukstanskii (2015)
CT	*c*. 1 μm (ideally), *c*. 130 μm (in practice)	Iodinates; nanoparticles of gold, silver, bismuth, tantalum, and other heavy transition metals	Burghardt et al. (2011)
PET and SPECT	Millimeter to centimeter scale	Radiotracers—water, salts, or biological molecules with radioactive atom substitutions. ^{11}C, ^{13}N, ^{15}O, and 18F are common replacements	Bateman et al. (2006), Ciernik et al. (2003)
Optical imaging and fluoroscopy	Abbe limit (ideally), millimeter scale (in practice, NIR)	Fluorophores, quantum dots, gold and silver nanoparticles, upconverting nanoparticles	Chang et al. (2003)
Ultrasound	*c*. 20 μm for high-frequency ultrasound (but low penetration), *c*. 100–150 μm typical	Gas microbubbles	Calliada et al. (1998), Stanga and Bird (2001), Wright et al. (2006)

Much work has been performed on potential methods to alleviate these deficiencies. The resolution limit of the MRI can be enhanced by using a stronger field, which requires more powerful (and often expensive) magnets (Lee et al., 2009). Likewise, the resolution of positron emission tomography (PET) can be increased by the use of thinner detector elements, the crystal components of which become exponentially more expensive to produce (Thompson et al., 1994). Consequently, high-resolution medical imaging devices are mostly limited to academic research, and the modification of the tracers and CAs is a more convenient means of circumventing the disadvantages of these methods. These modifications can be performed to allow the imaging agent to track a specific biological moiety, to decrease its toxicity, to prevent an immune response from being raised against the agent, to increase or decrease its renal clearance, or to grant it therapeutic functions in addition to its diagnostic capacity. In addition, the size and morphology of the agent may be modified, and the material may be designed so as to serve as an imaging agent for multiple complementary techniques (Lee et al., 2008; Xie et al., 2010).

Due to their ease of functionalization, relative lack of toxicity, suitable magnetic properties, and applicability to multiple imaging techniques, nanoparticles (NPs) in particular have been popular targets of CA research. Typically, these NPs possess a metal core that grants them their contrast-enhancing properties and are coated by one or more "shell" layers that decrease the toxicity of the metal used, increase the circulation time of the complex, or allow specific targeting of a specific tumor or tissue type. However, despite the variety of NPs developed for imaging enhancement, a broad overview of their properties, advantages, disadvantages, and potential uses in medical diagnosis has so far been lacking. In this chapter, we describe the characteristics of NP CAs proposed for use in medical imaging, detail the surface modification methods used to designate specific targets for their attachment, compare their effectiveness and toxicity compared to conventional methods of contrast enhancement, and discuss the contribution that nanoscience has had, and will have, on medical imaging and disease diagnosis at large.

9.2 NPs USED IN MRI

MRI is one of the most valuable noninvasive imaging modalities and is frequently used in both clinical and research studies to diagnose various disorders. In addition to its high spatial resolution, MRI utilizes nonionizing radiation and is therefore capable of gathering three-dimensional tomographic images with negligible damage to the imaged organ or tissue (Lodhia et al., 2010). Longitudinal (T_1) and transverse (T_2) proton relaxation times of water (or, less frequently, other molecules) are the two main parameters that affect the signal quality of MRI; consequently, contrast in MR images is a result of differences in proton relaxation times. Proton density differences between different tissues are another factor that affects the contrast in MRI: Since the altered physiological environment of a diseased region may yield a different intrinsic relaxation time compared to the rest of an organ or system, a specific contrast between healthy and aberrant tissues may be observed in MR images at more advanced stages of many disorders (Mulder et al., 2006). However, early diagnosis by MRI is feasible only for a select number of tissues and diseases; where in many other cases, the resolutions obtained are too low to conclusively determine the presence of disease. A variety of CAs have been designed to enhance the diagnostic value of MRI and lower the threshold at which the technique is able to differentiate between healthy and diseased tissue, which is accomplished by shortening T_1 or T_2 relaxation times or, in other words, increasing r_1 ($1/T_1$) or r_2 ($1/T_2$) relaxivities (Wang and Chen, 2009). Paramagnetic and superparamagnetic materials are typically used as T_1 and T_2 CAs, respectively, and applied in clinical imaging (Liu et al., 2012a).

9.2.1 T_1 CAs

Paramagnetic metals possess a large number of unpaired electrons, which create magnetic dipole moments when exposed to a magnetic field. Among all transition and lanthanide metals, gadolinium (Gd^{3+}), manganese (Mn^{2+}), and iron (Fe^{3+}) ions are known to have the strongest paramagnetic properties and show the most effective T_1 relaxation times, which increases both the signal intensity and the brightness of the image. This relaxation is enhanced as a result of dipolar interactions between the protons of water molecule inside the tissue and unpaired electrons of the metal when the paramagnetic compound is in close proximity to the tissue. The brightness of this signal makes anatomic details discernible in either pathogenic or ordinary conditions, and the paramagnetic nature of these compounds precludes them from altering the magnetic homogeneity of the body, effectively preventing them from disrupting the background of MR images. The r_2/r_1 ratio is the factor of interest for T_1 CAs (also called positive agents) and must ideally be as low as possible (i.e., a high r_1 value and a low r_2 value). T_1 agents have been used in either NP (Gd_2O_3, GdF_3, MnO, Mn_3O_4, Fe_2O_3, etc.) or complex form (Gd-DTPA, Mn-DOTA, etc.) (Hu and Zhao, 2012). In addition, there are several examples regarding the incorporation of Gd^{3+}

into nanostructured frameworks such as dendrimers, liposomes, polymers, and metal-organic frameworks (MOFs). Although they exhibit enhanced relaxivity and sensitivity compared to a single ionic chelate, the production of these nanostructures is laborious and costly, and their large sizes impose further limitations on their potential utility. Since NP-based T_1 CAs are superior to complex-based agents in terms of toxicity, sensitivity, relaxivity, contrast, retention time, and tailoring and targeting abilities, studies related to the development of the former have garnered great interest in recent years (Liang et al., 2013). As such, the present section will focus on inorganic NP-based systems designed for T_1 contrast enhancement.

9.2.1.1 Gd-Based CAs Recently, significant effort has been directed toward the utilization of gadolinium NPs, including gadolinium oxide (Gd_2O_3), gadolinium fluoride (GdF_3), and gadolinium phosphate ($GdPO_4$), in MRI. Synthetic approaches to produce these NPs show great variation: Reduction–coprecipitation, polyol, and hydrothermal processes are all utilized for Gd-based nanoparticle (GdNP) synthesis. The first technique results in the formation of polydisperse NPs and aggregates and is consequently not an efficient way to produce NPs for MRI. In polyol processes, crystal growth is controlled by a stabilizing agent in polyol, while extraction and exchange steps are performed with another, stronger stabilizing agent. In the last technique, hydrophobic NPs are prepared in high-boiling organic solvents, and the organic solvent is replaced by water in the presence of a hydrophilic ligand that solubilizes the NP in aqueous environment (Johnson et al., 2011). Following synthesis, the NPs must be functionalized in order to control their solubility, biocompatibility, stability, and retention times in the tissue of interest. Dextran, polyethylene glycol (PEG), polysiloxane, citrate, aminoethyl phosphate (AEP), and PEG-silica are some of the materials that impart these properties to metallic NPs and are commonly used in NP functionalization. Surface modification also enables tethering bioactive moieties onto the NP surface for targeting purposes (Na and Hyeon, 2009). Gd NPs can demonstrate different relaxivities depending on their size, shape, composition, and method of assembly (Hu et al., 2011). Park et al. prepared D-glucuronic acid-coated ultrasmall Gd_2O_3 NPs with an average diameter of 1 nm (Park et al., 2009). While the NP core provided a large longitudinal relaxivity of 9.9 $mM^{-1}{\cdot}s^{-1}$ (1.5 T) with a low r_1/r_2 ratio (1.1), the coating allowed *in vivo* T_1-weighted MRI of rat brain tumors. However, despite the effective visualization of subanatomic details of organs with Gd-based CAs, the toxicity of the metal remains as an obstacle to the use of these materials in clinical practice.

9.2.1.2 Mn-Based CAs Another paramagnetic ion used for T_1-weighted MRI is Mn^{2+}. Among manganese oxides, MnO and Mn_3O_4 are the two most commonly synthesized NPs and can be produced with great control over particle size. Since paramagnetic Mn^{2+} ions present on the NP surface is the main factor that shortens the longitudinal relaxation time T_1, the total surface area of the NP is a key factor to obtain high r_1 relaxivity values in Mn-based NPs. The pronounced effect of the NP surface-to-volume ratios on the r_1 value is exemplified by a study by Shin et al., in which solid MnO NPs were prepared with a hydrothermal method and stirred in an acidic buffer solution to create hollow structures. According to the results, r_1 values were found as 0.21 and 1.42 $mM^{-1}{\cdot}s^{-1}$ at 3 T for solid and hollow MnO nanoparticles (HMON), respectively (Shin et al., 2009). Since the inner surfaces of hollow NPs contain a higher concentration of Mn^{2+} ions compared to their solid counterparts, the enhancement probably stems from the fact that the former interacts more with water molecules. Other than HMON, mesoporous silica-coated MnO NPs were also synthesized to increase the accessibility of water to manganese ions with the help of the porous coating (Zhu et al., 2013). Different preparations of Mn-based NPs were used to visualize various organs and tissues, such as brain, liver, and kidney, using MRI. Although these agents effectively depict detailed structures in the brain and can detect neuronal activity in T_1-weighted MR, they are not applicable in humans because of the hepatic and cardiovascular toxicity effect of free manganese ions (Na and Hyeon, 2009).

9.2.1.3 Fe-Based CAs Both Gd-based and Mn-based T_1-weighted CAs show detrimental effects due to the dissociation and accumulation of these ions in the body. Consequently, much research has been performed on nontoxic formulations of these metals, as well as on nontoxic alternatives to traditional T_1 CAs. Since iron atoms are naturally found in the human blood and their excess can be stored as ferritin in the body, iron oxide is a promising candidate for use as a more biocompatible CA. Although Fe-based NPs are considered to be superparamagnetic T_2 CAs because of their high magnetic moments, the small-sized iron oxide NPs allow them to exhibit low r_2/r_1 ratios, which makes them appealing for T_1-weighted MRI. As the size of an NP decreases, so does the effects of magnetic anisotropy and spin disorders at the particle surface (Jun et al., 2008). However, uniformity and stability are two requirements for NPs to maintain a high T_1 contrast effect, and the synthesis of uniform NPs at very small sizes is problematic. Several approaches have been attempted to achieve the required sizes and uniformities in iron oxide NPs; these are principally classified under (i) hydrophobic and (ii) hydrophilic phase syntheses.

The former is performed at high temperatures (>200°C) to allow the decomposition of the iron precursor and the formation of homogeneous, extremely small iron oxide nanoparticles (ESPIONs), which then are made water soluble through ligand exchange (Tromsdorf et al., 2009). The latter is a simpler, one-step approach in which triethylene glycol (or diethylene glycol) acts as both a reductant and a surfactant to react with iron precursors at high temperatures, resulting in hydrophilic ESPIONs (Park et al., 2008). Hyeon's group synthesized PEG-derivatized phosphine oxide-capped ESPIONs with sizes down to 1.5 nm by a heat-up process and used them as T_1 MRI CAs for high-resolution angiography. They suggested that these NPs have the potential to be used in the diagnosis of several diseases, such as myocardial infarction, renal failure, atherosclerotic plaque, and thrombosis, because of their low toxicity, high r_1 relaxivity, long circulation times, and cost-effective synthesis (Kim et al., 2011). Very recently, Liu et al. also prepared glutathione-coated ESPIONs with 3.7 nm diameters using a single-step reaction in mild conditions, under atmospheric pressure and room temperature. The authors anticipate that these biocompatible high T_1 CAs ($3.63\,mM^{-1}\cdot s^{-1}$ at 4.7 T and 40°C) can be employed to great effect in vascular diagnosis, especially in stroke, venous thrombosis, renal disease, and urinary tract tumor diagnosis (Liu et al., 2014).

9.2.1.4 Hybrid Systems T_1-weighted MR contrast can be further enhanced using a variety of derivative methods, such as by embedding paramagnetic species into NPs (Zhou et al., 2013), developing surface-doped metal oxides (Sook Choi et al., 2010), or doping of Gd^{3+} to $NaYF_4$ NPs (Hou et al., 2013). Zhou et al. synthesized gadolinium-embedded iron oxide (GdIO) NPs (~5 nm diameter) and coated them with zwitterionic dopamine sulfonate molecules to create a hydrophilic surface, decrease nonspecific protein adsorption, prevent agglomeration, and provide a fast renal clearance. The embedding of Gd species in iron oxide NPs enhances the spin-canting effect, resulting in increased r_1 relaxivity ($7.85\,mM^{-1}\cdot s^{-1}$ at 7 T) and a lower r_2/r_1 ratio (5.24) for 4.8 nm diameter GdIO NPs compared to bare iron oxide NPs with similar sizes. Furthermore, the passive targeting ability of the NPs was investigated using a subcutaneous SKOV3 ovarian cancer model, and the NPs were found to create a marked improvement in T_1 contrast enhancement in the tumor site (Zhou et al., 2013). A similar system was reported by Choi et al., who doped Gd_2O_3 NPs with MnO and further functionalized the hybrid complex with lactobionic acid to increase biocompatibility and surface hydrophilicity. The complex was tested on the mouse model, and the r_1 relaxivity was found to increase from 9.9 to $12.8\,mM^{-1}\cdot s^{-1}$ after MnO doping; in addition, the r_2/r_1 ratio rose to 2.1 due to the dopant effect of MnO on r_2 relaxivity (Sook Choi et al., 2010). Among upconverting nanoparticles (UCNPs), Gd species-doped UCNPs demonstrate good MR contrast potential. Hou et al. synthesized $NaGdF_4$ NPs (~20 nm) exhibiting an r_1 relaxivity of 8.78 $mM^{-1}\cdot s^{-1}$, which is higher than that of Gd chelate (Gd-DTPA). In addition, a T_1 contrast enhancement was observed in tumor-bearing mice when these NPs were conjugated with an anti-EGFR monoclonal antibody (Hou et al., 2013).

9.2.2 T_2 CAs

Superparamagnetic materials exhibit a high susceptibility similar to ferromagnetic materials in the presence of an external magnetic field and a rapid demagnetization similar to paramagnetic molecules when the magnetic field is removed (Schaeffer, 1997). These materials, called T_2 or negative CAs, are typically associated with low T_2 relaxation times and the acquisition of darker images with low signal intensities, which stem from microscopic field inhomogeneities and the activation of proton dephasing. The spin–spin relaxivity r_2, which depends firmly on the magnetic moment and the relaxation processes of the magnetic spin, is a key factor for determining the degree of T_2 contrast effect on the NPs, with higher values resulting in greater contrast. The presence of T_2 agents creates darker MR images, and their signals may closely resemble the signatures associated with bleeding, calcification, or metal deposits; in addition, susceptibility artifacts that alter the background image are hard to distinguish from genuine signals with these CAs. Nonetheless, T_2 agents are still promising for use in clinical applications on account of their high r_2 and biocompatibility, as well as their prolonged circulation times in the body (Lee et al., 2012). T_2 CAs generally incorporate pure iron and cobalt metals; alloys such as $CoPt_3$, FePt, and FeZn; and iron oxides such as magnetite (Fe_3O_4) and maghemite (γ-Fe_2O_3). Unfortunately, problems associated with toxicity and susceptibility to oxidation limit the use of cobalt- and nickel-containing NPs in biomedical applications. Nonetheless, divalent cations such as Mn, Fe, Co, or Ni can be used as dopants for iron oxides to improve the magnetic properties of MFe_2O_4 structures (Veiseh et al., 2010).

9.2.2.1 Iron Oxide NPs Maghemite (γ-Fe_2O_3), magnetite (Fe_3O_4), and hematite (α-Fe_2O_3) are the main forms of iron oxide; and among these, magnetites are the most effective CAS and have received much attention as "next-generation" imaging agents due to their biocompatibility, biodegradability, and low levels of toxicity (Yu et al., 2008). In order to display superparamagnetic properties, a particle should have a suitable crystal structure, size, and shape (Jun et al., 2008). Several techniques allow the control of these features in iron oxide NPs, in addition to other factors such as solubility and size distribution. These synthetic approaches can be listed as (i) coprecipitation of iron salt solutions, (ii) thermal decomposition and/or reduction, (iii) hydrothermal synthesis, and (iv) polyol synthesis (Laurent et al., 2008). Among these techniques, thermal decomposition is the most widely used method for the production of NPs with tunable sizes (4–50 nm) and narrow size distributions, despite the limitations faced in large-scale production. In addition to scaling problems, a phase transfer into an aqueous solution is also required for these hydrophobic NPs to be used in biomedical applications (Barreto et al., 2011).

Since MRI CAs are administrated intravenously, the solubility, stability, and dispersity of NPs in water are also essential for NP CAs to be used effectively for any clinical application (Caravan, 2009). Carboxylates, phosphates, sulfonates, silicon compounds, gold, and polymers such as dextran, PEG, and polyvinyl alcohol (PVA) are frequently used as stabilizing agents to improve the stability of NPs in water (Barreto et al., 2011). Water-stabilized NPs can be further functionalized by incorporating targeting moieties (proteins, antibodies, peptides) and drug molecules to detect a specific disease or track cellular processes (Mahmoudi et al., 2011). In terms of MR contrast enhancement, NP size is the predominant parameter and should be taken into consideration in the design of both T_1 and T_2 CAs. Jun et al. investigated the effect of size on the spin–spin relaxivity (r_2) value and demonstrated that r_2 gradually increases with the size of Fe_3O_4 NPs, due to the enhanced magnetic moment produced by larger particles (Jun et al., 2005). Recently, Zhao et al. developed a new strategy to increase r_2 relaxivity by altering the morphology of iron oxide NPs. This method allows the production of size-controllable octapod iron oxide NPs with ultrahigh transverse relaxivity values ($679.3 \pm 30\,mM^{-1} \cdot s^{-1}$), which are of potential interest in applications involving *in vivo* imaging and small tumor detection (Zhao et al., 2013). Additionally, surface hydrophilicity and coating thickness are known to contribute greatly to MR contrast by affecting the proton relaxivities of iron oxide NPs (Huang et al., 2012).

9.2.2.2 Metal-Doped Iron Oxides Transition metal dopants (M^{2+}, where M = Co, Ni, or Mn) may be used instead of Fe^{2+} ions in MFe_2O_4 NPs to create significant MR contrast enhancement effects. Among all ferrites, $MnFe_2O_4$ NPs exhibit the highest mass magnetization value of 110 (emu/mass of magnetic atoms). In accordance with magnetization results, Mn-doped $MnFe_2O_4$ shows the strongest MR contrast effect, with an r_2 relaxivity value of $358\,mM^{-1} \cdot s^{-1}$ at 1.5 T. Thus, the material composition of NP complexes plays an important role in controlling the spin–spin relaxation processes of protons in the water molecules. Furthermore, herceptin-conjugated $MnFe_2O_4$ was shown to have a superior T_2 contrast effect compared to an undoped control in both *in vitro* cancer detection experiments and *in vivo* small tumor MRI (Lee et al., 2006).

9.2.2.3 Metal Alloy NPs Another class of MR probes, metal alloys, is exemplified by iron alloys such as FeCo and FePt. In these NPs, the parallel alignment of magnetic spins to the external magnetic field results in a higher magnetic moment compared to ferrimagnetic NPs. For instance, FeCo/graphitic shell nanocrystals, which were further coated with phospholipid–poly(ethylene glycol) (PL-PEG) to provide colloidal stability, were found to have a ultrahigh magnetization value of 215 emu/g metal and an r_2 relaxivity value of 644 $mM^{-1}{\cdot}s^{-1}$. Moreover, enhanced MR contrast was observed in FeCo NP-labeled mesenchymal stem cells, suggesting that these probes are able to provide excellent cellular MR signals. This study also shows that the toxic effect of highly reactive Co ions can be eliminated by using a suitable coating material, such as a graphite shell, to increase the safety of these heavy metal NPs in clinical practice (Seo et al., 2006).

9.2.3 Dual Modal Contrast Agents

In recent years, there has been a growing interest on the development of T_1–T_2 dual model strategies for MRI, which offers more precise diagnostic information by simultaneously utilizing the T_1 and T_2 imaging modes. While T_1 imaging provides high tissue resolution, T_2 imaging allows the detection of tumor sites with high confidence; as such, their combination allows both high-resolution imaging and the acquisition of biologically meaningful information. Since each imaging technique has a distinct penetration depth and spatiotemporal resolution, dual imaging within a single device may increase the diagnostic potential of MRI beyond what is expected from the individual imaging modes. However, the creation of an effective T_1–T_2 dual CA requires the prevention of direct contact between the two CAs, as the T_1 signal would otherwise be quenched due to the fact that the magnetic field generated by the T_2 CA perturbs the relaxation process of the T_1 CA. Cheon's group designed an inorganic core–shell model in which T_2 ($MnFe_2O_4$) and T_1 ($Gd_2O(CO_3)_2$) contrast materials were located in the core and the shell, respectively, and were separated by a SiO_2 layer with different thicknesses to modulate the degree of T_1 and T_2 coupling (Choi et al., 2010). Zhou and coauthors proposed a different core–shell model, in which superparamagnetic iron oxide nanoparticles

(SPIONs) and Gd species (e.g., Gd_2O_3) were integrated, and this system was used for the accurate detection of hepatic tumors in mice. This "core–shell" arrangement is also notable for facilitating a synergistic effect on r_1 and r_2 relaxivities by enhancing local magnetic field strengths under an external magnetic field (Zhou et al., 2012b).

9.3 NPs USED IN COMPUTED TOMOGRAPHY

Computed tomography (CT) imaging is another means by which high-resolution three-dimensional images of tissue structures can be acquired: While MRI scans change their focus by altering the applied magnetic field, a CT scanner physically moves around the patient to obtain a full "view" of the desired tissue. X-rays are used for imaging, and adverse effects caused by overexposure to radiation have long been a point of concern for CT scans (Chodick et al., 2007; Smith-Bindman et al., 2009). The images are interpreted on a similar basis as conventional X-ray radiography, in that the relative permittivity of the imaged area to the passage of X-rays is used to differentiate between tissues. This permittivity, called radiodensity, is evaluated using the Hounsfield scale, an arbitrary scale that sets the radiodensity of distilled water under standard conditions at zero Hounsfield units and the radiodensity of air under standard conditions at 1000 Hounsfield units (Forbes et al., 1978). X-ray attenuation, which is the combination of X-ray absorption and scattering, is also defined on the Hounsfield scale, and any given element has its own energy-dependent X-ray attenuation profile (Cormode et al., 2014). Damaged or diseased tissues typically possess different radiodensities, allowing their identification under CT, while natural differences between different tissue types allow the imaging of the local tissue structure. The latter property has led to the combination of the technique with PET scans, which visualize tissue activity rather than structure (von Schulthess et al., 2006; Wu et al., 2013).

The use of X-rays by this technique has resulted in frequent criticisms about its potential risks, especially on patients who may also benefit from less damaging imaging methods. Radiation exposure may be minimized by decreasing the exposure time, energy, or intensity of the X-ray beam; however, all three methods also decrease the resolution of the technique. Radiocontrast agents are typically administered both to increase the accuracy of the scan and to minimize the exposure to radiation. Presently, aromatic iodinated molecules are the clinical standard for *in vivo* contrast enhancement (Pasternak and Williamson, 2012). However, these CAs possess several limitations that restrict their use in clinical settings, mainly due to issues based on their (i) toxicity, (ii) high osmolality and viscosity, (iii) low payload capacity for targeted imaging, and (iv) low molecular weight. Although new approaches have been developed to eliminate these limitations, no marked improvement has so far been reported, especially with regard to the low attenuation coefficients and nontargeting natures of iodinated molecules (Jakhmola et al., 2012). Therefore, new radiocontrast agents with improved functionalities are required to increase the diagnostic capacity of X-ray based imaging methods and eliminate the toxicity issues associated with current contrast enhancers. These "next-generation" CAs should ideally fulfill several requirements: (i) High solubility, (ii) chemical stability and inertness in the physiological environment, (iii) opacity to X-rays, (iv) cost effectiveness, (v) biocompatibility, (vi) ease of functionalization, and (vii) long *in vivo* circulation times are all essential properties for a potential radiocontrast agent (Liu et al., 2012a). NP-based CAs fulfill many of these criteria and have therefore been developed to overcome the potential shortcomings of iodinated CT CAs.

9.3.1 Noble Metal-Based NPs

There has been a recent surge of interest in the use of noble metal NPs, especially gold and silver, in the field of CT imaging. These NPs are easy to synthesize and can be functionalized through well-established surface chemistry methods; as such, gold and silver are ideal materials for the production of surface-functionalized NPs in specific morphologies with little batch-to-batch variance.

9.3.1.1 Gold Nanoparticles Gold nanoparticles (AuNPs) are useful for X-ray CT imaging due to the high signal-to-noise ratios they exhibit, which allows the reduction of exposure times and the associated radiation damage to the patient. AuNP radiocontrast agents can also replace their iodine-based counterparts for *in vivo* imaging, as gold is a relatively inert element and its higher atomic number and electron density increase the resolution of *in vivo* X-ray imaging results. The X-ray absorption coefficient of gold is higher than that of iodine (5.16 and 1.94 cm^2/g, respectively, at 100 keV); as such, gold particles yield around 2.7 times higher contrast per unit weight than iodine (Hubbell and Seltzer, 1995).

AuNPs can be synthesized by the reduction of gold salts by a variety of reducing agents and can be further decorated with active molecules. Citrate reduction, also known as the Turkevich method, is the most popular technique for AuNP synthesis. Citrate is used both for the reduction of gold and as a stabilizing agent, and the size and shape of the resulting AuNPs can be altered by changing the reduction conditions (Kimling et al., 2006). The seeding growth method is another popular technique, producing size-adjustable particles within a size range of about 5–40 nm (Jana et al., 2001). Xu et al. examined the effect of size on the X-ray attenuation of NPs and compared the results with clinically administered Omnipaque. In this study, NPs synthesized by different synthetic approaches had a size range of 4–60 nm, and X-ray attenuation was found to be greater in smaller NPs compared to either larger NPs or Omnipaque, which is probably due to the increased surface area (Xu et al., 2008).

AuNPs can also be modified to exhibit additional functionalities, such as targeting capacity (Eck et al., 2010; Sun et al., 2009) or antifouling (Kim et al., 2007) properties, through facile surface chemistry. Sun et al. conjugated heparin–3,4-dihydroxy-L-phenylalanine (DOPA) functional group to AuNPs by a facile coating technique and found that intravenously injected AuNPs enhanced the resolution of liver-specific CT images, while no marked contrast signal was observed in the liver tissue of mice treated with the commercialized iodine-based CA eXIA160 (Sun et al., 2009). Eck et al. likewise demonstrated that anti-CD4 antibody-conjugated AuNPs facilitated the enhancement of CT contrast in the peripheral lymph nodes of mice (Eck et al., 2010). Kim et al. covalently anchored PEG to citrate-reduced AuNPs in an aqueous environment via thiol chemistry, thus preventing the nonspecific binding of plasma proteins and salts to AuNPs and alleviating the rapid clearance issues faced by NPs in the bloodstream. These PEG-coated AuNPs were still present in the bloodstream after 4 h without significant loss of contrast, exhibiting longer half-lives compared to the clinically used iodine-based CA Ultravist (<10 min) (Kim et al., 2007).

AuNPs can also be delivered as part of a therapeutic complex. Dendrimers have been used in the fabrication of organic/inorganic hybrid NPs either by entrapping or stabilizing AuNPs: While single or multiple NPs are entrapped within each dendrimer molecule in the formation of dendrimer-entrapped nanoparticles (DENPs), multiple dendrimer molecules surround the surface of a single NP in dendrimer-stabilized nanoparticles (DSNP). Dendrimer-entrapped gold nanoparticles (Au DENPs) are not only stable in different environments such as water, phosphate-buffered saline, and cell culture media but also retain their stability at various temperatures and pH (Guo et al., 2010). Fast reduction and nucleation chemistry have been used in the synthesis of Au DENPs, resulting in NPs smaller than 5 nm (Shen and Shi, 2010). Although the highly positive charges in the terminal amines of PAMAM dendrimers are known to result in high cytotoxicity and nonspecific binding to the cell membrane, these dendrimers are nonetheless used as templates due to their commercial availability. The surface of Au DENPs is commonly acetylated to minimize the risk of toxicity and nonspecific binding (Shi et al., 2008). Hybrid gold–dendrimer systems can also be modified with additional functionalities through the use of targeting molecules, such as folic acid (Wang et al., 2013) or the RGD peptide motif (Shukla et al., 2008). Au DSNPs can be synthesized in a similar manner as Au DENPs, as long as low-generation dendrimers are used. However, the production of Au DSNPs using higher-generation dendrimers requires complexation of gold salts with PAMAM dendrimers followed by physical or chemical reduction (Shen and Shi, 2010).

9.3.1.2 Silver NPs Silver NPs intended for use in biomedical applications generally suffer from toxicity and stability issues; however, these problems can be remedied using surface modifications. Liu et al., for example, formed size-tunable (8.8–23.2 nm) acetylated Ag DSNPs, which exhibited prolonged contrast enhancement in CT imaging performed on mice and did not show any toxic effects. Even though elemental silver has a lower X-ray absorption coefficient than iodine (due to the former's lower atomic weight), silver NPs with diameters of 16.1 nm were reported to show similar X-ray attenuation profiles with a clinically used iodine-based CA (Omnipaque), which may stem from the longer circulation times of silver NPs in the blood (Liu et al., 2010). The same group recently prepared spherical Au–Ag alloy DSNPs by using a G5 PAMAM dendrimer and folic acid as the stabilizer and targeting agent, respectively. Although the NPs have not yet been tried for targeted tumor imaging *in vivo*, the folic acid-modified particles were observed to be uptaken to a greater extent by cancer cells overexpressing folic acid receptor (Liu et al., 2013a).

9.3.2 Heavy Metal-Based NPs

Heavy metals possess good X-ray absorption properties because of their high density and atomic weights. More importantly, their X-ray attenuation is not significantly affected by changes in their X-ray energy, which is related to the K-edge values of heavy metals (Jakhmola et al., 2012). Unfortunately, the accumulation of certain heavy metals in the body may result in serious health problems, up to and including fatalities. Nevertheless, bismuth- and tantalum-based NPs have been produced and used for CT imaging without showing considerable toxic effects.

9.3.2.1 Bismuth-Based NPs The use of bismuth-based NPs in CT imaging has attracted great interest in recent years, due to high atomic number ($Z=83$) and large X-ray attenuation coefficient (5.74 cm^2/kg at 100 keV) of bismuth.

In addition, bismuth has low toxicity for a heavy metal and is readily cleared from the body without leaving residues. Since the synthesis of elemental bismuth NPs suffers from hydrolytic instability issues and the need for stringent purification steps, bismuth-based NPs often use compounds of bismuth, such as bismuth(III) sulfide (Bi_2S_3) (Brown and Goforth, 2012; Cormode et al., 2014). Rabin et al. reported the synthesis of polyvinylpyrrolidone (PVP)-coated Bi_2S_3 NPs with low cytotoxicity and good aqueous stability and demonstrated that the NPs display enhanced *in vivo* contrast efficacy and circulation times compared to commercialized iodine-based agents (Rabin et al., 2006). Ai et al. also synthesized PVP–Bi_2S_3 nanodots, which became water soluble after facile ligand exchange in dichloromethane. Their contrast efficacy was 1.9 times higher than that of the clinically used CA iobitridol, and X-ray CT imaging results on rats suggested that these particles had the potential to be used in vascular imaging and the detection of hepatic metastases (Ai et al., 2011). Sailor and his group reported the synthesis of Bi_2S_3 NPs conjugated to a cyclic peptide containing nine amino acid residues (CGNKRTRGC, LyP-1) and demonstrated that the NPs had a pronounced accumulation in tumor sites when administered to $4T_1$ breast tumors in mice (Kinsella et al., 2011).

9.3.2.2 Tantalum Oxide NPs Bioinertness, cost effectiveness, and a high X-ray attenuation coefficient value (4.3 cm^2/kg at 100 keV) all make elemental tantalum a good candidate for CT imaging. Bonitatibus et al. reported a two-step synthesis of Ta_2O_5 core–shell NPs, in which the core is prepared through the controlled hydrolysis of tantalum ethoxide, $Ta(OEt)_5$, and the surrounding shell is produced by coating with (2-diethylphosphatoethyl) triethoxysilane. The resulting NPs were water soluble and stable, had diameters of 6 nm, and showed enhanced CT contrast compared to iodine. Although no toxic effects were observed in rats at the concentrations administered, further analyses are required to establish whether the NPs are safe for use as CT CAs (Bonitatibus, 2010). Oh et al. reported large-scale synthesis of TaO_x NPs by a microemulsion method and further modified these core NPs with silane derivatives using an *in situ* sol–gel reaction. Following silane modification, the surface was modified with a secondary layer of PEG and/or Rhodamine B to allow prolonged circulation times and fluorescence imaging capacity, respectively. The addition of the fluorescent dye increased the *in vivo* circulation time, which is an important issue for CT agents intended for image-guided lymph node mapping and CT angiography (Oh et al., 2011).

Other elements, such as platinum, ytterbium, yttrium, gadolinium, palladium, holmium, and tungsten, have been used for CT imaging, albeit less frequently

(Cormode et al., 2014; Jakhmola et al., 2012). These will not be discussed in this chapter, as their methods of functionalization are similar to the ones that were outlined. From a biological standpoint, the major requirements for a radiocontrast agent are biocompatibility and solubility in biological environments, which are principally achieved by coating the core with a lipid, polymer, protein, silica, or any other suitable moiety and can be performed on both "conventional" and more uncommon metal NPs.

9.4 NPs USED IN OPTICAL AND FLUORESCENCE IMAGING

Although organs such as the eye and the outer ear are amenable to unaided visual inspection and endoscopic techniques such as colonoscopy and angioscopy can be used to visualize a broader range of tissues, the majority of the body is opaque to visible light. Nonetheless, it is possible to use light in the near-infrared (NIR) region, which displays a greater depth of penetration compared to the visible spectrum, to inspect the structure and function of a broader range of tissues. NIR light penetrates to a depth of up to 10 cm in the human body and, depending on the technique, can be used for the detection of tissue abnormalities at depths ranging from the superficial (such as burns and skin lesions) to deep tissue (such as brain and other solid organ tumors) (Huang et al., 2006; Still et al., 2001; Wang et al., 2004). However, tissues typically offer little NIR contrast by themselves, limiting the effectiveness of unenhanced NIR imaging and necessitating the use of CAs to increase imaging resolution (one exception is the detection of hemoglobin, which naturally absorbs in the NIR and shows different absorption patterns in the presence and absence of oxygen) (Pogue et al., 2001; Villringer and Chance, 1997). These CAs are typically fluorescent dyes that absorb in the NIR region and can be functionalized with one or more targeting molecules to ensure their delivery into a specific region (usually a tumor site).

Organic dyes and inorganic NPs are both used for contrast enhancement in *in vivo* NIR imaging applications. Organic fluorophores, as exemplified by the FDA-approved diagnostic agent indocyanine green, are limited in their effectiveness by their low quantum yields, tendency for rapid photobleaching, hard-to-control excitation and emission wavelengths, and inability to secondarily enhance targeting efficiency by functionalizing the CA with a large number of targeting molecules (Frangioni, 2003). The limitations of organic fluorophores can be overcome with the use of NP CAs, which display higher quantum yields, are resistant to photobleaching, and can be synthesized with great control over excitation and emission wavelengths. In addition, these NPs can be modified for *in vivo* applications by surface modification of the metal NP, which may be performed to introduce additional functionalities such as biocompatibility, enhanced clearance profiles, or targeting capacity.

It must be noted here that, in addition to NIR fluorescence imaging, a number of spectroscopic techniques have been repurposed for *in vivo* applications. The laser and detector elements of Raman and FTIR spectrometers, for example, can be mounted on the tip of a fiber-optic catheter and used in endoscopy in order to assist in diagnosis (Li et al., 2005; Qian et al., 2008). Although these techniques offer highly accurate diagnostic capacities and are sometimes used in tandem with NPs for contrast enhancement, they are not currently in clinical use and will not be covered by the present chapter (Qian et al., 2008).

9.4.1 Quantum Dots

Element pairs from the groups III–V, II–IV, or IV–VI of the periodic table can be used to produce fluorescent semiconductor nanocrystals called quantum dots (QDs), typically in sizes between 1 and 10 nm. These NPs are highly resistant to photobleaching and chemical degradation and display very high quantum yields, making them prime candidates for fluorescence imaging. They have been synthesized using different methodologies, resulting in either hydrophobic or hydrophilic QDs. The former necessitates further modification to increase their water solubility and biocompatibility and have been synthesized via organometallic and chalcogen precursor pyrolysis. Either ligand exchange with thiolated molecules or encapsulation by a variety of different moieties such as polymers, silica, or phospholipid micelles can be employed to allow the use of hydrophobic QDs in aqueous systems, such as in *in vivo* imaging. Hydrophilic QDs are synthesized using thiol-containing molecules as stabilizers, and the synthesis reactions are performed in aqueous conditions (Li and Zhu, 2013). In addition, higher band gap semiconductors, such as ZnS and ZnSe, are commonly used as a secondary layer around the core, not only to impede chemical degradation and oxidation but also to improve QD stability and photoluminescence (PL) (Barreto et al., 2011).

On account of the quantum confinement effect, absorption and PL spectra of QDs, as well as the color of emitted light, show size-dependent variance (Algar and Krull, 2009). In addition to size, the chemical compositions and core structures of QDs are the other two critical parameters to be considered for the control of the photophysical properties of QDs (Cooper and Nadeau, 2009). As opposed to organic fluorophores, QDs have a broad absorption spectra and a narrow emission peak, allowing multicolor fluorescence imaging by concurrently exciting multiple QDs with distinct emission wavelengths. Nie and coworkers devised a multifunctional NP system, in which encapsulation of QDs into ABC triblock copolymers was followed by the anchorage of tumor targeting ligands to an amphiphilic polymer, ultimately creating an NP complex for multicolor fluorescence imaging in live animals. The delivery and accumulation of QDs in tumor sites were investigated under passive and active targeting modes, the former with the help of the enhanced permeability and retention effect of the NPs and the latter through the specific interaction between antibody-conjugated QDs and cancer-specific cell surface biomarkers (Gao et al., 2004). Furthermore, Kobayashi et al. synthesized five different polymer-coated carboxyl-functionalized QDs (CdSe or CdTe) with similar sizes but different emission wavelengths (varying from 565 to 800 nm) for multiplexed lymphatic imaging. Intracutaneous injection of all QDs was performed into five different drainage sites of the lymphatic system in the upper body of mice, and the results revealed that no pronounced decrease would occur in fluorescence intensity for up to 3 h postinjection, which is longer than other NIR dye-labeled molecular systems (Kobayashi et al., 2007). Previously, Kim et al. showed that polydentate phosphine-coated CdTe/CdSe core–shell QDs could be used to detect sentinel lymph nodes in mice and pigs. The NPs had 15–20 nm diameters and high aqueous quantum yields, were stable in aqueous environments, and exhibited good retention in lymph nodes (Kim et al., 2003). Shuhendler and coauthors developed a hybrid system for deeper tissue penetration in optical imaging using NIR PbSe QDs embedded into solid fatty ester nanoparticles (QD-FENs). Following the intravenous administration, noticeable accumulation of QD-FENs was observed in the tumor sites of breast tumor-bearing mice after 30 min. Although the precise depth profile of the QD-FENs was difficult to measure under *in vivo* conditions, the system nonetheless displayed capacity for macroscopic deep tissue fluorescence imaging (Shuhendler et al., 2011).

In addition to QDs, considerable interest has been focused on the fabrication of anisotropic quantum wires (QWs) and quantum rods (QRs) with tailorable optical properties. More controlled synthetic approaches and surface modification techniques may allow the use of these one-dimensional semiconductors as effective probes for bioimaging applications. Prasad's group demonstrated successful *in vivo* tumor imaging of pancreatic cancer-bearing mice by using cyclic RGD (cRGD) peptide-conjugated CdSe/CdS/ZnS QRs. These QRs were first encapsulated in PEG-grafted phospholipid micelles, followed by surface conjugation of cRGD peptide for specific tumor targeting and imaging (Yong et al., 2009).

9.4.2 AuNPs

AuNPs are commonly used in optical cancer imaging due to their unique optical properties. As noble metal NPs, AuNPs exhibit strong light scattering at their localized surface plasmon resonance (SPR) frequency, which is related to the interaction between light and the conduction electrons of AuNPs. NP sizes and shapes as well as the dielectric constant of the medium change the SPR properties of AuNPs (Sharma et al., 2006). Spherical NPs are associated with a number of drawbacks, such as photochemical degradation and lack of deep tissue penetration because of strong absorption in the optical visible range; as such, AuNPs with alternative morphologies have been synthesized for tissue imaging purposes. AuNP rods, urchins, prisms, wires, disks, shells, and stars with more desirable SPR properties have been reported for use in imaging in the infrared region (Hutter and Maysinger, 2011). The surfaces of these NPs can also be modified to increase their aqueous dispersibility and to provide active sites for further biofunctionalization, which is crucial for targeting applications that are commonly employed in bioimaging.

Choi et al. used a seed-mediated growth approach to produce gold nanorods (GNRs) and functionalized the core NPs with cRGD peptide by EDC/NHS chemistry after heterobifunctional PEGylation. GNRs with aspect ratios of 4 were shown to exhibit maximum absorption at 780 nm and could be used to target and image glioblastoma under both *in vitro* and *in vivo* conditions. Gold accumulation in *in vivo* brain tumors was found to be higher for cRGD-tethered GNRs compared to control, while the opposite result was obtained in the case of liver uptake (Choi et al., 2011). In addition to GNRs, gold nanoclusters (GNCs) have also been used for the development of NP systems for optical imaging (Quek and Leong, 2012). Wu et al. produced ultrasmall GNCs with hydrodynamic sizes of 2.7 nm, which were tested on MDA-MB-45 and HeLa tumor xenograft models. Discernible differences in *ex vivo* fluorescence images were observed between tumor tissue and the muscle tissue surrounding the tumor, suggesting that the GNCs could be used as a CA in fluorescence tumor imaging (Wu et al., 2010).

Toxicity and the accumulation of the CA in the body are two major concerns that should be taken into consideration when dealing with NPs. Although factors such as size, shape, concentration, surface charge, surface coating, and aggregation status are known to alter the toxicity of AuNPs, special care must be taken while selecting targeted cell lines, exposure times, and determination methods for toxicity assessment. Toxicity is generally recognized to be unavoidable for AuNPs with very small sizes, negative charges, and functionalization with intrinsically toxic chemicals or with excessive amounts of coating materials (Ng et al., 2013). Surface modifiers have been developed to circumvent these issues by lowering the toxicity or preventing the long-term accumulation of NPs. Zheng's group synthesized glutathione-coated luminescent AuNPs and compared their results with a renal clearable organic fluorophore, IRDye 800CW. AuNPs were found not only to exhibit rapid normal tissue clearance but also to display prolonged tumor accumulation, which is a highly desirable result for clinical trials (Liu et al., 2013b).

9.4.3 UCNPs

UCNPs are lanthanide-doped nanocrystals and are composed of three major components: activators, sensitizers, and the host matrix. Lanthanide-based sensitizers (Yb^{3+}) and activators (Er^{3+}, Tm^{3+}, Ho^{3+}) are used as dopants to alter the emission properties of the nanocrystal composite. Host materials provide a suitable environment for upconversion by isolating the dopant from its surrounding and promoting energy transfers between dopant ions and are generally fluoride-based materials such as $NaYF_4$, LaF_3, and $NaYbF_4$ or oxide-based materials such as Y_2O_3, CeO_2, and YVO_4 (Nam et al., 2013).

In contrast to other fluorescent agents, UCNPs are excited at longer wavelengths (generally in the NIR region) by absorbing low-energy photons and emit at shorter wavelengths by releasing a high-energy photon through anti-Stokes emission. This upconversion process can be explained under three main mechanisms, which can be applicable alone or in combination: excited state absorption (ESA), photon avalanche (PA), and energy transfer upconversion (ETU) (Chen and Zhao, 2012). The upconversion behavior makes UCNPs enticing candidates for applications in bioimaging and provides superior properties over other fluorescent agents, such as decreased autofluorescence background, photobleaching resistivity, deeper penetration in tissues, and larger anti-Stokes shifts (Li et al., 2014).

Advances in the synthesis of lanthanide NPs have spurred the development of small UCNPs in the last decade (Bao et al., 2013). Thermal decomposition and hydrothermal synthesis are the most widely applied methods for the production of UCNPs with well-defined shapes and low variation in particle sizes. Since these two methods result in the production of hydrophobic NPs, further steps are necessary to render them water dispersible. Methods such as ligand exchange, organic ligand-free synthesis, ligand oxidation reaction, silanization, layer-by-layer methods, hydrophobic–hydrophobic interactions, or host–guest interactions can be used for this purpose (Wang et al., 2010; Zhou et al., 2012a). UCNPs can be further modified with specific functional groups by tethering bioactive moieties such as peptides, antibodies, DNA, and sugars (Zhou et al., 2012b). In terms of *in vivo* upconversion luminescence imaging, they can be used in whole-body imaging (Liu et al., 2011; Nyk et al., 2008), tumor imaging (Xiong et al., 2009), lymphatic imaging (Cao et al., 2011), and vascular imaging (Hilderbrand et al., 2009).

Since rapid progress has been made in the development of UCNPs, investigations on the short- and long-term cytotoxic effects of UCNPs are currently incomplete. Although these NPs have so far been reported to be biocompatible, further toxicological assessments are necessary to fully evaluate the risks associated with these novel CAs (Xiong et al., 2010).

9.5 THERANOSTIC APPROACHES AND MULTIMODAL SYSTEMS

Theranostics is a relatively new approach that proposes the integration of properties related to therapy and diagnostic imaging into a single system, and nanotechnology, as a developing area of science, is especially suitable for the development of such combined approaches. The coordination of therapeutic and imaging functionalities is an ideal means of increasing the efficacy of treatment, since doing so allows the medical practitioner to monitor and control each individual step of the therapy. However, the development of powerful imaging strategies is paramount to gain a better understanding of the nature of the disease at hand, which is of fundamental importance to theranostic approaches. Due to their ease of synthesis and functionalization, as well as their imaging potential, NPs are particularly suitable for the development of theranostic systems and have received much attention in this capacity (Brigger et al., 2002; Choi et al., 2012; Janib et al., 2010). A wide variety of NPs have been developed as powerful platforms to improve our understanding of the basic framework underlying NP-based diagnosis and therapy; however, clinical applications of these structures are limited. Compared to conventional methods of diagnosis and therapy, NPs present several advantages, such as (i) a high surface-to-volume ratio, (ii) adjustable size and surface properties, (iii) ease of functionalization, (iv) ability to form conjugates with specific targeting moieties, (v) prolonged circulation times in the blood, and (vi) enhanced uptake into the reticuloendothelial system. These properties allow multifunctional NPs to be produced as effective imaging and therapeutic systems with potentially promising applications in clinical settings (Brigger et al., 2002).

Theranostic NPs (Lee et al., 2012), such as iron oxide NPs, AuNPs, and GdNPs, can be specifically delivered to their target region by using passive and active targeting approaches. Passive targeting utilizes the pathophysiological characteristics of the target region to facilitate the accumulation of NPs. In particular, the enhanced permeability and retention (EPR) effect inherent to tumor sites allows NPs to selectively accumulate in these regions, as the high metabolic rates, elevated nutrient and oxygen demands, and acidic character of tumor tissues all enhance NP deposition. Uptake into specific cell types, however, usually necessitates the use of an active targeting strategy, which involves functionalization of the NP with a ligand molecule. This ligand may be a peptide, antibody, oligonucleotide, or small organic molecule and typically binds to a target moiety on the surface of the cell of interest (Danhier et al., 2010). In order to be applicable in theranostic approaches, NPs must be detectable through noninvasive imaging modalities, which include MRI, CT, PET, single-photon emission computed tomography (SPECT), and optical imaging. Each of these techniques has its own advantages and drawbacks, especially with regard to sensitivity and spatial resolution. Limits in the detection capacity of one method can be overcome by another; as such, combined imaging techniques such as PET/MRI (Pichler et al., 2010), PET/CT (Zhou et al., 2010), and MRI/CT (Chou et al., 2010) are particularly effective for disease diagnosis.

MRI is one of the most effective noninvasive imaging techniques in clinical diagnosis. Many types of NPs can be used as CAs in theranostic MRI. Of these, SPIONs have unique physical properties that make them particularly suitable for use as CAs in MRI, as well as in other techniques, and these NPs have been popular platforms for the development of theranostic methods (Cohen and Shoushan, 2013). One such example has been reported by Fang et al., who have produced a theranostic platform that combines SPIONs, a biodegradable and pH-sensitive poly (beta-amino ester) (PBAE) copolymer, and a chemotherapeutic agent (doxorubicin (Dox)) for an effective diagnostic–therapeutic delivery system. Their system demonstrated a pH-responsive drug release profile over 72 h of incubation: While Dox release was rapid between pH 5.5 and 6.4, the rate of release was slower at pH 7.4. In addition, theranostic NPs displayed a strong r_2 relaxivity of $146\,mM^{-1}{\cdot}s^{-1}$, which makes them suitable for MRI. *In vitro* studies with drug-resistant C6/ADR cells demonstrated that, compared to free Dox, the drug complex undergoes cellular internalization at significantly higher rates (Fang et al., 2012).

Magnetic NPs also display substantial potential for combining MRI with magnetic hyperthermia treatment (MHT), as they typically generate heat under alternating current (AC) magnetic fields (Kumar and Mohammad, 2011). In one such study, Hayashi et al. designed FA- and PEG-modified SPION nanoclusters (FA-PEG-SPION NCs) for cancer treatment. Twenty-four hours after the intravenous injection of NCs, the mice were exposed to an AC magnetic field with $H = 8$ kA/m and $f = 230$ kHz ($Hf = 1.8 \times 10^9$ A/m s) for 20 min, resulting in a temperature increase of 6°C above ambient at the tumor region. After 35 days of treatment, the tumor volume was found to be smaller compared to control mice (Hayashi et al., 2013).

Although the strong T_2 responses, versatile chemical functionalization capacity, and ability to track cells and tissues at low concentrations render SPIONs popular for a variety of biomedical applications, the practical use of these materials is limited by a number of drawbacks, especially with regard to their negative contrast characteristics and magnetic sensitivity. T_1 CAs, such as gadolinium (Gd) and manganese (Mn), are employed to bypass these limitations. Bae et al. have generated multifunctional HMON and functionalized them with a biocompatible adhesive ligand (DOPA) and a therapeutic monoclonal antibody specific to cancer cells (herceptin) for the targeted delivery of therapeutic siRNAs concurrent with MRI-based tumor monitoring. MRI and confocal microscopy results indicated that herceptin-conjugated HMON enabled the detection of HER2-overexpressing cancer cells in T_1-weighted MRI and facilitated effective intracellular delivery of siRNA to inhibit VEGF expression (Bae et al., 2011).

Gadolinium chelates are commonly used as MRI CAs but suffer from drawbacks such as rapid clearance from the body, nonspecific biodistribution, and interference with Ca(II) ion-mediated processes (Cabella et al., 2006). GdNPs are capable of circumventing many of these problems and are further advantageous in that they can be functionalized with a variety of ligands and delivery platforms for theranostic purposes (Liu and Zhang, 2012). One example of a GdNP-based combined imaging/delivery system was provided by Liu et al., who reported production of PEGylated Eu(III)-doped mesoporous gadolinium oxide nanorods as CAs for MRI/X-ray imaging/PL imaging and vehicles for chemotherapeutic drug delivery. These NPs were shown to possess high biocompatibility, good photoluminescent and magnetic properties, well-ordered mesopores, and rodlike morphology that may serve to increase internalization rates during cellular uptake. The relaxivity of these nanorods was 23.5 $mM^{-1} \cdot s^{-1}$, which is substantially higher than that of a commercially available CA Magnevist (4.0 $mM^{-1} \cdot s^{-1}$). The drug delivery potential of these nanorods was also evaluated by the use of the hydrophobic cancer drug camptothecin (CPT), the chemotherapeutic efficiency of which was greater when delivered with Gd nanorods compared to delivery by conventional CPT-loaded NPs (Liu et al., 2012b).

AuNPs are of considerable importance in many fields and are consequently one of the most well-studied nanomaterial types. Well-established synthesis methods make AuNPs attractive candidates for the design of novel theranostic delivery vehicles, and their unique physical characteristics lend them well to multimodal imaging and photodynamic therapy (Huang et al., 2007). A particularly interesting example of an Au-based delivery system has been reported by Liu et al., who generated a quintuple-modality nanoprobe by functionalizing gold nanostars with a surface-enhanced Raman scattering (SERS) reporter and Gd(III) ions for detection by SERS, MRI, CT, and two-photon luminescence (TPL) imaging, as well as tumor ablation by photothermal therapy (PTT). These multifunctional NPs may allow the use of multiple preoperative (MRI and CT) and intraoperative (SERS and TPL) imaging systems on the same tumor site for enhanced diagnostic capacity, in addition to actively contributing to therapeutic efforts by serving as absorptive materials for PTT (Liu et al., 2013c).

In addition to nanomaterials composed of pure metals, NP alloys can also be used for diagnostic and therapeutic efforts. These materials may offer enhanced resolution, sensitivity, and biocompatibility compared to their single metal counterparts. Amendola et al. reported the generation of magneto-plasmonic Au–Fe alloy nanoparticles (AuFeNPs) as multimodal CAs. These AuFeNPs displayed many advantages over conventional metal NPs, including a synthesis method that does not include harmful chemicals, a simple conjuration mechanism through the use of thiolated molecules, longer retention times in bloodstream, and better accumulation in tumor sites, making them promising candidates for simultaneous MRI, CT, and Raman imaging (Amendola et al., 2014). Also worthy of note are luminescent NPs, such as QDs, which have drawn considerable attention due to their unique optical properties such as narrow, symmetric, and size-tunable emission spectra, broad and continuous excitation spectra, high stability against photobleaching, and high florescence quantum yields. These properties, in tandem with their high functionalization capacity, make QDs attractive agents for biomedical imaging and drug and gene delivery (Sreenivasan et al., 2013). Bagalkot et al., for example, developed a QD–aptamer (Apt)–doxorubicin (Dox) conjugate system for targeted synchronous cancer imaging and drug delivery. The NP surfaces were functionalized with A10 RNA Apt, which recognize prostate-specific membrane antigen (PSMA) expressed in LNCaP cells, to ensure specific delivery of QDs into these cells. Doxorubicin (Dox), a widely used anthracycline drug, was intercalated into Apt for targeted delivery. The system mainly relied on a Bi-FRET mechanism between the QD/Dox and RNA Apt/Dox pairs. With this design, the fluorescence of the QD was "turned off" by the quenching effect of Dox, and the fluorescence of Dox was quenched in turn by the Apt. When Dox is released from the QD-Apt(Dox) complex, the quenching mechanism no longer functioned, resulting in fluorescence (Bagalkot et al., 2007). QDs can also be incorporated into other delivery vectors, such as liposomes, to increase their biocompatibility while also offering a strong fluorescent signal. Weng et al. reported that doxorubicin-loaded QD-immunoliposome complexes showed no cytotoxicity and displayed notably enhanced circulation times and anticancer activity (Weng et al., 2008).

Theranostic systems are as variable and versatile as the drug delivery and imaging components they incorporate, and this flexibility makes them attractive for the detection and treatment of many hard-to-diagnose, hard-to-treat diseases. There have been notable efforts to create multifunctional platforms for combined diagnostics and therapeutics, but challenges such as data reproducibility, NP biocompatibility and biodegradability, and targeting issues remain. If these limitations were to be overcome, theranostic systems can effectively combine several critical aspects of current diagnostic and therapeutic efforts to create a combined platform for the treatment of a wide range of disorders, from the common and mild to the rare and deadly.

9.6 OVERLOOK AND FUTURE DIRECTIONS

Despite their relatively recent addition to the roster of medical diagnostics tools, NPs are versatile and powerful imaging agents. The ability to modify a single NP with multiple targeting molecules enhances the targeting efficiency of the CA, and multiple steps of functionalization can be performed to produce an NP that can simultaneously perform several distinct functions (e.g., antibodies for targeting, PEG for avoiding complications with the immune system, and other molecules for lower toxicity and better clearance profiles). The contrast enhancements of NPs are typically greater than their conventional counterparts (e.g., iodinates for CT, bulk metal chelates for MRI, and organic fluorophores for NIR imaging). In addition to acting as stand-alone contrast enhancers, the larger-sized NPs also can deliver a therapeutic molecule alongside the CA, making use of the targeting capacity that these NPs are often modified to display. In short, NPs are not so much an individual class of CAs, but more an all-purpose platform that can be modified to a great extent to serve the specific needs of the tissue and disease to be imaged. This diversity of function makes them especially suitable for the era of next-generation medical diagnostics, where personalized treatment for each individual patient is necessary to tackle complex diseases such as cancer, diabetes, obesity, and neurodegenerative disorders.

Nevertheless, it must be kept in mind that NP diagnostics is still largely an area of research, rather than a standard of clinical use. The approval of any medical agent for human use requires very stringent conditions of testing, as well as a clear improvement over the previous standard of care. Despite their effectiveness in animal models and small-scale medical applications, issues such as long-term toxicity, environmental effects, and potential side effects on a large and genetically diverse human population should still be resolved for the widespread use of NPs for diagnostic purposes. The success of currently approved NP CAs, such as the SPIONs Feridex and Resovist (Wadajkar et al., 2013), stands as testament to the diagnostic potential of NPs, but much work remains to be done for NPs to truly realize this potential across all imaging methods and systems.

REFERENCES

Ai, K., Liu, Y., Liu, J., Yuan, Q., He, Y., and Lu, L. (2011). Large-scale synthesis of Bi_2S_3 nanodots as a contrast agent for in vivo X-ray computed tomography imaging. Advanced Materials, *23* (42), 4886–4891.

Algar, W.R. and Krull, U.J. (2009). Quantum Dots for the Development of Optical Biosensors Based on Fluorescence. In Biosensing Using Nanomaterials A. Merkoçi, ed. (Hoboken, NJ: John Wiley & Sons, Inc.), pp. 199–245.

Amendola, V., Scaramuzza, S., Litti, L., Meneghetti, M., Zuccolotto, G., Rosato, A., Nicolato, E., Marzola, P., Fracasso, G., and Anselmi, C. (2014). Magneto-plasmonic Au-Fe alloy nanoparticles designed for multimodal SERS-MRI-CT imaging. Small, *10* (12), 2476–2486.

Bae, K.H., Lee, K., Kim, C., and Park, T.G. (2011). Surface functionalized hollow manganese oxide nanoparticles for cancer targeted siRNA delivery and magnetic resonance imaging. Biomaterials, *32* (1), 176–184.

Bagalkot, V., Zhang, L., Levy-Nissenbaum, E., Jon, S., Kantoff, P.W., Langer, R., and Farokhzad, O.C. (2007). Quantum dot-aptamer conjugates for synchronous cancer imaging, therapy, and sensing of drug delivery based on bi-fluorescence resonance energy transfer. Nano Letters, *7* (10), 3065–3070.

Bao, G., Mitragotri, S., and Tong, S. (2013). Multifunctional nanoparticles for drug delivery and molecular imaging. Annual Review of Biomedical Engineering, *15* 253–282.

Barreto, J.A., O'Malley, W., Kubeil, M., Graham, B., Stephan, H., and Spiccia, L. (2011). Nanomaterials: applications in cancer imaging and therapy. Advanced Materials, *23* (12), H18–H40.

Bateman, T., Heller, G., McGhie, A., Friedman, J., Case, J., Bryngelson, J., Hertenstein, G., Moutray, K., Reid, K., and Cullom, S. (2006). Diagnostic accuracy of rest/stress ECG-gated Rb-82 myocardial perfusion PET: comparison with ECG-gated Tc-99m sestamibi SPECT. Journal of Nuclear Cardiology, *13* (1), 24–33.

Betz, C.S., Stepp, H., Janda, P., Arbogast, S., Grevers, G., Baumgartner, R., and Leunig, A. (2002). A comparative study of normal inspection, autofluorescence and 5-ALA-induced PPIX fluorescence for oral cancer diagnosis. International Journal of Cancer, *97* (2), 245–252.

Bonitatibus Jr, P.J. (2010). Synthesis, characterization, and computed tomography imaging of a tantalum oxide nanoparticle imaging agent. Chemical Communications, *46* (47), 8956–8958.

Brigger, I., Dubernet, C., and Couvreur, P. (2002). Nanoparticles in cancer therapy and diagnosis. Advanced Drug Delivery Reviews, *54* (5), 631–651.

Brown, A.L. and Goforth, A.M. (2012). pH-dependent synthesis and stability of aqueous, elemental bismuth glyconanoparticle colloids: potentially biocompatible X-ray contrast agents. Chemistry of Materials, *24* (9), 1599–1605.

Burghardt, A., Link, T., and Majumdar, S. (2011). High-resolution computed tomography for clinical imaging of bone microarchitecture. Clinical Orthopaedics and Related Research, *469* (8), 2179–2193.

Cabella, C., Crich, S.G., Corpillo, D., Barge, A., Ghirelli, C., Bruno, E., Lorusso, V., Uggeri, F., and Aime, S. (2006). Cellular labeling with Gd (III) chelates: only high thermodynamic stabilities prevent the cells acting as “sponges” of Gd^{3+} ions. Contrast Media & Molecular Imaging, *1* (1), 23–29.

Calliada, F., Campani, R., Bottinelli, O., Bozzini, A., and Sommaruga, M. (1998). Ultrasound contrast agents—basic principles. European Journal of Radiology, *27* S157–S160.

Cao, T., Yang, Y., Gao, Y., Zhou, J., Li, Z., and Li, F. (2011). High-quality water-soluble and surface-functionalized upconversion nanocrystals as luminescent probes for bioimaging. Biomaterials, *32* (11), 2959–2968.

Caravan, P. (2009). Protein-targeted gadolinium-based magnetic resonance imaging (MRI) contrast agents: design and mechanism of action. Accounts of Chemical Research, *42* (7), 851–862.

Chang, R.-F., Wu, W.-J., Moon, W.K., Chou, Y.-H., and Chen, D.-R. (2003). Support vector machines for diagnosis of breast tumors on US images. Academic Radiology, *10* (2), 189–197.

Chen, J. and Zhao, J.X. (2012). Upconversion nanomaterials: synthesis, mechanism, and applications in sensing. Sensors, *12* (3), 2414–2435.

Chodick, G., Ronckers, C.M., Shalev, V., and Ron, E. (2007). Excess lifetime cancer mortality risk attributable to radiation exposure from computed tomography examinations in children. The Israel Medical Association Journal, *9* (8), 584–587.

Choi, J.-S., Lee, J.-H., Shin, T.-H., Song, H.-T., Kim, E.Y., and Cheon, J. (2010). Self-confirming “AND” logic nanoparticles for fault-free MRI. Journal of the American Chemical Society, *132* (32), 11015–11017.

Choi, J., Yang, J., Park, J., Kim, E., Suh, J.-S., Huh, Y.-M., and Haam, S. (2011). Specific near-IR absorption imaging of glioblastomas using integrin-targeting gold nanorods. Advanced Functional Materials, *21* (6), 1082–1088.

Choi, K.Y., Liu, G., Lee, S., and Chen, X. (2012). Theranostic nanoplatforms for simultaneous cancer imaging and therapy: current approaches and future perspectives. Nanoscale, *4* (2), 330–342.

Chou, S.-W., Shau, Y.-H., Wu, P.-C., Yang, Y.-S., Shieh, D.-B., and Chen, C.-C. (2010). In vitro and in vivo studies of FePt nanoparticles for dual modal CT/MRI molecular imaging. Journal of the American Chemical Society, *132* (38), 13270–13278.

Ciernik, I., Dizendorf, E., Baumert, B., Reiner, B., Burger, C., Davis, J., Lutolf, U., Steinert, H., and Von Schulthess, G. (2003). Radiation treatment planning with an integrated positron emission and computer tomography (PET/CT): a feasibility study. International Journal of Radiation Oncology Biology Physics, *57* (3), 853–863.

Cohen, Y. and Shoushan, S.Y. (2013). Magnetic nanoparticles-based diagnostics and theranostics. Current Opinion in Biotechnology, *24* (4), 672–681.

Cooper, D.R. and Nadeau, J.L. (2009). Nanotechnology for in vitro neuroscience. Nanoscale, *1* (2), 183–200.

Cormode, D.P., Naha, P.C., and Fayad, Z.A. (2014). Nanoparticle contrast agents for computed tomography: a focus on micelles. Contrast Media & Molecular Imaging, *9* (1), 37–52.

Danhier, F., Feron, O., and Préat, V. (2010). To exploit the tumor microenvironment: passive and active tumor targeting of nanocarriers for anti-cancer drug delivery. Journal of Controlled Release, *148* (2), 135–146.

Dawson, P. (1985). Chemotoxicity of contrast media and clinical adverse effects: a review. Investigative Radiology, *20* (1 Suppl), S84–S91.

Eastwood, D.S., Bayley, P.M., Chang, H.J., Taiwo, O.O., Vila-Comamala, J., Brett, D.J., Rau, C., Withers, P.J., Shearing, P.R., Grey, C.P., et al. (2014). Three-dimensional characterization of electrodeposited lithium microstructures using synchrotron X-ray phase contrast imaging. Chemical communications (Cambridge, England), *51* (2), 266–268.

Eck, W., Nicholson, A.I., Zentgraf, H., Semmler, W., and Bartling, S.N. (2010). Anti-CD4-targeted gold nanoparticles induce specific contrast enhancement of peripheral lymph nodes in X-ray computed tomography of live mice. Nano Letters, *10* (7), 2318–2322.

Ekberg, O., Bergenfeldt, M., Aspelin, P., Genell, S., Lindholm, K., Nilsson, P., and Sigurjónsson, S. (1988). Reliability of ultrasound-guided fine-needle biopsy of pancreatic masses. Acta Radiologica, *29* (5), 535–539.

Fang, C., Kievit, F.M., Veiseh, O., Stephen, Z.R., Wang, T., Lee, D., Ellenbogen, R.G., and Zhang, M. (2012). Fabrication of magnetic nanoparticles with controllable drug loading and release through a simple assembly approach. Journal of Controlled Release, *162* (1), 233–241.

Forbes, G.S., Sheedy, P.F., Piepgras, D.G., and Houser, O.W. (1978). Computed tomography in the evaluation of subdural hematomas. Radiology, *126* (1), 143–148.

Frangioni, J.V. (2003). In vivo near-infrared fluorescence imaging. Current Opinion in Chemical Biology, *7* (5), 626–634.

Gao, X., Cui, Y., Levenson, R.M., Chung, L.W., and Nie, S. (2004). In vivo cancer targeting and imaging with semiconductor quantum dots. Nature Biotechnology, *22* (8), 969–976.

Guo, R., Wang, H., Peng, C., Shen, M., Pan, M., Cao, X., Zhang, G., and Shi, X. (2010). X-ray attenuation property of dendrimer-entrapped gold nanoparticles. The Journal of Physical Chemistry C, *114* (1), 50–56.

Hayashi, K., Nakamura, M., Sakamoto, W., Yogo, T., Miki, H., Ozaki, S., Abe, M., Matsumoto, T., and Ishimura, K. (2013). Superparamagnetic nanoparticle clusters for cancer theranostics combining magnetic resonance imaging and hyperthermia treatment. Theranostics, *3* (6), 366.

Hilderbrand, S.A., Shao, F., Salthouse, C., Mahmood, U., and Weissleder, R. (2009). Upconverting luminescent nanomaterials: application to in vivo bioimaging. Chemical Communications, (28), 4188–4190.

Hou, Y., Qiao, R., Fang, F., Wang, X., Dong, C., Liu, K., Liu, C., Liu, Z., Lei, H., and Wang, F. (2013). $NaGdF_4$ nanoparticle-based molecular probes for magnetic resonance imaging of intraperitoneal tumor xenografts in vivo. ACS Nano, *7* (1), 330–338.

Hu, F. and Zhao, Y.S. (2012). Inorganic nanoparticle-based T1 and T1/T2 magnetic resonance contrast probes. Nanoscale, *4* (20), 6235–6243.

Hu, F., Jia, Q., Li, Y., and Gao, M. (2011). Facile synthesis of ultrasmall PEGylated iron oxide nanoparticles for dual-contrast T1-and T2-weighted magnetic resonance imaging. Nanotechnology, *22* (24), 245604.

Huang, X., El-Sayed, I.H., Qian, W., and El-Sayed, M.A. (2006). Cancer cell imaging and photothermal therapy in the near-infrared region by using gold nanorods. Journal of the American Chemical Society, *128* (6), 2115–2120.

Huang, X., Jain, P.K., El-Sayed, I.H., and El-Sayed, M.A. (2007). Gold nanoparticles: interesting optical properties and recent applications in cancer diagnostics and therapy. Nanomedicine, *2* (5), 681–693.

Huang, J., Zhong, X., Wang, L., Yang, L., and Mao, H. (2012). Improving the magnetic resonance imaging contrast and detection methods with engineered magnetic nanoparticles. Theranostics, *2* (1), 86–102.

Hubbell, J.H. and Seltzer, S.M. (1995). Tables of X-ray mass attenuation coefficients and mass energy-absorption coefficients 1 keV to 20 MeV for elements Z = 1 to 92 and 48 additional substances of dosimetric interest (No. PB--95-220539/

XAB; NISTIR--5632). National Inst. of Standards and Technology-PL, Gaithersburg, MD. Ionizing Radiation Div. Available at http://www.nist.gov/pml/data/xraycoef/ (accessed on October 30, 2015).

Hutter, E. and Maysinger, D. (2011). Gold nanoparticles and quantum dots for bioimaging. Microscopy Research and Technique, *74* (7), 592–604.

Jakhmola, A., Anton, N., and Vandamme, T.F. (2012). Inorganic nanoparticles based contrast agents for X-ray computed tomography. Advanced Healthcare Materials, *1* (4), 413–431.

Jana, N.R., Gearheart, L., and Murphy, C.J. (2001). Seeding growth for size control of 5–40 nm diameter gold nanoparticles. Langmuir, *17* (22), 6782–6786.

Janib, S.M., Moses, A.S., and MacKay, J.A. (2010). Imaging and drug delivery using theranostic nanoparticles. Advanced Drug Delivery Reviews, *62* (11), 1052–1063.

Johnson, N.J., Oakden, W., Stanisz, G.J., Scott Prosser, R., and van Veggel, F.C. (2011). Size-tunable, ultrasmall $NaGdF_4$ nanoparticles: insights into their T1 MRI contrast enhancement. Chemistry of Materials, *23* (16), 3714–3722.

Jun, Y.-W., Huh, Y.-M., Choi, J.-S., Lee, J.-H., Song, H.-T., Kim, S., Yoon, S., Kim, K.-S., and Shin, J.-S. (2005). Nanoscale size effect of magnetic nanocrystals and their utilization for cancer diagnosis via magnetic resonance imaging. Journal of the American Chemical Society, *127* (16), 5732–5733.

Jun, Y.-W., Lee, J.-H., and Cheon, J. (2008). Chemical design of nanoparticle probes for high-performance magnetic resonance imaging. Angewandte Chemie International Edition, *47* (28), 5122–5135.

Kim, S., Lim, Y.T., Soltesz, E.G., De Grand, A.M., Lee, J., Nakayama, A., Parker, J.A., Mihaljevic, T., Laurence, R.G., and Dor, D.M. (2003). Near-infrared fluorescent type II quantum dots for sentinel lymph node mapping. Nature Biotechnology, *22* (1), 93–97.

Kim, D., Park, S., Lee, J.H., Jeong, Y.Y., and Jon, S. (2007). Antibiofouling polymer-coated gold nanoparticles as a contrast agent for in vivo X-ray computed tomography imaging. Journal of the American Chemical Society, *129* (24), 7661–7665.

Kim, B.H., Lee, N., Kim, H., An, K., Park, Y.I., Choi, Y., Shin, K., Lee, Y., Kwon, S.G., and Na, H.B. (2011). Large-scale synthesis of uniform and extremely small-sized iron oxide nanoparticles for high-resolution T1 magnetic resonance imaging contrast agents. Journal of the American Chemical Society, *133* (32), 12624–12631.

Kimling, J., Maier, M., Okenve, B., Kotaidis, V., Ballot, H., and Plech, A. (2006). Turkevich method for gold nanoparticle synthesis revisited. The Journal of Physical Chemistry B, *110* (32), 15700–15707.

Kinsella, J.M., Jimenez, R.E., Karmali, P.P., Rush, A.M., Kotamraju, V.R., Gianneschi, N.C., Ruoslahti, E., Stupack, D., and Sailor, M.J. (2011). X-ray computed tomography imaging of breast cancer by using targeted peptide-labeled bismuth sulfide nanoparticles. Angewandte Chemie International Edition, *50* (51), 12308–12311.

Kobayashi, H., Hama, Y., Koyama, Y., Barrett, T., Regino, C.A., Urano, Y., and Choyke, P.L. (2007). Simultaneous multicolor imaging of five different lymphatic basins using quantum dots. Nano Letters, *7* (6), 1711–1716.

Kumar, C.S.S.R. and Mohammad, F. (2011). Magnetic nanomaterials for hyperthermia-based therapy and controlled drug delivery. Advanced Drug Delivery Reviews, *63* (9), 789–808.

Laurent, S., Forge, D., Port, M., Roch, A., Robic, C., Vander Elst, L., and Muller, R.N. (2008). Magnetic iron oxide nanoparticles: synthesis, stabilization, vectorization, physicochemical characterizations, and biological applications. Chemical Reviews, *108* (6), 2064–2110.

Lee, J.-H., Huh, Y.-M., Jun, Y.-W., Seo, J.-W., Jang, J.-T., Song, H.-T., Kim, S., Cho, E.-J., Yoon, H.-G., and Suh, J.-S. (2006). Artificially engineered magnetic nanoparticles for ultra-sensitive molecular imaging. Nature Medicine, *13* (1), 95–99.

Lee, H.Y., Li, Z., Chen, K., Hsu, A.R., Xu, C., Xie, J., Sun, S., and Chen, X. (2008). PET/MRI dual-modality tumor imaging using arginine-glycine-aspartic (RGD)-conjugated radiolabeled iron oxide nanoparticles. Journal of Nuclear Medicine, *49* (8), 1371–1379.

Lee, A., Johnson, M., Policeni, B., and Smoker, W. (2009). Imaging for neuro-ophthalmic and orbital disease—a review. Clinical and Experimental Ophthalmology, *37* (1), 30–53.

Lee, D.-E., Koo, H., Sun, I.-C., Ryu, J.H., Kim, K., and Kwon, I.C. (2012). Multifunctional nanoparticles for multimodal imaging and theragnosis. Chemical Society Reviews, *41* (7), 2656–2672.

Li, J. and Zhu, J.-J. (2013). Quantum dots for fluorescent biosensing and bio-imaging applications. Analyst, *138* (9), 2506–2515.

Li, Q.-B., Xu, Z., Zhang, N.-W., Zhang, L., Wang, F., Yang, L.-M., Wang, J.-S., Zhou, S., Zhang, Y.-F., and Zhou, X.-S. (2005). In vivo and in situ detection of colorectal cancer using Fourier transform infrared spectroscopy. World Journal of Gastroenterology, *11* (3), 327–330.

Li, Z., Sun, Q., Zhu, Y., Tan, B., Xu, Z.P., and Dou, S.X. (2014). Ultra-small fluorescent inorganic nanoparticles for bioimaging. Journal of Materials Chemistry B, *2* (19), 2793–2818.

Liang, G., Cao, L., Chen, H., Zhang, Z., Zhang, S., Yu, S., Shen, X., and Kong, J. (2013). Ultrasmall gadolinium hydrated carbonate nanoparticle: an advanced T1 MRI contrast agent with large longitudinal relaxivity. Journal of Materials Chemistry B, *1* (5), 629–638.

Liu, Y. and Zhang, N. (2012). Gadolinium loaded nanoparticles in theranostic magnetic resonance imaging. Biomaterials, *33* (21), 5363–5375.

Liu, H., Wang, H., Guo, R., Cao, X., Zhao, J., Luo, Y., Shen, M., Zhang, G., and Shi, X. (2010). Size-controlled synthesis of dendrimer-stabilized silver nanoparticles for X-ray computed tomography imaging applications. Polymer Chemistry, *1* (10), 1677–1683.

Liu, Q., Sun, Y., Yang, T., Feng, W., Li, C., and Li, F. (2011). Sub-10 nm hexagonal lanthanide-doped $NaLuF_4$ upconversion nanocrystals for sensitive bioimaging in vivo. Journal of the American Chemical Society, *133* (43), 17122–17125.

Liu, Y., Ai, K., and Lu, L. (2012a). Nanoparticulate X-ray computed tomography contrast agents: from design validation to in vivo applications. Accounts of Chemical Research, *45* (10), 1817–1827.

Liu, Z., Liu, X., Yuan, Q., Dong, K., Jiang, L., Li, Z., Ren, J., and Qu, X. (2012b). Hybrid mesoporous gadolinium oxide nanorods: a platform for multimodal imaging and enhanced insoluble anticancer drug delivery with low systemic toxicity. Journal of Materials Chemistry, *22* (30), 14982–14990.

Liu, H., Shen, M., Zhao, J., Zhu, J., Xiao, T., Cao, X., Zhang, G., and Shi, X. (2013a). Facile formation of folic acid-modified dendrimer-stabilized gold–silver alloy nanoparticles for potential cellular computed tomography imaging applications. Analyst, *138* (7), 1979–1987.

Liu, J., Yu, M., Zhou, C., Yang, S., Ning, X., and Zheng, J. (2013b). Passive tumor targeting of renal-clearable luminescent gold nanoparticles: long tumor retention and fast normal tissue clearance. Journal of the American Chemical Society, *135* (13), 4978–4981.

Liu, Y., Chang, Z., Yuan, H., Fales, A.M., and Vo-Dinh, T. (2013c). Quintuple-modality (SERS-MRI-CT-TPL-PTT) plasmonic nanoprobe for theranostics. Nanoscale, *5* (24), 12126–12131.

Liu, C.-L., Peng, Y.-K., Chou, S.-W., Tseng, W.-H., Tseng, Y.-J., Chen, H.-C., Hsiao, J.-K., and Chou, P.-T. (2014). One-step, room-temperature synthesis of glutathione-capped iron-oxide nanoparticles and their application in in vivo T1-weighted magnetic resonance imaging. Small, *10* (19), 3962–3969.

Lodhia, J., Mandarano, G., Ferris, N., Eu, P., and Cowell, S. (2010). Development and use of iron oxide nanoparticles (Part 1): synthesis of iron oxide nanoparticles for MRI. Biomedical Imaging and Intervention Journal, *6* (2), e12.

Mahmoudi, M., Sant, S., Wang, B., Laurent, S., and Sen, T. (2011). Superparamagnetic iron oxide nanoparticles (SPIONs): development, surface modification and applications in chemotherapy. Advanced Drug Delivery Reviews, *63* (1), 24–46.

Mulder, W.J., Strijkers, G.J., van Tilborg, G.A., Griffioen, A.W., and Nicolay, K. (2006). Lipid-based nanoparticles for contrast-enhanced MRI and molecular imaging. NMR in Biomedicine, *19* (1), 142–164.

Na, H.B. and Hyeon, T. (2009). Nanostructured T1 MRI contrast agents. Journal of Materials Chemistry, *19* (35), 6267–6273.

Nam, J., Won, N., Bang, J., Jin, H., Park, J., Jung, S., Jung, S., Park, Y., and Kim, S. (2013). Surface engineering of inorganic nanoparticles for imaging and therapy. Advanced Drug Delivery Reviews, *65* (5), 622–648.

Ng, V.W., Berti, R., Lesage, F., and Kakkar, A. (2013). Gold: a versatile tool for in vivo imaging. Journal of Materials Chemistry B, *1* (1), 9–25.

Nyk, M., Kumar, R., Ohulchanskyy, T.Y., Bergey, E.J., and Prasad, P.N. (2008). High contrast in vitro and in vivo photoluminescence bioimaging using near infrared to near infrared up-conversion in Tm^{3+} and Yb^{3+} doped fluoride nanophosphors. Nano Letters, *8* (11), 3834–3838.

Oh, M.H., Lee, N., Kim, H., Park, S.P., Piao, Y., Lee, J., Jun, S.W., Moon, W.K., Choi, S.H., and Hyeon, T. (2011). Large-scale synthesis of bioinert tantalum oxide nanoparticles for X-ray computed tomography imaging and bimodal image-guided sentinel lymph node mapping. Journal of the American Chemical Society, *133* (14), 5508–5515.

Pan, D., Caruthers, S., Senpan, A., Schmieder, A., Wickline, S., and Lanza, G. (2011). Revisiting an old friend: manganese-based MRI contrast agents. Wiley Interdisciplinary Reviews-Nanomedicine and Nanobiotechnology, *3* (2), 162–173.

Park, J.Y., Daksha, P., Lee, G.H., Woo, S., and Chang, Y. (2008). Highly water-dispersible PEG surface modified ultra small superparamagnetic iron oxide nanoparticles useful for target-specific biomedical applications. Nanotechnology, *19* (36), 365603.

Park, J.Y., Baek, M.J., Choi, E.S., Woo, S., Kim, J.H., Kim, T.J., Jung, J.C., Chae, K.S., Chang, Y., and Lee, G.H. (2009). Paramagnetic ultrasmall gadolinium oxide nanoparticles as advanced T1 MRI contrast agent: account for large longitudinal relaxivity, optimal particle diameter, and in vivo T1 MR images. ACS Nano, *3* (11), 3663–3669.

Pasternak, J.J. and Williamson, E.E. (2012). Clinical pharmacology, uses, and adverse reactions of iodinated contrast agents: a primer for the non-radiologist. In *Mayo Clinic Proceedings* (Vol. 87, No. 4, pp. 390–402, Elsevier).

Pichler, B.J., Kolb, A., Nägele, T., and Schlemmer, H.-P. (2010). PET/MRI: paving the way for the next generation of clinical multimodality imaging applications. Journal of Nuclear Medicine, *51* (3), 333–336.

Pogue, B.W., Poplack, S.P., McBride, T.O., Wells, W.A., Osterman, K.S., Osterberg, U.L., and Paulsen, K.D. (2001). Quantitative hemoglobin tomography with diffuse near-infrared spectroscopy: pilot results in the breast. Radiology, *218* (1), 261–266.

Qian, X., Peng, X.-H., Ansari, D.O., Yin-Goen, Q., Chen, G.Z., Shin, D.M., Yang, L., Young, A.N., Wang, M.D., and Nie, S. (2008). In vivo tumor targeting and spectroscopic detection with surface-enhanced Raman nanoparticle tags. Nature Biotechnology, *26* (1), 83–90.

Quek, C.-H. and Leong, K.W. (2012). Near-infrared fluorescent nanoprobes for in vivo optical imaging. Nanomaterials, *2* (2), 92–112.

Rabin, O., Perez, J.M., Grimm, J., Wojtkiewicz, G., and Weissleder, R. (2006). An X-ray computed tomography imaging agent based on long-circulating bismuth sulphide nanoparticles. Nature Materials, *5* (2), 118–122.

Schaeffer, M. (1997). Properties of paramagnetic metals in MRI. Metal-Based Drugs, *4* 159–172.

von Schulthess, G.K., Steinert, H.C., and Hany, T.F. (2006). Integrated PET/CT: current applications and future directions. Radiology, *238* (2), 405–422.

Seo, W.S., Lee, J.H., Sun, X., Suzuki, Y., Mann, D., Liu, Z., Terashima, M., Yang, P.C., McConnell, M.V., and Nishimura, D.G. (2006). FeCo/graphitic-shell nanocrystals as advanced magnetic-resonance-imaging and near-infrared agents. Nature Materials, *5* (12), 971–976.

Sharma, P., Brown, S., Walter, G., Santra, S., and Moudgil, B. (2006). Nanoparticles for bioimaging. Advances in Colloid and Interface Science, *123*, 471–485.

Shen, M. and Shi, X. (2010). Dendrimer-based organic/inorganic hybrid nanoparticles in biomedical applications. Nanoscale, *2* (9), 1596–1610.

Shi, X., Sun, K., and Baker Jr, J.R. (2008). Spontaneous formation of functionalized dendrimer-stabilized gold nanoparticles. The Journal of Physical Chemistry C, *112* (22), 8251–8258.

Shin, J., Anisur, R.M., Ko, M.K., Im, G.H., Lee, J.H., and Lee, I.S. (2009). Hollow manganese oxide nanoparticles as multifunctional agents for magnetic resonance imaging and drug delivery. Angewandte Chemie International Edition, *48* (2), 321–324.

Shuhendler, A.J., Prasad, P., Chan, H.-K.C., Gordijo, C.R., Soroushian, B., Kolios, M., Yu, K., O'Brien, P.J., Rauth, A.M., and Wu, X.Y. (2011). Hybrid quantum dot–fatty ester stealth nanoparticles: toward clinically relevant in vivo optical imaging of deep tissue. ACS Nano, *5* (3), 1958–1966.

Shukla, R., Hill, E., Shi, X., Kim, J., Muniz, M.C., Sun, K., and Baker, J.R. (2008). Tumor microvasculature targeting with dendrimer-entrapped gold nanoparticles. Soft Matter, *4* (11), 2160–2163.

Smith-Bindman, R., Lipson, J., Marcus, R., Kim, K.-P., Mahesh, M., Gould, R., de González, A.B., and Miglioretti, D.L. (2009). Radiation dose associated with common computed tomography examinations and the associated lifetime attributable risk of cancer. Archives of Internal Medicine, *169* (22), 2078–2086.

Sook Choi, E., Young Park, J., Ju Baek, M., Xu, W., Kattel, K., Hyun Kim, J., Jun Lee, J., Chang, Y., Jeong Kim, T., and Eun Bae, J. (2010). Water-soluble ultra-small manganese oxide surface doped gadolinium oxide (Gd_2O_3@ MnO) nanoparticles for MRI contrast agent. European Journal of Inorganic Chemistry, *2010* (28), 4555–4560.

Sreenivasan, V.K., Zvyagin, A.V., and Goldys, E.M. (2013). Luminescent nanoparticles and their applications in the life sciences. Journal of Physics: Condensed Matter, *25* (19), 194101.

Stanga, P.E. and Bird, A.C. (2001). Optical coherence tomography (OCT): principles of operation, technology, indications in vitreoretinal imaging and interpretation of results. International Ophthalmology, *23* (4–6), 191–197.

Still, J., Law, E., Klavuhn, K., Island, T., and Holtz, J. (2001). Diagnosis of burn depth using laser-induced indocyanine green fluorescence: a preliminary clinical trial. Burns, *27* (4), 364–371.

Strong, M.S., Vaughan, C.W., and Incze, J.S. (1968). Toluidine blue in the management of carcinoma of the oral cavity. Archives of Otolaryngology, *87* (5), 527–531.

Sun, I.-C., Eun, D.-K., Na, J.H., Lee, S., Kim, I.-J., Youn, I.-C., Ko, C.-Y., Kim, H.-S., Lim, D., and Choi, K. (2009). Heparin-coated gold nanoparticles for liver-specific CT imaging. Chemistry--A European Journal, *15* (48), 13341–13347.

Thompson, C.J., Murthy, K., Weinberg, I.N., and Mako, F. (1994). Feasibility study for positron emission mammography. Medical Physics, *21* (4), 529–538.

Tromsdorf, U.I., Bruns, O.T., Salmen, S.C., Beisiegel, U., and Weller, H. (2009). A highly effective, nontoxic T1 MR contrast agent based on ultrasmall PEGylated iron oxide nanoparticles. Nano Letters, *9* (12), 4434–4440.

Veiseh, O., Gunn, J.W., and Zhang, M. (2010). Design and fabrication of magnetic nanoparticles for targeted drug delivery and imaging. Advanced Drug Delivery Reviews, *62* (3), 284–304.

Villringer, A. and Chance, B. (1997). Non-invasive optical spectroscopy and imaging of human brain function. Trends in Neurosciences, *20* (10), 435–442.

Wadajkar, A.S., Menon, J.U., Kadapure, T., Tran, R.T., Yang, J., and Nguyen, K.T. (2013). Design and application of magnetic-based theranostic nanoparticle systems. Recent Patents on Biomedical Engineering, *6* (1), 47–57.

Wang, H. and Chen, X. (2009). Applications for site-directed molecular imaging agents coupled with drug delivery potential. Expert Opinion on Drug Delivery, *6* (7), 745–768.

Wang, X., Ku, G., Wegiel, M.A., Bornhop, D.J., Stoica, G., and Wang, L.V. (2004). Noninvasive photoacoustic angiography of animal brains in vivo with near-infrared light and an optical contrast agent. Optics Letters, *29* (7), 730–732.

Wang, F., Banerjee, D., Liu, Y., Chen, X., and Liu, X. (2010). Upconversion nanoparticles in biological labeling, imaging, and therapy. Analyst, *135* (8), 1839–1854.

Wang, H., Zheng, L., Peng, C., Shen, M., Shi, X., and Zhang, G. (2013). Folic acid-modified dendrimer-entrapped gold nanoparticles as nanoprobes for targeted CT imaging of human lung adencarcinoma. Biomaterials, *34* (2), 470–480.

Weng, K.C., Noble, C.O., Papahadjopoulos-Sternberg, B., Chen, F.F., Drummond, D.C., Kirpotin, D.B., Wang, D., Hom, Y.K., Hann, B., and Park, J.W. (2008). Targeted tumor cell internalization and imaging of multifunctional quantum dot-conjugated immunoliposomes in vitro and in vivo. Nano Letters, *8* (9), 2851–2857.

Wright, C.I., Kroner, C.I., and Draijer, R. (2006). Non-invasive methods and stimuli for evaluating the skin's microcirculation. Journal of Pharmacological and Toxicological Methods, *54* (1), 1–25.

Wu, X., He, X., Wang, K., Xie, C., Zhou, B., and Qing, Z. (2010). Ultrasmall near-infrared gold nanoclusters for tumor fluorescence imaging in vivo. Nanoscale, *2* (10), 2244–2249.

Wu, P., Zhang, Y., Sun, Y., Shi, X., Li, F., and Zhu, Z. (2013). Clinical applications of ^{18}F-FDG PET/CT in monitoring anti-cancer therapies. Current Pharmaceutical Biotechnology, *14* (7), 658–668.

Xie, J., Chen, K., Huang, J., Lee, S., Wang, J., Gao, J., Li, X., and Chen, X. (2010). PET/NIRF/MRI triple functional iron oxide nanoparticles. Biomaterials, *31* (11), 3016–3022.

Xiong, L.-Q., Chen, Z.-G., Yu, M.-X., Li, F.-Y., Liu, C., and Huang, C.-H. (2009). Synthesis, characterization, and *in vivo* targeted imaging of amine-functionalized rare-earth up-converting nanophosphors. Biomaterials, *30* (29), 5592–5600.

Xiong, L., Yang, T., Yang, Y., Xu, C., and Li, F. (2010). Long-term in vivo biodistribution imaging and toxicity of polyacrylic acid-coated upconversion nanophosphors. Biomaterials, *31* (27), 7078–7085.

Xu, C., Tung, G.A., and Sun, S. (2008). Size and concentration effect of gold nanoparticles on X-ray attenuation as measured on computed tomography. Chemistry of Materials, *20* (13), 4167–4169.

Yablonskiy, D.A. and Sukstanskii, A.L. (2015). Generalized Lorentzian Tensor Approach (GLTA) as a biophysical background for quantitative susceptibility mapping. Magnetic Resonance in Medicine, *73*(2), 757–764.

Yong, K.-T., Hu, R., Roy, I., Ding, H., Vathy, L.A., Bergey, E.J., Mizuma, M., Maitra, A., and Prasad, P.N. (2009). Tumor targeting and imaging in live animals with functionalized semiconductor quantum rods. ACS Applied Materials & Interfaces, *1* (3), 710–719.

Yu, M.K., Jeong, Y.Y., Park, J., Park, S., Kim, J.W., Min, J.J., Kim, K., and Jon, S. (2008). Drug-loaded superparamagnetic iron oxide nanoparticles for combined cancer imaging and therapy in vivo. Angewandte Chemie International Edition, *47* (29), 5362–5365.

Zhao, Z., Zhou, Z., Bao, J., Wang, Z., Hu, J., Chi, X., Ni, K., Wang, R., Chen, X., and Chen, Z. (2013). Octapod iron oxide nanoparticles as high-performance T2 contrast agents for magnetic resonance imaging. Nature Communications, *4* 2266.

Zhou, M., Zhang, R., Huang, M., Lu, W., Song, S., Melancon, M.P., Tian, M., Liang, D., and Li, C. (2010). A chelator-free multifunctional [^{64}Cu] CuS nanoparticle platform for simultaneous micro-PET/CT imaging and photothermal ablation therapy. Journal of the American Chemical Society, *132* (43), 15351–15358.

Zhou, Z., Huang, D., Bao, J., Chen, Q., Liu, G., Chen, Z., Chen, X., and Gao, J. (2012a). A synergistically enhanced T1–T2 dual-modal contrast agent. Advanced Materials, *24* (46), 6223–6228.

Zhou, J., Liu, Z., and Li, F. (2012b). Upconversion nanophosphors for small-animal imaging. Chemical Society Reviews, *41* (3), 1323–1349.

Zhou, Z., Wang, L., Chi, X., Bao, J., Yang, L., Zhao, W., Chen, Z., Wang, X., Chen, X., and Gao, J. (2013). Engineered iron-oxide-based nanoparticles as enhanced T1 contrast agents for efficient tumor imaging. ACS Nano, *7* (4), 3287–3296.

Zhu, D., Liu, F., Ma, L., Liu, D., and Wang, Z. (2013). Nanoparticle-based systems for T1-weighted magnetic resonance imaging contrast agents. International Journal of Molecular Sciences, *14* (5), 10591–10607.

10

BIOSENSORS FOR EARLY DISEASE DIAGNOSIS

AHMET E. TOPAL, ALPER DEVRIM OZKAN, AYKUTLU DANA, AYSE B. TEKINAY, AND MUSTAFA O. GULER

Institute of Materials Science and Nanotechnology, National Nanotechnology Research Center (UNAM), Bilkent University, Ankara, Turkey

10.1 INTRODUCTION

Early diagnosis is an important aspect of modern medicine and significantly improves the prognosis of a wide range of life-threatening disorders (Ping et al., 1997; Wang, 2006). However, early diagnosis is often difficult or impossible without an appropriate means of detecting small differences in the patient's physiology. As such, a great diversity of both general and specialized methods has been developed to identify the presence of genetic disorders, cancers, viral or bacterial infections, and other diseases prior to the point they present their pathological effects (Saravolatz et al., 2003; Wulfkuhle et al., 2003). These methods principally involve either medical imaging, in which the patient's tissues are checked *in situ* with or without the assistance of a dye or contrast agent (see Chapter 9); histopathology, in which a sample of the patient's tissues is removed by biopsy for visual inspection, usually with the help of specific cell and tissue stains; or a variety of diagnostic assays, in which the patient's bodily fluids (or tissue samples) are biochemically analyzed for disease-specific *biomarkers*.

Therapeutic Nanomaterials, First Edition. Edited by Mustafa O. Guler and Ayse B. Tekinay.

While powerful, current medical diagnostic techniques nonetheless suffer from clinical limitations. Histopathology is the gold standard for the diagnosis of many diseases but also requires an expert practitioner and introduces an element of subjectivity to disease diagnosis (Uckermann et al., 2014). Likewise, medical imaging is commonly used in disease diagnosis but cannot easily detect early-stage disorders, which often affect tissues on scales too small to be resolved by the imaging technique. In contrast, biochemical assays combine low detection limits with objective criteria of quantification and are therefore more suitable for the diagnosis of hard-to-detect diseases. Standard biochemical assays are now employed for the diagnosis of tuberculosis, HIV, hepatitis, *Escherichia coli* enteritis, and many other pathological conditions.

Nonetheless, these assays frequently require a laboratory and a human practitioner, which may not be readily available in rural areas where infectious diseases are common (Peeling and Mabey, 2010). In addition, complex diseases often require testing of multiple biomarkers in tandem, and an automated system would be of considerable advantage for the parallel testing (or "multiplexing") of a broad range of molecules. The amount of analytes and buffer solutions used in these assays can also be reduced if the protocols of these assays could be replicated on a small device, allowing the development of cheaper diagnostic methods. The benefits of such fast, repeatable "autoassay" devices are therefore obvious, and numerous attempts have been made to perform each step of a diagnostic assay on a small device that ultimately produces an output that can either be confirmed by the naked eye or quantified by a spectrometer or similar, commonly available equipment. These devices are typically referred to as *biosensors*.

A biosensor is a compact analytical device that is capable of selectively identifying biological signals, such as proteins, nucleic acids, small molecules, or secondary metabolites, which are collectively called *analytes*. The presence of a specific biological moiety can either be detected directly ("label-free") or through the assistance of a *label*. Label-bearing biosensors usually have a *recognition element*, which specifically interacts with the target analyte, and a *signal transducer element*, which transforms that interaction into an optical, electrochemical, or mechanical signal. Recognition elements are called *bioreceptors* if they consist of biological materials with recognition capability, such as antibodies and complementary DNA or RNA sequences (due to their specificity, most biosensors employ bioreceptors as their recognition element). Transducers, in turn, convert the analyte/recognition element interaction into a measurable signal and are composed of one or more *interface elements*, which are device elements such as thin films and field-effect transistor (FET) devices or nanomaterials such as nanoparticles and nanowire arrays. The output of the biosensor is either confirmed visually (especially in the case of colorimetric biosensors) or quantified by a readout system (Medley et al., 2008).

The first biosensor, designed for the detection of glucose, was introduced in 1962 by Clark and Lyons (Rapp et al., 2010). Also called an "enzyme electrode," it was a biosensor of the amperometric type. Since then, the number and diversity of biosensors have grown enormously, and the design of biosensors has become an important field in medical diagnostics. As such, this chapter will focus on biosensor types, their detection limits, analysis times, and the diseases they are suitable for detecting. In addition, as nanomaterials are an effective means of producing small-scale diagnostic devices, nanostructures have been commonly employed in biosensor design. Consequently, a section will be devoted to the types of nanomaterials currently under use in biosensor design.

10.2 BIOSENSOR ELEMENTS

Biosensors can be classified according to their recognition element (e.g., enzymes, antibodies, nucleic acids), output type (e.g., optical, electrical, mechanical), detection principle (e.g., surface plasmon resonance (SPR) based, surface-enhanced Raman spectroscopy (SERS) based, quartz crystal microbalance (QCM) based), or intended use (*in vivo* or *ex vivo*). These factors all play vital roles in determining the sensitivity and selectivity of a biosensor and will be considered separately.

10.2.1 Recognition Elements

Recognition elements are almost uniformly biological, since enzymes, antibodies, and complementary nucleic acid sequences display specificities unparalleled by almost any nonbiological material (molecular imprinting, however, is also capable of creating highly specific binding sites for biological analytes and can be employed in sensor design). Biological sensing materials used for analyte recognition include enzymes, antibodies, and DNA or RNA constructs, which may either be used in purified form or expressed on the surface of bacteria or viruses (live-cell detection) (Van Dorst et al., 2010).

Enzyme-based biosensors typically detect the breakdown products of an enzymatic reaction between the analyte and the enzyme (i.e., catalytic recognition) (Iles and Kallichurn, 2012). They are often employed for the detection of small molecules, as antibody- and nucleic acid-based methods are either unfeasible (if the analyte isn't recognized as an antigen) or unnecessary (if easy-to-produce enzymes are able to yield accurate measurements) for these materials. As enzyme targets are often found in both healthy and diseased tissues, diagnosis often relies on the concentration, rather than the presence, of the analyte. Enzymes have been used in the design of biosensors for glucose and other sugars, nerve gas agents, heavy metals, urea, ascorbic acid, acetylcholine, malate, and other small molecules (Mulchandani et al., 1999; Tsai and Doong, 2005; Wang, 2001). In addition to acting as the recognition element, enzymes can also be used as a means of visualizing the analyte–recognition element, such as by the use of HRP-tagged secondary antibodies.

Antibodies are often considered the gold standard for the detection of proteins and are commonly used in protein-detecting biosensors. Antibodies may either be monoclonal or polyclonal; monoclonal antibodies target a specific recognition site, while polyclonal antibodies bind to different recognition sites on the same antigen. As such, monoclonal antibodies are generally more specific, although they are costlier to produce and the recognition specificity of even polyclonal antibodies is considerable. Antibodies are also selective enough to quantify very small differences in the concentration of a specific protein, which is of considerable advantage in situations where small changes in expression patterns are indicative of early-stage disease. In addition to antibodies, binding peptides with similarly high affinities (down to the picomolar range) to specific analytes can be used as detector elements (Sidhu et al., 2000). Antibody-based biosensors have been designed for the detection of a wide range of proteins, varying from cancer markers to viral antigens and bacterial cell membrane components (El-Sayed et al., 2005; Pathirana et al., 2000; Torrance et al., 2006).

The most prominent technique for antibody-based detection is the enzyme-linked immunosorbent assay (ELISA). Digital ELISA is capable of detecting analyte concentrations at as low as the femtomolar scale, for example, prostate-specific antigen (PSA) could be detected at 14 fg/ml (0.4 fM) from patients who had undergone radical prostatectomy (Rissin et al., 2010). However, ELISA is relatively costly as a routine diagnostic technique and, in some cases, not sensitive enough for use in diagnosis, especially for the detection of early biomarkers for cancer (Tothill, 2009). As such, one main focus in biosensor development is on surpassing the detection limit of conventional ELISA (Park et al., 2009).

While antibodies can be raised against specific nucleic acid sequences, a complementary DNA or RNA strand can also be used for nucleic acid detection. These can be used not only for the detection of bacterial or viral nucleic acids but also to rapidly measure gene expression at the RNA level or to detect mutations in genomic DNA (Dell'Atti et al., 2006). However, DNA and RNA may face stability issues when exposed to serum or other biological media; as such, nucleic acids with alternative backbones, such as peptide nucleic acids (PNAs), have been developed for use as recognition elements with increased stability (Ray and Norden, 2000). In addition, oligonucleotide aptamers are promising recognition agents that, much like antibodies, can be raised against specific protein targets (Du et al., 2013). Aptamer production does not require a cell line or animal to serve as a source, which is an advantage compared to antibodies (Iliuk et al., 2011). Nucleic acid-based biosensor probes have been reported for the detection of mutations, human and animal viruses, and heavy metals, as well as for use in gene expression studies.

Recognition elements are typically of biological origin, as it is difficult to match the efficiency of detection mechanisms that have been continuously improved by natural selection since the emergence of the earliest immune systems. However, polymers can also be etched to create highly specific binding sites using a technique called molecular imprinting, which has been employed in the detection of amino acids, sugars, antibiotics, and simple organic molecules (Kriz et al., 1997). Electronic noses, devices that contain no biological elements and instead rely on the differential binding of gas or solute molecules to the device surface, have also been used in the development of breath test biosensors for diabetes, pneumonia, fungal toxins, and blood in urine (Di Natale et al., 1999; Hanson and Thaler, 2005; Logrieco et al., 2005; Ping et al., 1997).

10.2.2 Output Type and Detection Techniques

The signal created by the binding of the analyte to the recognition element is transformed into a detectable form by a *transducer*. The transducer may detect either the binding event itself or, in the case of an enzymatic reaction, the products that are formed in the aftermath of catalytic activity. Alternatively, the recognition element itself might yield a detectable signal, such as a change in absorption properties, after binding to the analyte. No matter the case, the resulting output must be measured and quantified for the biosensor to function. This output may be optical, electrical, electrochemical, gravimetric/piezoelectric, mechanical, or magnetic and may be amplified and processed to increase signal quality prior to diagnostic assessment. Secondary equipment, such as spectrometers, is commonly used in data evaluation in this manner.

The output signal and the method used for its detection are important factors in determining the sensitivity of the biosensor. A table of biosensor output types, detection techniques, and the associated sensitivity and analysis time comparisons is provided in Table 10.1. It is worth noting that most biosensors provide optical, electrical/electrochemical, or mechanical output and other detection methods are relatively rare.

TABLE 10.1 Output Types, Recognition Elements, Sensitivities, and Analysis Times for Recent Biosensors

	Technology, Technique or Phenomenon	Nanomaterial, Bioreceptor, or Component	Analyte	Sensitivity	Analysis Duration	References
Electrical	Amperometric	Fructosyl amino acid oxidase immobilized onto zinc oxide nanoparticle–polypyrrole film	Hemoglobin A1c	50 μM	5–25 s current response time	Chawla and Pundir (2012)
		Bionanocomposite film consisting of glucose oxidase/Pt/functional graphene sheets/chitosan (GOD/Pt/FGS/chitosan)	Glucose	0.6 μM	~200 s	Wu et al. (2009)
	Capacitance	Citrate-capped gold nanoparticles (AuNPs) on a screen-printed electrode	Cardiac troponin I	0.2 ng/ml	~25 min	Bhalla et al. (2012)
	Field-effect transistor (FET) biosensors	Thiamine-immobilized, glutaraldehyde-modified SiO_2 gate on p-type Si	Prion proteins	2 nM	>1 h	Wustoni et al. (2014)
		Borrelia burgdorferi (Lyme) flagellar antibodies on SWNT FET	Lyme disease antigen	1 ng/ml	20 min	Lerner et al. (2013)
		Silicon nanowires (SiNWs)	Prostate-specific antigen (PSA)	10^{-6} ng/ml	Real time	Luo and Davis (2013)
Electrochemical	Electrochemical impedance spectroscopy (EIS)	Antihuman alpha-synuclein-based detection; device consists of PEG-thiol monolayer on gold electrode	Alpha-synuclein	55 + 3 pM	A few minutes	Bryan et al. (2012)

Amperometric	Fructosyl amino acid oxidase (FAO) immobilized on core–shell magnetic Fe–Si bionanoparticles (with chitosan); device uses modified gold electrode	Glycosylated hemoglobin	0.1 mM	<4 s	Chawla and Pundir (2011)
	Porous redox-active Cu_2O–SiO_2 nanoparticles (NPs)	Ferritin	0.4 ng/ml	30 min	Yang et al. (2009)
Squarewave voltammetry	CdSe QDs functionalized with streptavidin- (SA) labeled DNA; detection based on endonuclease activity; AuNPs used for signal amplification	*Mycobacterium* DNA	8.7×10^{-15} M	10 s for pH = 2	Zhang et al. (2015)
Cyclic and differential pulse voltammetry	Chitosan-modified glassy carbon electrode (GCE)	SCWL-diseased plant ssDNA	4.709 ng/μl	>10 min	Wongkaew and Poosittisak (2014)
	SiNW/AuNP-modified indium tin oxide (ITO)	DNA related to dengue virus	3.5 ng/ml	Within seconds	Rashid et al. (2014)
Electrochemiluminescence (ECL)	Biotin-anti-cTnI-luminol-AuNPs and SA-AuNPs	Troponin I	0.06 ng/ml	A few minutes	(Li et al., 2013a)
	Gold electrode labeled with hairpin DNA incorporating a ruthenium complex	Target ssDNA	9×10^{-11} M	2–4 s	Bertoncello and Forster (2009)
Electrochemical immunosensor	Antitestosterone antibody on hybrid gold NP/CNT–Teflon composite electrodes	Testosterone in human serum	85 ± 6 pg/ml		Serafín et al. (2011)
	Interdigitated electrodes (IDE) with nanoislands; device functions through protein immobilization to a parylene-A surface	Hepatitis B virus surface antigen (HBsAg)	<100 pg/ml		Jung et al. (2014)

(*Continued*)

TABLE 10.1 *(Continued)*

	Technology, Technique or Phenomenon	Nanomaterial, Bioreceptor, or Component	Analyte	Sensitivity	Analysis Duration	References
	Electrochemically enhanced LSPR	Nanoscale Lycurgus cup arrays	BSA	13 μg/ml		Zhang et al. (2014)
Optical	Labeled SPR	ssDNA aptamer-functionalized gold chip	DNA	75 nM (1.58 μg/ml) RBP4		Lee et al. (2008)
		Oligoethylene glycol linker conjugated testosterone covalently immobilized to gold; detection enhanced using a secondary antibody with an attached AuNP	Testosterone	3.7 pg/ml with standard in running buffer 15.4 pg/ml in a stripped human saliva matrix	Real-time in 13 min	Mitchell and Lowe (2009)
	Fluorescence	Hepatitis B virus (HBV) capsid-derived chimeric NPs on nickel nanohairs; detection achieved by antibody–antigen binding	Troponin I	0.1 nM		Park et al. (2009)
		Graphene oxide (GO) array deposited on amino-modified glass; antibody–DNA–AuNP complexes used for detection	Rotavirus	10^3–10^5 pfu/ml		Jung et al. (2010)
		PDMS microchannels with gold thin-film sensing layers coated with anti-CTX-II antibodies covalently conjugated to fluoro-microbeads	C-telopeptide fragments of type II collagen (CTX-II) serum (sCTX-II; homodimers) and urine (uCTX-II; monomers or variant monomers)	50 ng/mmol	<70 min	Park et al. (2014)

Transmission spectroscopy	Fluorescent Qdot-antinitrotyrosine conjugate	Nitrated ceruloplasmin, a significant biomarker for cardiovascular disease	1 ng/ml	10 min	Li et al. (2010)
	Plasmonic gold nanorods	Single nucleotide polymorphism (SNP) in DNA	310 nm/RIU, 10 nM		Dodson et al. (2015)
	100 nm Au thin film with plasmonic nanoholes	Vesicular stomatitis virus (VSV), pseudotyped Ebola (PT-Ebola), and Vaccinia virus	10^6 PFU/ml	>90 min	Yanik et al. (2010)
SERS	Detection of antibody conjugated to gold nanostar or spheres bound to malachite green isothiocyanate (MGITC) Raman signals is enhanced by a silica sandwich nanostructure	Human immunoglobulin G (human IgG), VEGF	7 fg/ml	10 s	Li et al. (2013b)
	Iodide-modified AgNPs in fluid	Lysozyme, avidin, bovine serum albumin (BSA), cytochrome c, and hemoglobin	3 μg/ml	Within minutes	Xu et al. (2014)
Colorimetric, naked eye	AuNP-based plasmonic ELISA	PSA HIV-1 capsid antigen p24	1×10^{-18} g/ml		de la Rica and Stevens (2012)

(*Continued*)

TABLE 10.1 *(Continued)*

	Technology, Technique or Phenomenon	Nanomaterial, Bioreceptor, or Component	Analyte	Sensitivity	Analysis Duration	References
		Water-soluble cationic conjugated polymers (CCP)	DNA mutation	fM, 2% mutant DNAs from total DNA	A few hours	Song et al. (2012a)
		Mutation-specific single-base extension (SBE) primers	PIK3CA mutations for breast cancer			
		Trinder's reaction-based enzymatic assay	Glucose	1.3 ng/ml		Duk Han et al. (2014)
		Competitive enzyme-linked immunosorbent assay	Urinary collagen type II C-telopeptide fragments (uCTX-II)			
		PDMS microchannel bonded with polydopamine coating and/or G4 PAMAM dendrimer				
		Cell-SELEX aptamers and gold colloid NPs	Cancer cells including T-cells, leukemia, and lymphoma	90 cells	>30 min	Medley et al. (2008)
Mechanical	Suspended microchannel resonator (SMR)	U-shaped microchannels or microcantilevers with anti-ALCAM monoclonal IgGs coated on silicon	Activated leukocyte cell adhesion molecule (ALCAM)	300 pM, 10 ng/ml	1 min	Arlett et al., (2011), von Muhlen et al. (2010)

Atomic force microscopy (AFM)	Functionalization with 3-aminopropyltriethoxysilane (APTES) followed by another functionalization step with a layer of very small peptides, which have a high affinity to the Fc region of the antibody	Antibody–antigen complexes	10 ng/ml		Ierardi et al. (2013)
Surface acoustic wave (SAW) biosensor	MUC1 aptamer-modified gold interdigital transducers for leaky SAW on lithium tantalite ($LiTaO_3$) surface	MCF-7 human breast cancer cells	32 cells/ml	Real time	Chang et al. (2014)
Quartz crystal microbalance (QCM) biosensor (acoustic-based biosensor)	ssDNA cross-linked polymeric hydrogel immobilized on the gold surface	Avian influenza virus (AIV) H5N1	0.0128 HAU (HA unit)	30 min	Wang and Li (2013)
	Biotin-labeled probe DNA immobilized on gold surface after immobilizing avidin on carboxyl chip prior to biotin	Fish pathogenic viral hemorrhagic septicemia (VHS) virus	About 0.1 nM of real-target viral RNA	~25 min	Hong et al. (2010)
Microcantilever biosensor	Biotinylated polyclonal antibody-functionalized silicon microcantilever arrays Multifunctional Fe_3O_4 at SiO_2 core TiO_2 shell magnetic–photocatalytic NPs	Interleukin-6, interferon-γ, and alpha-fetoprotein	<0.1 pg/ml	~1 h	Joo et al. (2012)
	FhuA-containing proteoliposomes immobilized on the chemically activated gold-coated surface	Bacterial virus T5	3 pM T5 virus	>20 min	Braun et al. (2009)

(Continued)

TABLE 10.1 (*Continued*)

	Technology, Technique or Phenomenon	Nanomaterial, Bioreceptor, or Component	Analyte	Sensitivity	Analysis Duration	References
Lab-on-a-chip (combined approaches)	Chemiluminometric ELISA on a chip	HRP-labeled biotinylated anti-HRP antibody biotin–SA capture on plastic plates	Troponin I	0.027 ng/ml	30 s	Cho et al. (2009)
Magnetic biosensors		Core–shell Fe_3O_4 at poly (dopamine) magnetic nanoparticles (MNPs)	H_2O_2	182 nM	10 s	Martín et al. (2014)
	Superconducting quantum interference devices (SQUIDs)	Antibody-coated magnetic particles	*Listeria monocytogenes* bacteria	5×10^4 NPs for a substrate-based assay and 1.1×10^5 bacteria in a 20 ml sample volume	<0.1 s	Grossman et al. (2004)
		Dextran-coated iron oxide MNPs attached to antibodies	Amyloid ß 40 and 42	10 pg/ml	Real time	Chiu et al. (2011)
		Freely floating SA-functionalized MNPs with biotinylated monoclonal antibodies	PSA	0.7 nM or 100 ng/ml in a 2 µl sample volume	Real time	Oisjöen et al. (2010)
		Antibody-coated magnetic particles	*L. monocytogenes* bacteria	$(5.6\pm1.1)\times10^6$ bacterial cells in sample volume of 20 µl	<0.1 s	Grossman et al. (2004)

Whole cell	Microbial cell-based biosensor	Bacterium *Sphingomonas* sp. on the inner epidermis of onion bulb scale	p-Nitrophenol (PNP)	4–80 μM of methyl parathion	>5 min	Michelini and Roda (2012)
	Cell-based biosensor	TurboRFP-based DNA damage reporter cell line in NIH-3T3 cells that fluoresce in response to stress	Genotoxic stress caused by a wide variety of agents, from chemical genotoxic agents to UV-C radiation	>80%	After 12 h of exposure to UV-C	Fendyur et al. (2015)
Microfluidics	Paper-based chemiluminescence biosensor	Paper-based, μPCAD, glucose oxidase (GOx), urate oxidase (UOx) enzymes	Glucose and uric acid	0.14 mmol/l for glucose and 0.52 mmol/l for uric acid	40–50 s	Yu et al. (2011)
	Electrophoretic microfluidic immunoassay with magnetic-beads detection	Biotinylated magnetic beads combined with anti-SA-IgG antibody array	SA	20 aM or 2×10^{-17} M	Within 2–3 min	Morozov et al. (2007)
Other types	Nanopore-based detection	Bionanopore phi29 DNA-packaging motor	Colon cancer antigen EpCAM antibody	1 nM at single-molecule level	Real time	Wang et al. (2013)

10.2.3 Optical Biosensors

Optical biosensors yield outputs that can be confirmed by the naked eye, by changes in absorption, fluorescence peaks, or refractive index or by using other optical or spectroscopic techniques such as fluorescence, FTIR, and Raman spectroscopies. The widespread availability of optical detection equipment makes optical biosensor techniques attractive for the development of low-cost lab-on-a-chip devices for use in areas where rapid diagnosis of potential disease is essential and more sophisticated diagnostic methods are not readily available, such as during pathogen outbreaks in rural regions.

Colorimetric biosensors change color when exposed to their target analyte, and quantification is performed through changes in absorption at a specific wavelength. Despite their simplicity, these biosensors may have considerable sensitivity and selectivities: An absorption wavelength shift-based biosensor for vesicular stomatitis virus (VSV), pseudotyped Ebola (PT-Ebola), and vaccinia virus was able to reach a sensitivity figure of merit (FOM) of 40, a detection limit of below 10^5 PFU/ml virus and a resolution of 0.05 nm (Yanik et al., 2010). Likewise, HIV RNA molecules could be colorimetrically detected using PNAs bearing different numbers of cyclopentane chemical groups, with sufficient sensitivity to detect 20–30,000 copies/ml plasma of this virus (Zhao et al., 2014a).

Changes in chemical composition can also be detected through spectroscopic methods, such as Raman or FTIR spectroscopy. These methods quantify the absorbance, reflectance, or fluorescence of a material following exposure to light at a specific wavelength or range of wavelengths and yield chemical information in the form of molecular interactions, ionic and covalent bonds, and vibrational and rotational motions. They are useful for label-free biosensing efforts, as IR absorption and Raman scattering can be used to detect conformational changes of proteins and structural variations between materials and molecules, allowing the detailed analysis of chemical bonds without an intermediary reporter or dye. Alzheimer's disease, for example, could be detected by quantifying amyloid β (Aβ) peptide titers using an attenuated total reflection Fourier transform infrared (ATR-FTIR) spectroscopy biosensor (Kleiren et al., 2010).

Raman signals can be enhanced through a method called SERS, which utilizes the fact that the Raman signal becomes much more prominent (up to a factor of 10^{14}, although typical values are in the 10^6–10^8 range) when the analyte molecules are situated between gold and silver surfaces set apart by distances around 20–30 μm (Kneipp et al., 1997). The interstructural distances required for SERS are usually created using nanoparticles or nanopatterned surfaces, and gold and silver are often the materials of choice because their plasmon resonances (and, therefore, Raman enhancements) are in the near-infrared and visible spectral ranges, respectively. SERS-active substrates for protein detection may also be designed to display a tunable resonance in the infrared range in order to produce a signal enhancement effect around the spectral positions of amide bands; or the samples may be labeled by a Raman-active dye for easier detection (Han et al., 2009). A detection level of 7 fg/ml could be achieved in a cancer biomarker detection study from plasma using SERS with 3D hierarchical plasmonic nanoarchitectures, including Au nanospheres and Au stars, as well as with a Raman dye (malachite green isothiocyanate, MGITC) and silica nanoparticles (Li et al., 2013b).

Surface plasmons can be used for the enhancement of other spectroscopic detection methods, often using noble metal nanoparticles or metallic thin films. SPR reflectivity measurements in particular are useful for the detection of molecular interactions and may reach very high sensitivities. SPR-based sensing systems measure the coupling of light with the plasmons (electron cloud oscillations) present at the surface of a nanoscale thin film of gold or silver. Recognition agents for SPR sensing and imaging are generally antibodies. Picogram per milliliter-level sensitivities can be obtained using SPR; for example, testosterone detection limits in one SPR-based system was found to be 3.7 pg/ml in standard running buffer and 15.4 pg/ml in a human saliva matrix (Mitchell and Lowe, 2009). Biological assays, nonetheless, may prove superior in detection capacity: In another study, the detection capacity of SPR for retinol binding protein 4 (RBP4, a useful marker for type 2 diabetes) was found to be greater than ELISA but lower than Western blot (Lee et al., 2008).

A variant technique of SPR, called localized SPR (LSPR), uses metal nanoparticles (usually gold or silver) of specific sizes and geometries to further improve the detection limits of this technique (Swierczewska et al., 2012). LSPR nanostructures may be employed for biomolecule detection through the measurement of refractive index changes and colorimetric SPR imaging, and it has been shown in a theoretical study that the detection capacity of LSPR-based methods can exceed that of systems based on traditional Kretschmann geometry systems in terms of wavelength shift sensitivity outperforming the detection system with Kretschmann geometry (Kaya et al., 2014). LSPR-based detection of colon cancer was also achieved using Au nanorods, and the width- and length-dependent changes in the plasmonic and photonic properties of nanorod antennae were investigated to identify optimal geometries for nanorod arrays for use in cancer biosensors (Dodson et al., 2015).

10.2.4 Electrical and Electrochemical Biosensors

Like optical signals, electrical signals are generally easy to detect and quantify. Electrical and electrochemical biosensors consequently see common use in both research and commercial purposes. Electrical biosensors typically use an electrode as their transducer, and the recognition element (usually an enzyme) is immobilized onto this electrode. The enzyme–ligand interaction creates a change in the electrical properties of the electrode (this may be electrical potential, resistivity, impedance, conductivity, or capacitance, as well as the current running between the recognition element-bound electrode and a reference electrode), which is measured and quantified to determine the concentration of the analyte. Cyclic voltammetry, chronoamperometry, chronopotentiometry, impedance spectroscopy, and various FET-based methods are some common tools that are used in the measurement of the output signal. More sensitive measurements may be performed with the use of nanomaterials such as nanowires, nanotubes, and nanoparticles.

As one of the most popular biosensor types, amperometric biosensors quantify changes in current, which is usually linear with the change in analyte concentration within the detection range (Wang, 1999). Amperometric biosensors typically display relatively modest detection capabilities; for example, a biosensor based on a multiwalled carbon nanotube (CNT)–gold nanoparticle composite yielded a limit of detection of 0.01 mM for uric acid (Chauhan and Pundir, 2011). Likewise, a detection range of 0.001–5 mM was obtained for glucose detection using an early biosensor based on Prussian blue-functionalized electrodes (Wu et al., 2012). In addition to gold, nanoparticles of other metals, such as zinc and platinum, can also be used in biosensor design (Chawla and Pundir, 2012; Wu et al., 2009).

The transducer electrode can be modified to produce amperometric sensors with much higher sensitivities. Sotiropoulou and Chaniotakis have reported dichlorvos detection limits at the picomolar level using a biosensor with a nanoporous carbon electrode, which enhances enzyme–ligand interactions by adsorbing the enzyme on its surface (Sotiropoulou and Chaniotakis, 2005). Salimi et al. also reported picomolar-level detection capacity using a modified electrode; their system used a glassy carbon electrode modified by guanine and nickel oxide nanoparticles with the help of cyclic voltammetry (Salimi et al., 2008). Likewise, an electrical detection-based FET biosensor was used to detect prion proteins through the use of thiamine molecules immobilized on a glutaraldehyde-modified SiO_2 gate surface on p-Si (Wustoni et al., 2014). Another biosensor design, using an antibody-immobilized single-walled carbon nanotube (SWNT)

FET, was hypothesized to be able to detect the Lyme disease antigen at concentrations up to 1 ng/ml in buffer (Lerner et al., 2013).

Potentiometric biosensors measure changes in the electrical potential of the electrode and are less common than amperometric biosensors. Nonetheless, potentiometry-based biosensors using modified electrodes have been described for biomaterials such as urea and polyglycerides, with working ranges between 10^{-2} and 30 mM (Lakard et al., 2004; Saurina et al., 1998). Due to the relative rarity of this detection type, high-sensitivity studies involving potentiometric biosensors are lacking, although a combined SPR–potentiometric analysis method for nerve gas detection at nanomolar-to-picomolar detection ranges was reported (Taranekar et al., 2006). Biosensors that utilize changes in conductivity, capacitance, admittance, and impedance have also been recorded in the literature, although nonamperometric output types are somewhat uncommon (Berggren et al., 2001; Gerard et al., 2002; Varshney and Li, 2007).

10.2.5 Mechanical Biosensors

The binding interaction between the recognition element and the analyte imposes an additional weight on the surface on which the former is immobilized. This change can be quantified using specialized devices and measurement methods, such as microcantilevers, QCM, microring resonators (MRR), suspended microchannel resonators (SMR), or atomic force microscopy (AFM). Many of these methods have exceptionally high sensitivities; QCM biosensors, for example, can detect femtomolar concentrations of DNA. However, they also have long analysis times and require uncommon and highly sensitive equipment for analysis (Arlett et al., 2011). As such, these techniques are generally suitable for research purposes rather than point-of-care diagnostics.

Mechanical biosensors can broadly be divided into two types, those that experience a surface deflection following analyte binding (surface stress mechanical biosensors) and those that change their oscillation frequency in the presence of the analyte (dynamic-mode mechanical biosensors) (Arlett et al., 2011). Cantilever biosensors can be of both types and have been used for sensing cancer biomarkers by using the change in cantilever frequency or deflection following antigen binding (Choi et al., 2010). Microcantilever array biosensors can be used with protein–ligand interactions to detect viruses at sensitivity levels down to subpicomolar concentrations (Braun et al., 2009). QCM biosensors can likewise reach picomolar sensitivities and have been developed for the detection of biomaterials such as human antibodies and bacterial toxins (Alfonta et al., 2001; Das et al., 2003; Yao et al., 2010).

Other types of mechanical biosensors include nanoelectromechanical systems (NEMS), such as nanomechanical resonators, which may possess single-molecule detection capacities. NEMS devices may use mass, elastic modulus, and surface stress changes as sensing parameters and can possess complex device designs that render them ideal for lab-on-a-chip applications that require more sophisticated assay conditions than the detection of a recognition element–analyte binding interaction. In addition, while not generally employed for the analysis of biosensor outputs, AFM is an exceptionally sensitive material characterization and manipulation tool that has seen extensive use in medical proteomics and potential diagnosis of cancer and infectious disease (Archakov and Ivanov, 2007). AFM is not an ideal technique for commercial biosensors; however, its subatomic detection capacity is effective for mechanically characterizing the recognition element–analyte interactions that drive biosensor development, and the similarity of AFM probes to mechanical biosensor cantilevers allows this technique to yield potentially valuable information about the design of novel mechanical biosensors (Baselt et al., 1998).

10.2.6 Other Biosensor Types

In addition to optical, electrical, and mechanical detection methods, changes in the magnetic field of an analyte-bound material can be quantified for biological detection. One example in the literature involves quantification of magnetic responsivity with a high-density sensor array that exhibits giant magnetoresistance (GMR), and the method was shown reach very high antigen–antibody binding sensitivities. A solute sensitivity of 20 zeptomoles, which is difficult to reach even with SPR and mechanical detection methods, was obtained using this strategy (Gaster et al., 2011).

Acoustic biosensors are another uncommon type and can be regarded as a subtype of mechanical biosensors. In these biosensors, changes in the properties of an acoustic wave are used to gain information about the binding interaction, usually by measuring the mass increase that results when the recognition element binds to the analyte. QCM and other types of acoustic resonators are employed in biosensors of this type and such devices have been shown to detect bacteria at sensitivities up to 0.4 cells/μl (Ferreira et al., 2009; Rocha-Gaso et al., 2009).

Entire cells can also be used for biosensor applications, especially for the detection of pollutants and other chemicals. In these biosensors, the physiological response of the cell is used as an indicator of analyte presence, and the complex sensory systems of cells are employed in place of advanced device design. The cells used may be bacteria, yeasts, and fungi and are often genetically modified to yield an easily detectable signal in the presence of the analyte. These "one-cell biosensors" may be exposed to a water or soil sample to yield a bioluminescent or fluorescent response that is subsequently quantified in an assay-like process or the cells may be immobilized in an integrated system that truly qualifies as a biosensor (Belkin, 2003).

While analytes are typically presented in a solution, volatile organic compounds (VOCs) in gas form can also be quantified through mass spectroscopy methods for disease diagnosis. Gas chromatography–mass spectrometry (GC-MS), for example, was used to detect VOCs in human breath for the diagnosis of lung cancer (Dragonieri et al., 2009). Sensors for VOCs typically display sensitivities in the ppm range (Adiguzel and Kulah, 2014).

10.3 THE IMPACT OF NANOTECHNOLOGY AND NANOMATERIALS IN BIOSENSOR DESIGN

Although the analytes and recognition elements involved are often in the nanoscale, nanomaterials in the strict sense are not required for the design of biosensors. Nonetheless, the large total surface areas associated with nanomaterials, as well as the small amounts of analytes and buffers required for nanoscale device-based assays, render them attractive for biosensor applications. In addition, nanostructures may present ideal surface properties for analyte or recognition element immobilization or offer increased detection capacity through phenomena such as self-assembly or plasmon enhancement. Nanospheres, nanorods, nanowires, graphene- and CNT-based structures, quantum dots, magnetic nanoparticles, NEMS systems, and other nanoscale materials are therefore often used in biosensor design.

Nanospheres are the most common nanoscale biosensor components and may be functionalized with surface groups such as thiols to better bind to an enzyme, antibody, or other recognition molecules. In addition, they may be organized into nanostructure arrays to better present their enzyme or antibody load (Xu and Han, 2004). The total surface area presented by nanoparticles is larger relative to the bulk material of equal volume; as such, nanospheres and similar nanostructures can immobilize a greater amount of recognition molecule per volume compared to their macroscale counterparts (Li et al., 2008). In addition, the size, composition, or material properties of the nanospheres might allow enhanced detection capability, as is the case with magnetic nanoparticles: Cross-linked magnetic nanoparticles (CLIOs), manganese-doped magnetic nanoparticles, and core–shell structures such as elemental iron-ferrite nanoparticles are among the magnetic materials used in biosensor applications (Haun et al., 2010). Quantum dots can also be used to improve the detection capacity of conventional recognition elements, and QD sensors that react to a broad range of factors, such as pH change, protein or nucleic acid cleavage, or DNA synthesis, have been reported in the literature (Suzuki et al., 2008). A quantum dot–aptamer conjugate was also shown to possess a detection limit in the attomolar scale, while conventional aptamer-based biosensors generally display nanomolar-level sensitivity (Hansen et al., 2006).

While nanospheres are commonly used for their ease of production, nanorods and other nonspherical nanoparticles may present advantages over the nanosphere morphology. Thiol groups, for example, prefer to bind to the tips of gold nanorods, and this effect can be used to produce self-assembled nanoparticle chains by attaching thiolated oligonucleotides to gold nanorods and allowing complementary sequences to link individual nanoparticles (Sepulveda et al., 2009). Nonspherical morphologies also alter the optical response of nanoparticles; both theoretical and experimental reports suggest that triangular silver nanoparticles enhance sensitivity in SPR-based biosensors (Haes and Van Duyne, 2002; Peng and Miller, 2011; Xu and Kall, 2002). Another study on a ZnO-based biosensor found that biotin-bound ZnO nanorods could detect streptavidin at concentrations lower than that previously reported for ZnO nanospheres (Kim et al., 2006).

Nanowires of silicon and noble metals can also be incorporated into sensor design. An Au nanowire waveguide, for example, was successfully used for plasmonic waveguide sensing in water and other liquids (Wang et al., 2014). Nanowires can also be organized into larger assemblies using techniques such as flow-assisted, Langmuir–Blodgett, bubble-blown, electric-field, smearing-transfer, roll-printing, and PDMS-transfer assembly processes (Chen et al., 2011). In addition to metal or metal oxide nanowires, single- or multiwalled CNTs can also be used for biosensor applications. In these biosensors, the carbon structure is used to modify the electrode of an amperometric biosensor and may be functionalized to carry the recognition molecule of interest. Different configurations can be used for CNT-based biosensors: CNTs may be coated onto the electrode or synthesized in aligned networks around it or the CNT itself may serve as the electrode. Submicromolar detection limits have been obtained in CNT-based biosensors for glucose monitoring (Wang, 2005).

Nanoporous silica and similar materials have also been used in the design of biosensors. Light reflected off a thin, porous layer of silicon creates interference patterns (Fabry–Perot fringes) that change following the binding of the target analyte to a recognition element immobilized on the surface of the silicon film. The extent of this change depends on the change in the refractive index of the silicon layer and can be quantified to detect DNA, proteins, and other organic molecules at picomolar-to-femtomolar scales (Lin et al., 1997). Molecular imprinting can also be employed to create polymer matrices with "holes" that serve as recognition sites for specific biomolecules and may reach detection levels of 0.1 μM for small organic molecules (Yano and Karube, 1999). Lipid membranes can also be employed as biosensors, either serving as a platform to support an enzyme or antibody or acting as the recognition element itself; however, these sensors are more suited to membrane transport-related research than clinical diagnosis (Nikolelis and Krull, 1992; Reimhult and Kumar, 2008). Pore-bearing proteins that mimic the structure and selectivity of biological pore components can also be used in biosensor applications (Braha et al., 1997).

In addition to nanoparticle-, thin film-, and nanoporous matrix-based detection methods, the techniques used in biosensor design and quantification are often deeply rooted in nanotechnology. As such, advances in nanomaterial-based detection methods, such as QCM (the sensitivity of which depends on crystal thickness) and SERS, or nanofabrication methods, such as laser, AFM, or electron beam lithographies (which are used for the fabrication of cantilevers used in mechanical biosensors), will indirectly improve the sensitivity of current biosensors. However, while nanostructures offer several advantages over conventional biosensors, it must be kept in mind the *in vivo* use of nanoparticles in disease diagnosis is largely still in preclinical stages of development (Thakor and Gambhir, 2013). Further advances are no doubt necessary for the use of these devices in clinical settings.

10.4 EARLY DIAGNOSIS AND BIOSENSOR-BASED DISEASE DETECTION

The greatest potential of biosensors lies in their medical applications. While the detection of a broad spectrum of biomaterials is feasible using biosensors, the sheer incidence rates, physiological diversity, and difficulty of treatment that is characteristic of cancer, heart disease, and similar disorders ensure that a substantial amount of biosensor-related research is directed toward their diagnosis. Likewise, the detection of widespread foodborne pathogens such as *Salmonella*, *Klebsiella*, and *Staphylococcus* or viruses such as HIV and dengue virus is vital for the control of the associated diseases. As such, a list of diseases and target molecules that have been the subject of high-sensitivity biosensor development efforts is provided in the present section (Table 10.2).

TABLE 10.2 Target Molecules, Detection Methods, and Sensitivities of Biosensor-based Diagnostic Methods

Disease		Markers	Method	Sensitivity	Reference
Cancer	Prostate cancer	Prostate-specific antigen (PSA)	MEMS/NEMS Silicon nitride cantilever Optical deflection	0.2 ng/ml	Waggoner and Craighead (2007)
			Digital ELISA	14 fg/ml (0.4 fM).	Rissin et al. (2010)
	Hepatocellular carcinoma	Alpha-fetoprotein (AFP)	Resonant microcantilevers	Nanogram/ml	Liu et al. (2009)
			Immunosensor with single-walled CNT inside mesoporous silica	0.06 ng/ml	Lin et al. (2013)
	Ovarian	Human epididymis protein 4 (HE4)	Anti-HE4 antibody covalently attached to silver nanoparticles (synthesized through NSL technology) on glass substrate	4 pM	Yuan et al. (2012)
		Cancer antigen 125 (CA125) (most frequently tested)	Anti-CA125 was immobilized on gold surface through a self-assembled monolayer SPR biosensor and capacitive label-free immunosensor, separately	0.05 U/ml with a capacitive biosensor system, 0.1 U/ml with SPR	Suwansa-ard et al. (2009)
	Breast	PIK3CA mutations	FRET-based visual detection and fluorescence measurement	fM, 2% mutant DNAs from total DNA	Song et al. (2012a)
	Lung	Urine, CD59 glycoprotein, transthyretin (TTR), GM2 activator protein (GM2AP), and Ig-free light chain			Arya and Bhansali (2011)
	Nonsmall cell	Fragile histidine triad (FHIT), mucin1 (MUC1), β-catenin (CTNNB1)			Altintas and Tothill (2013)

	Mutations in the EGFR gene	Circulating tumor cell detection using a microfluidic device	Median of 74 cells per milliliter (mean, 133; range, 5–771)	Collura et al. (2008)
Cervical cancer		Electrochemical impedance spectroscopy, impedimetric biosensor based on PEGylated arginine-functionalized magnetic nanoparticles	10 cells/ml	Chandra et al. (2011)
	Squamous cell carcinoma antigen (SCCa)	LSPR biosensor	0.125 pM	Zhao et al. (2014b)
Myocardial infarction	Troponin I, T			Steuer et al. (2009)
	Fatty acid binding protein			Kakoti and Goswami (2013)
Rheumatoid arthritis (RA)	Rheumatoid factor (RF), anticyclic citrullinated peptide antibody			Chandrashekara (2014)
	Activated CD8 cells			
	IL-6, IL-7, IL-10			
	Antimutated citrullinated vimentin (anti-MCV) antibody			
Systemic lupus erythematosus (SLE)	Autoantibodies against C1q	ELISA		Andrejevic et al. (2013), Vanhecke et al. (2012)
Osteoarthritis	C-telopeptide fragments of type II collagen (CTX-II)	Fluorescence-based biosensor	200 ng/mmol	Kim et al. (2013)
	Cartilage oligomeric matrix protein (COMP)	ELISA with fluoro-microbeads	0.8 ng/ml	Song et al. (2012b)

Cancer markers are some of the most common targets for biosensor design, as early detection is vital for successful treatment in many cancers. Early diagnosis of oral cancer, for example, increases survival rates from 50 to 80%, while pancreatic cancer, which is often diagnosed after the initial tumor has either spread locally or metastasized into another tissue, has a 5-year survival rate of about 5% (Pannala et al., 2009; Silverman, 1988). Diagnosis of cervical cancer, likewise, must be performed at an early stage or "the cancer metastasizes to the rest of the uterus, bladder, rectum and abdominal wall and eventually reaches pelvic lymph nodes, thereby invading other organs and leading to death" (Chandra et al., 2011). As such, biosensors using protein markers, DNA sequences, membrane glycans, and cancer-associated estrogen derivatives have been developed for the early detection of cancer, and the responses of cancer cells, such as drug resistance after taxane treatment in breast cancer cells, hydrogen peroxide production in human hepatoma cells, or formaldehyde presence in glioblastoma cells following treatment with a formaldehyde-releasing drug, have been observed using various biosensor designs (Bareket et al., 2010; Braunhut et al., 2005; Cavalieri et al., 2006; Feng et al., 2006; Myung et al., 2011; Rui et al., 2010; Zhang et al., 2010).

Advancements in nanoscience have also created the prospect of developing small biosensors that are implanted to the body of a prospective patient and trigger at the onset of disease state or other changes in physiological conditions. This type of *in vivo* diagnostic method is useful for the detection of diseases that are nonsymptomatic in their early stages and for the monitoring of chronic conditions, such as diabetes, that require regular treatment. In addition to their direct healthcare benefits, these biosensors may also serve as a means of collecting physiological data from a large set of patients, which would in turn enable the development of better treatment options. These sensors typically detect glucose and are designed for the management of diabetes; although biosensors that use fluorescein (introduced by the biosensor) or lactate for the detection of internal bleeding have also been developed (Kotanen et al., 2012; Mo and Smart, 2004; Ryou et al., 2011). Some glucose sensors of this type are approved for use by the FDA (Klonoff, 2007).

10.5 CONCLUSION AND FUTURE DIRECTIONS

Biosensor research combines the investigation of biochemical recognition processes, signal transduction systems, and output-specific detection methods, which makes it a highly multidisciplinary research field. As such, the development of highly specific recognition elements and advancements in various signal detection techniques will indirectly result in the development of more accurate biosensors. Recent discoveries in gene regulation and biological signaling mechanisms have revealed the presence of various regulatory elements that are highly specific to their targets and may be used in biosensor applications. In addition, methods for the synthesis of proteins and nucleic acids incorporating nonstandard chemical groups have advanced greatly in the recent years; as such, combination systems incorporating signal detecting, amplifying, and transducing elements can now be produced with considerable ease.

As resolution limits are bypassed and more selective biomolecular recognition agents are discovered, the detection efficiency of future biosensors will no doubt continue to increase. Furthermore, advances in nanotechnology and material fabrication methods may both allow novel detection techniques that employ nanoscale phenomena to be used in biosensor design and decrease costs by lowering biosensor dimensions and the quantities of biological materials incorporated into the sensor structure. Consequently, it would appear that the multifaceted nature of biosensor design, as well as the advantages it offers in terms of assay sensitivity, selectivity, and detection time and costs, will continue attracting commercial and research efforts and may allow the early detection of diseases that currently are undetectable during periods for which their treatment would have been the most effective.

REFERENCES

Adiguzel, Y. and Kulah, H. (2014). Breath sensors for lung cancer diagnosis. Biosens. Bioelectron. *65*, 121–138.

Alfonta, L., Willner, I., Throckmorton, D., and Singh, A. (2001). Electrochemical and quartz crystal microbalance detection of the cholera toxin employing horseradish peroxidase and GM1-functionalized liposomes. Anal. Chem. *73*, 5287–5295.

Altintas, Z. and Tothill, I. (2013). Biomarkers and biosensors for the early diagnosis of lung cancer. Sens. Actuators B Chem. *188*, 988–998.

Andrejevic, S., Jeremic, I., Sefik-Bukilica, M., Nikolic, M., Stojimirovic, B., and Bonaci-Nikolic, B. (2013). Immunoserological parameters in SLE: High-avidity anti-dsDNA detected by ELISA are the most closely associated with the disease activity. Clin. Rheumatol. *32*, 1619–1626.

Archakov, A.I. and Ivanov, Y.D. (2007). Analytical nanobiotechnology for medicine diagnostics. Mol. Biosyst. *3*, 336–342.

Arlett, J., Myers, E., and Roukes, M. (2011). Comparative advantages of mechanical biosensors. Nat. Nanotechnol. *6*, 203–215.

Arya, S.K. and Bhansali, S. (2011). Lung cancer and its early detection using biomarker-based biosensors. Chem. Rev. *111*, 6783–6809.

Bareket, L., Rephaeli, A., Berkovitch, G., Nudelman, A., and Rishpon, J. (2010). Carbon nanotubes based electrochemical biosensor for detection of formaldehyde released from a cancer cell line treated with formaldehyde-releasing anticancer prodrugs. Bioelectrochemistry *77*, 94–99.

Baselt, D., Lee, G., Natesan, M., Metzger, S., Sheehan, P., and Colton, R. (1998). A biosensor based on magnetoresistance technology. Biosens. Bioelectron. *13*, 731–739.

Belkin, S. (2003). Microbial whole-cell sensing systems of environmental pollutants. Curr. Opin. Microbiol. *6*, 206–212.

Berggren, C., Bjarnason, B., and Johansson, G. (2001). Capacitive biosensors. Electroanalysis *13*, 173–180.

Bertoncello, P. and Forster, R.J. (2009). Nanostructured materials for electrochemiluminescence (ECL)-based detection methods: Recent advances and future perspectives. Biosens. Bioelectron. *24*, 3191–3200.

Bhalla, V., Carrara, S., Sharma, P., Nangia, Y., and Raman Suri, C. (2012). Gold nanoparticles mediated label-free capacitance detection of cardiac troponin I. Sens. Actuators B Chem. *161*, 761–768.

Braha, O., Walker, B., Cheley, S., Kasianowicz, J., Song, L., Gouaux, J., and Bayley, H. (1997). Designed protein pores as components for biosensors. Chem. Biol. *4*, 497–505.

Braun, T., Ghatkesar, M.K., Backmann, N., Grange, W., Boulanger, P., Letellier, L., Lang, H.-P., Bietsch, A., Gerber, C., and Hegner, M. (2009). Quantitative time-resolved measurement of membrane protein-ligand interactions using microcantilever array sensors. Nat. Nanotechnol. *4*, 179–185.

Braunhut, S., McIntosh, D., Vorotnikova, E., Zhou, T., and Marx, K. (2005). Detection of apoptosis and drug resistance of human breast cancer cells to taxane treatments using quartz crystal microbalance biosensor technology. Assay Drug Dev. Technol. *3*, 77–88.

Bryan, T., Luo, X., Forsgren, L., Morozova-Roche, L.A., and Davis, J.J. (2012). The robust electrochemical detection of a Parkinson's disease marker in whole blood sera. Chem. Sci. *3*, 3468.

Cavalieri, E., Chakravarti, D., Guttenplan, J., Hart, E., Ingle, J., Jankowiak, R., Muti, P., Rogan, E., Russo, J., Santen, R., et al. (2006). Catechol estrogen quinones as initiators of breast and other human cancers: Implications for biomarkers of susceptibility and cancer prevention. Biochim. Biophys. Acta *1766*, 63–78.

Chandra, S., Barola, N., and Bahadur, D. (2011). Impedimetric biosensor for early detection of cervical cancer. Chem. Commun. (Camb.) *47*, 11258–11260.

Chandrashekara, S. (2014). Current studies of biomarkers for the early diagnosis of rheumatoid arthritis. Curr. Biomark. Find. *4*, 107–119.

Chang, K., Pi, Y., Lu, W., Wang, F., Pan, F., Li, F., Jia, S., Shi, J., Deng, S., and Chen, M. (2014). Label-free and high-sensitive detection of human breast cancer cells by aptamer-based leaky surface acoustic wave biosensor array. Biosens. Bioelectron. *60*, 318–324.

Chauhan, N. and Pundir, C.S. (2011). An amperometric uric acid biosensor based on multiwalled carbon nanotube-gold nanoparticle composite. Anal. Biochem. *413*, 97–103.

Chawla, S. and Pundir, C.S. (2011). An electrochemical biosensor for fructosyl valine for glycosylated hemoglobin detection based on core-shell magnetic bionanoparticles modified gold electrode. Biosens. Bioelectron. *26*, 3438–3443.

Chawla, S. and Pundir, C.S. (2012). An amperometric hemoglobin A1c biosensor based on immobilization of fructosyl amino acid oxidase onto zinc oxide nanoparticles-polypyrrole film. Anal. Biochem. *430*, 156–162.

Chen, K.-I., Li, B.-R., and Chen, Y.-T. (2011). Silicon nanowire field-effect transistor-based biosensors for biomedical diagnosis and cellular recording investigation. Nano Today *6*, 131–154.

Chiu, M.J., Horng, H.E., Chieh, J.J., Liao, S.H., Chen, C.H., Shih, B.Y., Yang, C.C., Lee, C.L., Chen, T.F., Yang, S.Y., et al. (2011). Multi-channel SQUID-based ultra-high-sensitivity in-vitro detections for bio-markers of Alzheimer's disease via immunomagnetic reduction. IEEE Trans. Appl. Supercond. *21*, 477–480.

Cho, I.-H., Paek, E.-H., Kim, Y.-K., Kim, J.-H., and Paek, S.-H. (2009). Chemiluminometric enzyme-linked immunosorbent assays (ELISA)-on-a-chip biosensor based on cross-flow chromatography. Anal. Chim. Acta *632*, 247–255.

Choi, Y.-E., Kwak, J.-W., and Park, J.W. (2010). Nanotechnology for early cancer detection. Sensors (Basel) *10*, 428–455.

Collura, C.V., Inserra, E., Diederichs, S., Ph, D., Iafrate, A.J., Bell, D.W., Digumarthy, S., Muzikansky, A., Irimia, D., Settleman, J., et al. (2008). Detection of mutations in EGFR in circulating lung-cancer cells. N. Engl. J. Med. *359*, 366–377.

Das, K., Penelle, J., and Rotello, V. (2003). Selective picomolar detection of hexachlorobenzene in water using a quartz crystal microbalance coated with a molecularly imprinted polymer thin film. Langmuir *19*, 3921–3925.

De la Rica, R. and Stevens, M.M. (2012). Plasmonic ELISA for the ultrasensitive detection of disease biomarkers with the naked eye. Nat. Nanotechnol. *7*, 821–824.

Dell'Atti, D., Tombelli, S., Minunni, M., and Mascini, M. (2006). Detection of clinically relevant point mutations by a novel piezoelectric biosensor. Biosens. Bioelectron. *21*, 1876–1879.

Di Natale, C., Mantini, A., Macagnano, A., Antuzzi, D., Paolesse, R., and D'Amico, A. (1999). Electronic nose analysis of urine samples containing blood. Physiol. Meas. *20*, 377–384.

Dodson, S.L., Cao, C., Zaribafzadeh, H., Li, S., and Xiong, Q. (2015). Engineering plasmonic nanorod arrays for colon cancer marker detection. Biosens. Bioelectron. *63*, 472–477.

Dragonieri, S., Annema, J.T., Schot, R., van der Schee, M.P.C., Spanevello, A., Carratú, P., Resta, O., Rabe, K.F., and Sterk, P.J. (2009). An electronic nose in the discrimination of patients with non-small cell lung cancer and COPD. Lung Cancer *64*, 166–170.

Du, Y., Li, B., and Wang, E. (2013). "Fitting" makes "sensing" simple: Label-free detection strategies based on nucleic acid aptamers. Acc. Chem. Res. *46*, 203–213.

Duk Han, Y., Jin Chun, H., and Yoon, H.C. (2014). The transformation of common office supplies into a low-cost optical biosensing platform. Biosens. Bioelectron. *59*, 259–268.

El-Sayed, I., Huang, X., and El-Sayed, M. (2005). Surface plasmon resonance scattering and absorption of anti-EGFR antibody conjugated gold nanoparticles in cancer diagnostics: Applications in oral cancer. Nano Lett. *5*, 829–834.

Fendyur, A., Varma, S., Lo, C.T., and Voldman, J. (2015). Cell-based biosensor to report DNA damage in micro- and nanosystems. Anal. Chem. *86*, 7598–7605.

Feng, K., Yang, Y., Wang, Z., Jiang, J., Shen, G., and Yu, R. (2006). A nanoporous CeO_2/chitosan composite film as the immobilization matrix for colorectal cancer DNA sequence-selective electrochemical biosensor. Talanta *70*, 561–565.

Ferreira, G., Da-Silva, A., and Tome, B. (2009). Acoustic wave biosensors: Physical models and biological applications of quartz crystal microbalance. Trends Biotechnol. *27*, 689–697.

Gaster, R.S., Xu, L., Han, S.-J., Wilson, R.J., Hall, D.A., Osterfeld, S.J., Yu, H., and Wang, S.X. (2011). Quantification of protein interactions and solution transport using high-density GMR sensor arrays. Nat. Nanotechnol. *6*, 314–320.

Gerard, M., Chaubey, A., and Malhotra, B. (2002). Application of conducting polymers to biosensors. Biosens. Bioelectron. *17*, 345–359.

Grossman, H.L., Myers, W.R., Vreeland, V.J., Bruehl, R., Alper, M.D., Bertozzi, C.R., and Clarke, J. (2004). Detection of bacteria in suspension by using a superconducting quantum interference device. Proc. Natl. Acad. Sci. U.S.A. *101*, 1–6.

Haes, A. and Van Duyne, R. (2002). A nanoscale optical blosensor: Sensitivity and selectivity of an approach based on the localized surface plasmon resonance spectroscopy of triangular silver nanoparticles. J. Am. Chem. Soc. *124*, 10596–10604.

Han, X.X., Zhao, B., and Ozaki, Y. (2009). Surface-enhanced Raman scattering for protein detection. Anal. Bioanal. Chem. *394*, 1719–1727.

Hansen, J., Wang, J., Kawde, A., Xiang, Y., Gothelf, K., and Collins, G. (2006). Quantum-dot/aptamer-based ultrasensitive multi-analyte electrochemical biosensor. J. Am. Chem. Soc. *128*, 2228–2229.

Hanson, C. and Thaler, E. (2005). Electronic nose prediction of a clinical pneumonia score: Biosensors and microbes. Anesthesiology *102*, 63–68.

Haun, J., Yoon, T., Lee, H., and Weissleder, R. (2010). Magnetic nanoparticle biosensors. Wiley Interdiscip. Rev. Nanomed. Nanobiotechnol. *2*, 291–304.

Hong, S.-R., Jeong, H.-D., and Hong, S. (2010). QCM DNA biosensor for the diagnosis of a fish pathogenic virus VHSV. Talanta *82*, 899–903.

Ierardi, V., Ferrera, F., Millo, E., Damonte, G., Filaci, G., and Valbusa, U. (2013). Bioactive surfaces for antibody-antigen complex detection by atomic force microscopy. J. Phys. Conf. Ser. *439*, 012001.

Iles, R.K. and Kallichurn, H. (2012). What will be the future development of electrochemical biosensors for the detection and quantification of biomarkers? J Bioeng. Biomed. Sci. *2*, e110.

Iliuk, A.B., Hu, L., and Tao, W.A. (2011). Aptamer in bioanalytical applications. Anal. Chem. *83*, 4440–4452.

Joo, J., Kwon, D., Yim, C., and Jeon, S. (2012). Highly sensitive diagnostic assay for the detection of protein biomarkers using microresonators and multifunctional nanoparticles. ACS Nano *6*, 4375–4381.

Jung, J., Cheon, D., Liu, F., Lee, K., and Seo, T. (2010). A graphene oxide based immuno-biosensor for pathogen detection. Angew. Chem. Int. Ed. Engl. *49*, 5708–5711.

Jung, H.-W., Chang, Y.W., Lee, G., Cho, S., Kang, M.-J., and Pyun, J.-C. (2014). A capacitive biosensor based on an interdigitated electrode with nanoislands. Anal. Chim. Acta *844*, 27–34.

Kakoti, A. and Goswami, P. (2013). Heart type fatty acid binding protein: Structure, function and biosensing applications for early detection of myocardial infarction. Biosens. Bioelectron. *43*, 400–411.

Kaya, Y., Ayas, S., Topal, A.E., Guner, H., and Dana, A. (2014). Sensitivity comparison of localized plasmon resonance structures and prism coupler. Sens. Actuators B Chem. *191*, 516–521.

Kim, J., Park, W.I., Lee, C., and Yi, G. (2006). ZnO nanorod biosensor for highly sensitive detection of specific protein binding. J. Kor. Phys. Soc. *49*, 1635–1639.

Kim, S.J., Park, Y.M., Min, B.-H., Lee, D.-S., and Yoon, H.C. (2013). Optical immunosensor for quantifying C-telopeptide fragments of type II collagen as an osteoarthritis biomarker in urine. BioChip J. *7*, 399–407.

Kleiren, E., Ruysschaert, J., Goormaghtigh, E., and Raussens, V. (2010). Development of a quantitative and for Alzheimer' s disease : The effect of deuteration on the detection of the A β peptide. Spectroscopy *24*, 61–66.

Klonoff, D.C. (2007). The benefits of implanted glucose sensors. J. Diabetes Sci. Technol. *1*, 797–800.

Kneipp, K., Wang, Y., Kneipp, H., Perelman, L., Itzkan, I., Dasari, R., and Feld, M. (1997). Single molecule detection using surface-enhanced Raman scattering (SERS). Phys. Rev. Lett. *78*, 1667–1670.

Kotanen, C., Moussy, F., Carrara, S., and Guiseppi-Elie, A. (2012). Implantable enzyme amperometric biosensors. Biosens. Bioelectron. *35*, 14–26.

Kriz, D., Ramstrom, O., and Mosbach, K. (1997). Molecular imprinting—New possibilities for sensor technology. Anal. Chem. *69*, A345–A349.

Lakard, B., Herlem, G., Lakard, S., Antoniou, A., and Fahys, B. (2004). Urea potentiometric biosensor based on modified electrodes with urease immobilized on polyethylenimine films. Biosens. Bioelectron. *19*, 1641–1647.

Lee, S.J., Youn, B.-S., Park, J.W., Niazi, J.H., Kim, Y.S., and Gu, M.B. (2008). ssDNA aptamer-based surface plasmon resonance biosensor for the detection of retinol binding protein 4 for the early diagnosis of type 2 diabetes. Anal. Chem. *80*, 2867–2873.

Lerner, M.B., Dailey, J., Goldsmith, B.R., Brisson, D., and Johnson, A.T.C. (2013). Detecting Lyme disease using antibody-functionalized single-walled carbon nanotube transistors. Biosens. Bioelectron. *45*, 163–167.

Li, C., Liu, Y., Li, L., Du, Z., Xu, S., Zhang, M., Yin, X., and Wang, T. (2008). A novel amperometric biosensor based on NiO hollow nanospheres for biosensing glucose. Talanta *77*, 455–459.

Li, Z., Wang, Y., Wang, J., Tang, Z., Pounds, J.G., and Lin, Y. (2010). Rapid and sensitive detection of protein biomarker using a portable fluorescence biosensor based on quantum dots and a lateral flow test strip. Anal. Chem. *82*, 7008–7014.

Li, F., Yu, Y., Cui, H., Yang, D., and Bian, Z. (2013a). Label-free electrochemiluminescence immunosensor for cardiac troponin I using luminol functionalized gold nanoparticles as a sensing platform. Analyst *138*, 1844–1850.

Li, M., Cushing, S.K., Zhang, J., Suri, S., Evans, R., Petros, W.P., Gibson, L.F., Ma, D., Liu, Y., and Wu, N. (2013b). Three-dimensional hierarchical plasmonic nano-architecture enhanced surface-enhanced Raman scattering immunosensor for cancer biomarker detection in blood plasma. ACS Nano *7*, 4967–4976.

Lin, V., Motesharei, K., Dancil, K., Sailor, M., and Ghadiri, M. (1997). A porous silicon-based optical interferometric biosensor. Science *278*, 840–843.

Lin, J., Wei, Z., Zhang, H., and Shao, M. (2013). Sensitive immunosensor for the label-free determination of tumor marker based on carbon nanotubes/

mesoporous silica and graphene modified electrode. Biosens. Bioelectron. *41*, 342–347.

Liu, Y., Li, X., Zhang, Z., Zuo, G., Cheng, Z., and Yu, H. (2009). Nanogram per milliliter-level immunologic detection of alpha-fetoprotein with integrated rotating-resonance microcantilevers for early-stage diagnosis of hepatocellular carcinoma. Biomed. Microdevices *11*, 183–191.

Logrieco, A., Arrigan, D., Brengel-Pesce, K., Siciliano, P., and Tothill, I. (2005). DNA arrays, electronic noses and tongues, biosensors and receptors for rapid detection of toxigenic fungi and mycotoxins: A review. Food Addit. Contam. *22*, 335–344.

Luo, X. and Davis, J.J. (2013). Electrical biosensors and the label free detection of protein disease biomarkers. Chem. Soc. Rev. *42*, 5944–5962.

Martín, M., Salazar, P., Villalonga, R., Campuzano, S., Pingarrón, J.M., and González-Mora, J.L. (2014). Preparation of core–shell Fe_3O_4@poly(dopamine) magnetic nanoparticles for biosensor construction. J. Mater. Chem. B *2*, 739.

Medley, C.D., Smith, J.E., Tang, Z., Wu, Y., Bamrungsap, S., and Tan, W. (2008). Gold nanoparticle-based colorimetric assay for the direct detection of cancerous cells. Anal. Chem. *80*, 1067–1072.

Michelini, E. and Roda, A. (2012). Staying alive: New perspectives on cell immobilization for biosensing purposes. Anal. Bioanal. Chem. *402*, 1785–1797.

Mitchell, J.S. and Lowe, T.E. (2009). Ultrasensitive detection of testosterone using conjugate linker technology in a nanoparticle-enhanced surface plasmon resonance biosensor. Biosens. Bioelectron. *24*, 2177–2183.

Mo, J. and Smart, W. (2004). Lactate biosensors for continuous monitoring. Front. Biosci. *9*, 3384–3391.

Morozov, V.N., Groves, S., Turell, M.J., and Bailey, C. (2007). Three minutes-long electrophoretically assisted zeptomolar microfluidic immunoassay with magnetic-beads detection. J. Am. Chem. Soc. *129*, 12628–12629.

Mulchandani, A., Pan, S., and Chen, W. (1999). Fiber-optic enzyme biosensor for direct determination of organophosphate nerve agents. Biotechnol. Prog. *15*, 130–134.

Myung, S., Solanki, A., Kim, C., Park, J., Kim, K., and Lee, K. (2011). Graphene-encapsulated nanoparticle-based biosensor for the selective detection of cancer biomarkers. Adv. Mater. *23*, 2221–2225.

Nikolelis, D. and Krull, U. (1992). Establishment and control of artificial ion-conductive zones for lipid-membrane biosensor development. Anal. Chim. Acta *257*, 239–245.

Oisjöen, F., Schneiderman, J.F., Astalan, A.P., Kalabukhov, A., Johansson, C., and Winkler, D. (2010). A new approach for bioassays based on frequency- and time-domain measurements of magnetic nanoparticles. Biosens. Bioelectron. *25*, 1008–1013.

Pannala, R., Basu, A., Petersen, G., and Chari, S. (2009). New-onset diabetes: A potential clue to the early diagnosis of pancreatic cancer. Lancet Oncol. *10*, 88–95.

Park, J.-S., Cho, M.K., Lee, E.J., Ahn, K.-Y., Lee, K.E., Jung, J.H., Cho, Y., Han, S.-S., Kim, Y.K., and Lee, J. (2009). A highly sensitive and selective diagnostic assay based on virus nanoparticles. Nat. Nanotechnol. *4*, 259–264.

Park, Y.M., Kim, S.J., Lee, K.J., Yang, S.S., Min, B.-H., and Yoon, H.C. (2014). Detection of CTX-II in serum and urine to diagnose osteoarthritis by using a fluoro-microbeads guiding chip. Biosens. Bioelectron. *67*, 192–199.

Pathirana, S., Barbaree, J., Chin, B., Hartell, M., Neely, W., and Vodyanoy, V. (2000). Rapid and sensitive biosensor for Salmonella. Biosens. Bioelectron. *15*, 135–141.

Peeling, R. and Mabey, D. (2010). Point-of-care tests for diagnosing infections in the developing world. Clin. Microbiol. Infect. *16*, 1062–1069.

Peng, H. and Miller, B. (2011). Recent advancements in optical DNA biosensors: Exploiting the plasmonic effects of metal nanoparticles. Analyst *136*, 436–447.

Ping, W., Yi, T., Xie, H., and Shen, F. (1997). A novel method for diabetes diagnosis based on electronic nose. Biosens. Bioelectron. *12*, 1031–1036.

Rapp, B.E., Gruhl, F.J., and Länge, K. (2010). Biosensors with label-free detection designed for diagnostic applications. Anal. Bioanal. Chem. *398*, 2403–2412.

Rashid, J.I.A., Yusof, N.A., Abdullah, J., Hashim, U., and Hajian, R. (2014). The utilization of SiNWs/AuNPs-modified indium tin oxide (ITO) in fabrication of electrochemical DNA sensor. Mater. Sci. Eng. C *45*, 270–276.

Ray, A. and Norden, B. (2000). Peptide nucleic acid (PNA): Its medical and biotechnical applications and promise for the future. FASEB J. *14*, 1041–1060.

Reimhult, E. and Kumar, K. (2008). Membrane biosensor platforms using nano- and microporous supports. Trends Biotechnol. *26*, 82–89.

Rissin, D.M., Kan, C.W., Campbell, T.G., Howes, S.C., Fournier, D.R., Song, L., Piech, T., Patel, P.P., Chang, L., Rivnak, A.J., et al. (2010). Single-molecule enzyme-linked immunosorbent assay detects serum proteins at subfemtomolar concentrations. Nat. Biotechnol. *28*, 595–599.

Rocha-Gaso, M., March-Iborra, C., Montoya-Baides, A., and Arnau-Vives, A. (2009). Surface generated acoustic wave biosensors for the detection of pathogens: A review. Sensors *9*, 5740–5769.

Rui, Q., Komori, K., Tian, Y., Liu, H., Luo, Y., and Sakai, Y. (2010). Electrochemical biosensor for the detection of H_2O_2 from living cancer cells based on ZnO nanosheets. Anal. Chim. Acta *70*, 57–62.

Ryou, M., Netniroski, A., Azagury, D., Shaikh, S., Ryan, M., Westervelt, R., and Thompson, C. (2011). An implantable wireless biosensor for the immediate detection of upper GI bleeding: A new fluorescein-based tool for diagnosis and surveillance. Gastrointest. Endosc. *74*, 189–194.

Salimi, A., Noorbakhash, A., Sharifi, E., and Semnani, A. (2008). Highly sensitive sensor for picomolar detection of insulin at physiological pH, using GC electrode modified with guanine and electrodeposited nickel oxide nanoparticles. Biosens. Bioelectron. *24*, 792–798.

Saravolatz, L., Manzor, O., VanderVelde, N., Pawlak, J., and Belian, B. (2003). Broad-range bacterial polymerase chain reaction for early detection of bacterial meningitis. Clin. Infect. Dis. *36*, 40–45.

Saurina, J., Hernandez-Cassou, S., Fabregas, E., and Alegret, S. (1998). Potentiometric biosensor for lysine analysis based on a chemically immobilized lysine oxidase membrane. Anal. Chim. Acta *371*, 49–56.

Sepulveda, B., Angelome, P., Lechuga, L., and Liz-Marzan, L. (2009). LSPR-based nanobiosensors. Nano Today *4*, 244–251.

Serafín, V., Eguílaz, M., Agüí, L., Yáñez-Sedeño, P., and Pingarrón, J.M. (2011). An electrochemical immunosensor for testosterone using gold nanoparticles—Carbon nanotubes composite electrodes. Electroanalysis *23*, 169–176.

Sidhu, S., Lowman, H., Cunningham, B., and Wells, J. (2000). Phage display for selection of novel binding peptides. In: Thorner, J., Emr, S.D., Abelson, J.N. (eds.), Methods in Enzymology, vol. *328*, Applications of chimeric genes and hybrid proteins, Pt. C, pp. Academic Press, San Diego, pp. 333–363.

Silverman, S. (1988). Early diagnosis of oral cancer. Cancer *62*, 1796–1799.

Song, J., Yang, Q., Lv, F., Liu, L., and Wang, S. (2012a). Visual detection of DNA mutation using multicolor fluorescent coding. ACS Appl. Mater. Interfaces *4*, 2885–2890.

Song, S.Y., Han, Y.D., Hong, S.Y., Kim, K., Yang, S.S., Min, B.-H., and Yoon, H.C. (2012b). Chip-based cartilage oligomeric matrix protein detection in serum and synovial fluid for osteoarthritis diagnosis. Anal. Biochem. *420*, 139–146.

Sotiropoulou, S. and Chaniotakis, N. (2005). Lowering the detection limit of the acetylcholinesterase biosensor using a nanoporous carbon matrix. Anal. Chim. Acta *530*, 199–204.

Reichlin, T., Hochholzer, W., Bassetti, S., Steuer, S., Stelzig, C., Hartwiger, S., Biedert, S., Schaub, N., Buerge, C., Potocki, M., Noveanu, M., Breidthardt, T., Twerenbold, R., Winkler, K., Bingisser, R., Mueller, C. (2009). Early diagnosis of myocardial infarction with sensitive cardiac troponin assays. N. Engl. J. Med. *361*, 858–867.

Suwansa-ard, S., Kanatharana, P., Asawatreratanakul, P., Wongkittisuksa, B., Limsakul, C., and Thavarungkul, P. (2009). Comparison of surface plasmon resonance and capacitive immunosensors for cancer antigen 125 detection in human serum samples. Biosens. Bioelectron. *24*, 3436–3441.

Suzuki, M., Husimi, Y., Komatsu, H., Suzuki, K., and Douglas, K. (2008). Quantum dot FRET biosensors that respond to pH, to proteolytic or nucleolytic cleavage, to DNA synthesis, or to a multiplexing combination. J. Am. Chem. Soc. *130*, 5720–5725.

Swierczewska, M., Liu, G., Lee, S., and Chen, X. (2012). High-sensitivity nanosensors for biomarker detection. Chem. Soc. Rev. *41*, 2641–2655.

Taranekar, P., Baba, A., Park, J., Fulghum, T., and Advincula, R. (2006). Dendrimer precursors for nanomolar and picomolar real-time surface plasmon resonance/potentiometric chemical nerve agent sensing using electrochemically crosslinked ultrathin films. Adv. Funct. Mater. *16*, 2000–2007.

Thakor, A.S. and Gambhir, S.S. (2013). Nanooncology: The future of cancer diagnosis and therapy. CA Cancer J. Clin. *63*, 395–418.

Torrance, L., Ziegler, A., Pittman, H., Paterson, M., Toth, R., and Eggleston, I. (2006). Oriented immobilisation of engineered single-chain antibodies to develop biosensors for virus detection. J. Virol. Methods *134*, 164–170.

Tothill, I.E. (2009). Biosensors for cancer markers diagnosis. Semin. Cell Dev. Biol. *20*, 55–62.

Tsai, H. and Doong, R. (2005). Simultaneous determination of pH, urea, acetylcholine and heavy metals using array-based enzymatic optical biosensor. Biosens. Bioelectron. *20*, 1796–1804.

Uckermann, O., Galli, R., Tamosaityte, S., Leipnitz, E., Geiger, K., Schackert, G., Koch, E., Steiner, G., and Kirsch, M. (2014). Label-free delineation of brain tumors by coherent anti-stokes Raman scattering microscopy in an orthotopic mouse model and human glioblastoma. PLoS One *9*, e107115.

Van Dorst, B., Mehta, J., Bekaert, K., Rouah-Martin, E., De Coen, W., Dubruel, P., Blust, R., and Robbens, J. (2010). Recent advances in recognition elements of food and environmental biosensors: A review. Biosens. Bioelectron. *26*, 1178–1194.

Vanhecke, D., Roumenina, L.T., Wan, H., Osthoff, M., Schaller, M., and Trendelenburg, M. (2012). Identification of a major linear C1q epitope allows detection of systemic lupus erythematosus anti-C1q antibodies by a specific peptide-based enzyme-linked immunosorbent assay. Arthritis Rheum. *64*, 3706–3714.

Varshney, M. and Li, Y. (2007). Interdigitated array microelectrode based impedance biosensor coupled with magnetic nanoparticle-antibody conjugates for detection of *Escherichia coli* O157: H7 in food samples. Biosens. Bioelectron. *22*, 2408–2414.

Von Muhlen, M.G., Brault, N.D., Knudsen, S.M., Jiang, S., and Manalis, S.R. (2010). Label-free biomarker sensing in undiluted serum with suspended microchannel resonators. Anal. Chem. *82*, 1905–1910.

Waggoner, P.S. and Craighead, H.G. (2007). Micro- and nanomechanical sensors for environmental, chemical, and biological detection. Lab Chip *7*, 1238–1255.

Wang, J. (1999). Amperometric biosensors for clinical and therapeutic drug monitoring: A review. J. Pharm. Biomed. Anal. *19*, 47–53.

Wang, J. (2001). Glucose biosensors: 40 years of advances and challenges. Electroanalysis *13*, 983–988.

Wang, J. (2005). Carbon-nanotube based electrochemical biosensors: A review. Electroanalysis *17*, 7–14.

Wang, J. (2006). Electrochemical biosensors: Towards point-of-care cancer diagnostics. Biosens. Bioelectron. *21*, 1887–1892.

Wang, R. and Li, Y. (2013). Hydrogel based QCM aptasensor for detection of avian influenza virus. Biosens. Bioelectron. *42*, 148–155.

Wang, S., Haque, F., Rychahou, P.G., Evers, B.M., and Guo, P. (2013). Engineered nanopore of Phi29 DNA-packaging motor for real-time detection of single colon cancer specific antibody in serum. ACS Nano *7*, 9814–9822.

Wang, Y., Guo, X., Tong, L., and Lou, J. (2014). Modeling of Au-nanowire waveguide for plasmonic sensing in liquids. J. Lightwave Technol. *32*, 3631–3636.

Wongkaew, P. and Poosittisak, S. (2014). Diagnosis of sugarcane white leaf disease using the highly sensitive DNA based voltammetric electrochemical determination. Am. J. Plant Sci., *5*, 2256–2268.

Wu, H., Wang, J., Kang, X., Wang, C., Wang, D., Liu, J., Aksay, I.A., and Lin, Y. (2009). Glucose biosensor based on immobilization of glucose oxidase in platinum nanoparticles/graphene/chitosan nanocomposite film. Talanta *80*, 403–406.

Wu, S., Liu, G., Li, P., Liu, H., and Xu, H. (2012). A high-sensitive and fast-fabricated glucose biosensor based on Prussian blue/topological insulator Bi_2Se_3 hybrid film. Biosens. Bioelectron. *38*, 289–294.

Wulfkuhle, J., Liotta, L., and Petricoin, E. (2003). Proteomic applications for the early detection of cancer. Nat. Rev. Cancer *3*, 267–275.

Wustoni, S., Hideshima, S., Kuroiwa, S., Nakanishi, T., Hashimoto, M., Mori, Y., and Osaka, T. (2014). Sensitive electrical detection of human prion proteins using field effect transistor biosensor with dual-ligand binding amplification. Biosens. Bioelectron. *67*, 256–262.

Xu, S. and Han, X. (2004). A novel method to construct a third-generation biosensor: Self-assembling gold nanoparticles on thiol-functionalized poly(styrene-*co*-acrylic acid) nanospheres. Biosens. Bioelectron. *19*, 1117–1120.

Xu, H. and Kall, M. (2002). Modeling the optical response of nanoparticle-based surface plasmon resonance sensors. Sens. Actuators B Chem. *87*, 244–249.

Xu, L.-J., Zong, C., Zheng, X.-S., Hu, P., Feng, J.-M., and Ren, B. (2014). Label-free detection of native proteins by surface-enhanced Raman spectroscopy using iodide-modified nanoparticles. Anal. Chem. *86*, 2238–2245.

Yang, X., Yuan, R., Chai, Y., Zhuo, Y., Hong, C., Liu, Z., and Su, H. (2009). Porous redox-active Cu_2O-SiO_2 nanostructured film: Preparation, characterization and application for a label-free amperometric ferritin immunosensor. Talanta *78*, 596–601.

Yanik, A.A., Huang, M., Kamohara, O., Artar, A., Geisbert, T.W., Connor, J.H., and Altug, H. (2010). An optofluidic nanoplasmonic biosensor for direct detection of live viruses from biological media. Nano Lett. *10*, 4962–4969.

Yano, K. and Karube, I. (1999). Molecularly imprinted polymers for biosensor applications. Trends Anal. Chem. *18*, 199–204.

Yao, C., Zhu, T., Qi, Y., Zhao, Y., Xia, H., and Fu, W. (2010). Development of a quartz crystal microbalance biosensor with aptamers as bio-recognition element. Sensors *10*, 5859–5871.

Yu, J., Ge, L., Huang, J., Wang, S., and Ge, S. (2011). Microfluidic paper-based chemiluminescence biosensor for simultaneous determination of glucose and uric acid. Lab Chip *11*, 1286–1291.

Yuan, J., Duan, R., Yang, H., Luo, X., Xi, M. (2012). Detection of serum human epididymis secretory protein 4 in patients with ovarian cancer using a label-free biosensor based on localized surface plasmon resonance. Int. J. Nanomed. *7*, 2921–2928.

Zhang, X., Teng, Y., Fu, Y., Xu, L., Zhang, S., He, B., Wang, C., and Zhang, W. (2010). Lectin-based biosensor strategy for electrochemical assay of glycan expression on living cancer cells. Anal. Chem. *82*, 9455–9460.

Zhang, D., Lu, Y., Jiang, J., Zhang, Q., Yao, Y., Wang, P., Chen, B., Cheng, Q., Liu, G.L., and Liu, Q. (2014). Nanoplasmonic biosensor: Coupling electrochemistry to localized surface plasmon resonance spectroscopy on nanocup arrays. Biosens. Bioelectron. *140*, 506–511.

Zhang, C., Lou, J., Tu, W., Bao, J., and Dai, Z. (2015). Ultrasensitive electrochemical biosensing for DNA using quantum dots combined with restriction endonuclease. Analyst *140*, 506–511.

Zhao, C., Hoppe, T., Setty, M.K.H.G., Murray, D., Chun, T.-W., Hewlett, I., and Appella, D.H. (2014a). Quantification of plasma HIV RNA using chemically engineered peptide nucleic acids. Nat. Commun. *5*, 5079.

Zhao, Q., Duan, R., Yuan, J., Quan, Y., Yang, H., and Xi, M. (2014b). A reusable localized surface plasmon resonance biosensor for quantitative detection of serum squamous cell carcinoma antigen in cervical cancer patients based on silver nanoparticles array. Int. J. Nanomed. *9*, 1097–1104.

11

SAFETY OF NANOMATERIALS

NURAY GUNDUZ, ELIF ARSLAN, MUSTAFA O. GULER, AND AYSE B. TEKINAY

Institute of Materials Science and Nanotechnology, National Nanotechnology Research Center (UNAM), Bilkent University, Ankara, Turkey

11.1 INTRODUCTION

Advances in the design and fabrication of nanomaterials have created novel opportunities for the development of effective therapeutic agents against hard-to-treat disorders. However, the long-term biocompatibility of these materials remains an important concern for their use in biomedical applications. A thorough assessment of potential toxicity is therefore necessary for any material intended for use in biological environments; nonetheless standard methods for the evaluation of nanomaterial toxicity are often lacking. In this chapter, we will overview the attempts in understanding the biocompatibility of nanomaterials and provide an account of how these views changed in light of recent findings, with emphasis on the methodology used in nanotoxicology studies.

Nanoparticles (NPs) have commonly been used in the fields of engineering, medicine, pharmacology, and biotechnology over the last 20 years (Granqvist et al., 1976). NPs of iron and noble metals are especially popular in biomedical research, as they are largely inert toward biological agents, display optical properties ideal for *in vivo* imaging, and can be

Therapeutic Nanomaterials, First Edition. Edited by Mustafa O. Guler and Ayse B. Tekinay.

functionalized with a variety of targeting molecules. These properties render NPs and other engineered nanomaterials (ENMs) attractive for use in clinical studies, and ENM-based methods have been proposed for applications such as targeted gene and drug delivery, computed tomography, Raman spectroscopy, photoacoustic imaging, and photothermal therapy (Choi and Wang, 2011; Neuberger et al., 2005; Prasad, 2009). However, the popularity of ENMs in biomedical and industrial applications has led to questions regarding their safety and especially on their long-term toxic effects on living organisms. As with any other foreign element, nanomaterials can affect the functioning of the body by disrupting any of a large number of interconnected pathways, and the nanotoxicology of ENMs should be evaluated in a broad and comprehensive manner by taking account of the individual and synergistic aspects of ENM biology. The physicochemical features, exposure routes, uptake mechanisms, cellular localization, and biodistribution, clearance, adverse molecular responses (inflammation and ROS generation), and genotoxicity of ENMs should therefore all be considered to provide a coherent account of their potential toxic effects.

A large number of techniques exist for the evaluation of these features under *in vitro* or *in vivo* conditions, and some (but not all) are typically employed to test the safety of ENMs intended for biomedical use. However, despite the considerable volume of research regarding the adverse effects of nanostructures, a comprehensive toxicity assessment of ENMs is yet to be attempted in the literature, and guidelines for the safe use and manufacturing of these materials are likewise wanting. Nonetheless, institutional organizations such as the Environmental Protection Agency (EPA), World Health Organization (WHO), National Institute for Occupational Safety and Health (NIOSH), and the European Commission frequently perform toxicology assessments on nanostructures to better understand the risks involved with this class of materials (Park, 2012).

11.2 CHARACTERIZATION, DESIGN, AND SYNTHESIS OF NANOMATERIALS

11.2.1 Chemical Identity and Physicochemical Properties

NPs are commonly used in various *in vitro* and *in vivo* toxicity studies, and computational approaches have also been used for the *in silico* toxicity assessment of ENMs with minimal cost and no harm to living organisms (Clark et al., 2011; García-Remesal et al., 2013; Han et al., 2011; Sayes and Ivanov, 2010). The determination of the chemical identity of ENMs is a vital aspect of their characterization, and potential impurities or batch-to-batch variances may play a major role in the toxicity of NPs. ENMs can be synthesized with various shapes, structures, sizes, and chemical and physical properties. Methods for the synthesis and assembly of ENM are likewise variable and may involve a large number of physical and chemical processes using solid, liquid, or gas precursors (Saravanan et al., 2008). Synthesis methods are broadly classified as either bottom-up or top-down approaches (Saravanan et al., 2008): Bottom-up techniques assemble a larger structure from smaller units or "building blocks" (such as the synthesis of NPs from a solvent in a sol-gel method), while top-down techniques reduce a bulk material to an array of smaller structures through microfabrication processes such as grinding, polishing, and lithography (Ghosh Chaudhuri and Paria, 2011).

The design approach chosen for ENM synthesis may heavily influence its toxicity; as such, the biocompatibility of an ENM can be considered to be a major design characteristic. Biocompatibility for ENMs is defined as the ability to show an intended biomedical function without any adverse effects (Williams, 2008) and depends heavily on the size, shape, surface area, composition, solubility, surface charge, and purity of the nanomaterial. In addition, it must be noted that ENMs are typically functionalized with a variety of biological moieties to modulate their toxic effects or impart additional properties to the material. These moieties represent another critical aspect of ENM fabrication, as most NPs are designed and developed according to their intended biological functionality. It is therefore vital to preserve this functionality through the path of the nanomaterial from its site of introduction to the site of action, and the failure to do so may result in adverse effects due to the nonspecific activity of the nanomaterial.

Surface coating is the most common method for introducing a functional moiety onto an NP. An outer layer of biofunctional material not only provides bioactivity to the NP but also increases its stability in serum or storage. Surface conjugation can be performed for purposes such as cell adhesion or targeting of a specific tissue, organ, or disease (Li et al., 2008; Matsuo et al., 2003; Xie et al., 2005). Targeting moieties also eliminate potential side effects by preventing the NPs from binding to cells or tissues outside their site of interest. As such, another important role of coating is to reduce the toxic or adverse effects of ENMs, and it is known that different surface coatings may modulate the physicochemical properties and cytotoxic effects of NPs (Suresh et al., 2012).

The primary structural features of ENMs are their sizes and surface areas. The size of the ENM not only affects its physicochemical characteristics but also plays roles in its uptake and interaction with cells, tissues, organs, and, in cases where the NP is not administered but surreptitiously encountered as a contaminant, whole organisms. Nanomaterials with smaller sizes have greater surface areas per mass, which allows a better interaction between the nanomaterial surface and biological materials (Ghosh Chaudhuri and Paria, 2011). For example, the uptake mechanism of ENMs was dramatically influenced by the sizes of the nanomaterials in case of silver nanoparticles (NPs) with different sizes. Smaller NPs were also more active in increasing the generation of ROS and expression of inflammatory markers (Park et al., 2011). Size characterization is therefore critical for ENMs that are intended for use in biomedical applications and can be performed by TEM analysis (visual characterization) and UV/vis spectroscopy (quantified characterization) (Browning et al., 2013). In addition, while ENMs may be stable in water or standard buffers, biological environments are often complex and contain proteins that may trigger the aggregation of NPs. As such, size characterization should also be performed in a medium that adequately simulates the biological environment at the site of action (Powers et al., 2007).

The biological activity of an ENM is also influenced by its morphology, the importance of which is addressed in several studies. One such paper demonstrates that AuNR–PEG (Au nanorods functionalized with PEG) nanostructures were more cytotoxic to human skin cells than AuNS–MPS (Au nanospheres functionalized with mercaptopropane sulfonate) and suggests that differences in shape and agglomeration responses play a crucial role in NP toxicity (Schaeublin et al., 2012). The aspect ratio of an ENM is often a good predictor for the potential effects of its morphology, as high-aspect ratio ENMs (nanotubes and nanowires) and low-aspect ratio ENMs (nanospheres and similar structures) display distinct behaviors in uptake and cytotoxicity.

Chemical composition is another critical parameter that is frequently characterized for the nanotoxicology assessment of ENMs: Prahalad et al., for example, have shown that material composition, solubility, and acidity are important factors for determining the responses of cells to transition metal-based ENMs (Prahalad et al., 1999). In addition to the three major parameters of size, shape, and composition, properties such as surface chemistry, surface structure and charge, purity, solubility, crystal structure, catalytic activity, acidity/alkalinity, and agglomeration behavior also affect the toxicity of ENMs (Arora et al., 2012; Donaldson and Poland, 2013; Kroll et al., 2009; Park, 2012). All these parameters may alter cellular responses to ENMs by changing uptake mechanisms (through associations with endo- or exocytosis pathways) or facilitating interactions with proteins and other biological macromolecules (Lynch et al., 2009; Nel et al., 2009). As such, their evaluation is key for predicting the potential toxicity of an NP.

The chemistry, surface charge, and nanotopology of an ENM can change greatly through the process of its synthesis and application, especially due to the moieties used in surface modification and the structural alterations they undergo in biological environments. Therefore, the characterization of the surface chemistry of ENMs is an essential step for nanotoxicology studies (Johnston et al., 2010a). The charge of the ENM is another critical physicochemical aspect and should be characterized to determine its potential interactions with molecules, cells, and especially membranes, which are often charged (Johnston et al., 2010a). The solubility of the ENM is important for its biocompatibility (as biological environments are mostly aqueous) and for the evaluation of biopersistence (Johnston et al., 2010a). Both of these parameters must therefore be investigated in detail to provide an adequate explanation for the toxicity (or the lack thereof) of a material, as well as to design novel ENMs with minimal toxic effects. The reliability and applicability of the analysis method used is also crucial, as experimental protocols may potentially alter the physicochemical features of ENMs, making it difficult to predict or explain their *in vivo* behavior.

11.2.2 Biological Identity

Biological responses toward ENMs are strongly affected by the interactions between the ENM and cells, body fluids, proteins, and other macromolecules, which are also important for the distribution of ENMs in the target organism (Nel et al., 2006). The adsorption capacity of an ENM correlates with its size and influences material toxicity. NPs with smaller sizes possess larger surface areas per mass, which increases their adsorption capacity and biological interactions. These interactions trigger the formation of a protein coat, called the corona, when NPs are exposed to a biological environment (Kroll et al., 2009). The corona affects the effective size, charge, and accumulation/aggregation behaviors of ENMs and is in turn influenced by the surface chemistry of the NP (Kroll et al., 2009; Park, 2012). After the ENM assembles its corona of aggregating proteins, it may also gain the ability to interact with cells and tissues that it would normally fail to recognize (Nel et al., 2006). In addition, these interactions may result in the degradation or denaturation of proteins.

Blood is the first biological medium encountered by ENMs delivered by intravenous injection and contains a large variety of proteins, lipids, and other macromolecules that may interact with NPs (Rahman et al., 2013). These interactions generate two types of coronae: hard coronae and soft coronae. The hard corona is formed by those proteins that display a greater affinity to NPs than to other proteins. These proteins make strong bonds with the ENM and are not easily detached. The soft corona, in contrast, is formed by proteins with low affinity toward NPs and is a more dynamic layer (Rahman et al., 2013). It is also suggested that soft corona proteins interact or associate with hard corona proteins rather than the ENM itself.

The structure and composition of the protein corona are determined by the physicochemical characteristics of NPs, such as size, shape, curvature, surface charge, surface modification, and solubility. In addition, the type of biological environment, route of administration, and exposure time also affect the corona formation process (Rahman et al., 2013; Walkey and Chan, 2012). As such, both the inherent properties of NPs and the influence of the biological environment must be characterized to perform an adequate toxicology assessment of NPs.

One of the most important features of the protein corona is its thickness, which is affected by protein concentration, and ENM's size and surface properties. For instance, blood proteins generate a hydrodynamic diameter of approximately 3–15 nm, which triggers the formation of a thick, multilayer protein corona (Rahman et al., 2013). The surface charge of the ENM has been indicated as a critical parameter for protein corona formation, as higher surface charges are generally associated with greater protein adsorption (Park, 2012). Surface modifications on ENMs also alter the adsorption of proteins and thus the formation of the corona, which can be used to engineer the corona composition by changing the surface chemistry of an ENM.

Cell culture media used in *in vitro* studies also possess protein components that can form coronae around material surfaces, which should be considered for the nanotoxicological assessment of ENMs. It has been demonstrated that different media compositions have different impacts on both the formation of the protein corona and the cellular responses given to the NPs (Rahman et al., 2013). Nonetheless, culture media are not fully adequate substitutes for the tissue microenvironment, and corona compositions and cell responses may differ significantly under *in vitro* and *in vivo* conditions. In addition, the composition of the protein corona is also altered when the ENM is transported from one tissue microenvironment to another, which in turn triggers a different set of cellular responses (García-Remesal et al., 2013). As such, a standardized toxicology assessment of ENMs should ideally be performed through the integration of both *in vitro* and *in vivo* results in various media that the ENM is expected to encounter.

11.3 INTERACTIONS AT THE CELL–MATERIAL INTERFACE

The high costs and ethical concerns associated with large-scale animal research have necessitated the development of reliable, high-throughput *in vitro* screening assays and the use of computational models in toxicology studies (Gangwal et al., 2011; Krewski et al., 2010). Although various *in vitro* assays have been developed for the toxicity analysis of ENMs, these methods are not widely employed and may yield results inconsistent with *in vivo* studies (Demokritou et al., 2013; Han et al., 2012; Oberdorster, 2012). This is hardly surprising, as *in vivo* systems are directed through a complex multicellular network that connects the actions of cells, tissues, and organs, while these interactions are largely lacking in *in vitro* models (DeLoid et al., 2014). However, the toxicity of ENMs is dependent on the physicochemical properties of nanomaterials, including size (Dykman and Khlebtsov, 2014), morphology (Agarwal et al., 2013; Barua et al., 2013; Gitrowski et al., 2014), surface chemistry (Yang et al., 2014), agglomeration state (Grassian et al., 2007) and solubility (Xia et al., 2008), biological milieu (Zhu et al., 2012), and cell type (dos Santos et al., 2011; Kroll et al., 2011; Kuhn et al., 2014), which are important for both *in vivo* and *in vitro* systems.

Investigations on the toxicity of ENMs often fail to consider that the cell itself is a bionano cosmos assembled from a great variety of nanoscale machines. For instance, the ribosome is an impressive example of a nanomaterial, with a diameter around approximately 20 nm. Another example of a nanoscale biological machine is the adenovirus capsid (~30 nm), which selectively transports its nucleic acid cargo to specific target organs and has been utilized for gene therapy due to this property. The targeted delivery mechanism of adenoviruses is in turn stopped by the immunological detection and capture methods employed by macrophages in the liver. As such, the human body is already subject to a large number of nanoscale interactions, and a better understanding of these interactions will assist greatly in the understanding of the effects of ENMs (Feliu and Fadeel, 2010).

The biological milieu plays an important role during cell–nanomaterial interactions. Lipids and proteins cover the surfaces of ENMs immediately after their introduction into biological systems (Nel et al., 2009), and this effect, in tandem with other physicochemical properties, affects the entry mechanisms of ENMs in all cell types. The localization and intracellular fate of ENMs depend on their internalization pathways, which in turn are affected by surface chemistry (Yang et al., 2014), size (Dykman and Khlebtsov, 2014), and shape (Agarwal et al., 2013; Barua et al., 2013; Gitrowski et al., 2014) of ENMs, in addition to biological milieu (Zhu et al., 2012), cell type (dos Santos et al., 2011; Kuhn et al., 2014), and other factors. Internalization pathways and interactions at the cell–ENM interface are dependent on the biomolecular corona, which partially blocks the bare material surface from interacting with the cell membrane and therefore reduces the surface free energy of ENMs, resulting in a decrease in their cellular uptake (Lesniak et al., 2013). Other studies have also reported that the thickness of the biocorona, as directed by changing serum concentrations, may affect the cellular uptake of ENMs (Zhu et al., 2012).

11.3.1 Intracellular Activity

After cellular entry, ENMs are transported within the cell and may concentrate in various organelles. However, owing to the lack of information regarding the intracellular transport mechanisms used by ENMs, there is no uniform consensus regarding the precise series of events that occur following internalization. Thus, the intracellular transport of ENMs cannot be generalized, especially in view of the structural and biochemical variability of the ENMs themselves (Wu et al., 2014). Once inside the cell, ENMs may interact with cellular compartments such as the cytosol, endosome, mitochondria, lysosome, nucleus, and Golgi apparatus (Fig. 11.1). The cytoskeleton is another common target for cytotoxic ENMs and plays significant roles in cellular events such as intracellular transport, cell division, and motility, all of which can be disrupted by NP activity. It has been reported that the aggregation state of AuNPs strongly influences their toxicity against human dermal fibroblast cells: NP clusters were fourfold more toxic than their nonaggregated counterparts and triggered the disruption of F-actin fibers. In addition, the aggregation states of these NPs were closely related to their surface chemistry and the conditions under which the NPs were introduced to the biological milieu (Yang et al., 2014). Despite their lack of acute cytotoxicity, single-walled carbon nanotubes (SWCNTs) were also shown to greatly reduce the proliferation of HeLa cells, which was attributed to actin-related cell division defects (Holt et al., 2010). Similar effects were observed in neuronal progenitor cells and endothelial cells treated with iron oxide NPs, resulting in concentration-dependent alterations in cellular physiology. High levels of iron oxide NPs affected the actin cytoskeleton through the formation and maturation of focal adhesion complexes (FACs) and resulted in a reduction in cell proliferation (Soenen et al., 2010). ENMs might also localize in other organelles after escaping from endosomes: Rhodamine-loaded poly(lactic-*co*-glycolic acid) (PLGA) NPs, for example, were reported to localize in various

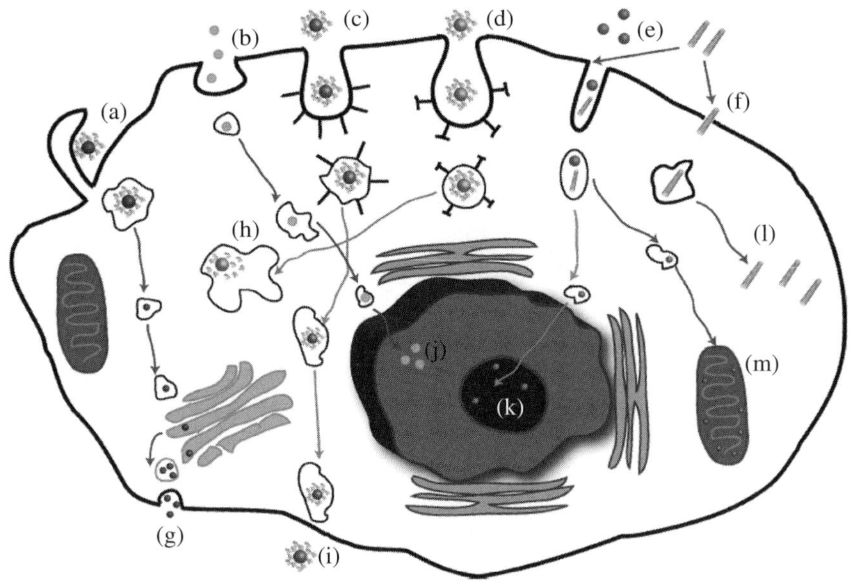

FIGURE 11.1 Uptake mechanisms and the intracellular transport of ENMs. Uptake may occur through micropinocytosis (a), pinocytosis (b), caveolin-mediated endocytosis (c), clathrin-mediated endocytosis (d), unknown endocytic pathways (e), or diffusion/penetration (f). After cellular entry, ENMs may localize into different organelles and cellular compartments. ENMs that enter to the Golgi apparatus can cause organelle damage and/or leave the cell via vesicles involved in the regular secretion system (g). ENMs remaining in early endosomes move slowly along cytoskeletal elements, fuse with late endosomes, and finally accumulate in the lysosome (h), nucleus (j), nucleolus (k), and mitochondria (m). In addition to these possible endpoints, ENMs can also accumulate in the cytoplasm (l) after escaping from late endosomes or lysosomes. Ultimately, ENMs can be excreted from the cell by transcytosis (i).

intracellular compartments such as the Golgi apparatus, cytosol, and endoplasmic reticulum (Cartiera et al., 2009).

Mitochondria are another potential endpoint for ENM transport, as the high-membrane potential (~ −200 mV) of mitochondria would strongly attract positively charged NPs (Szabo et al., 2012). The inner membrane of the mitochondrion might be especially suitable for the accumulation of ENMs, as a result of its hydrophobic, anionic, and high-membrane potential characteristics. Mitochondriotropic moieties have been developed to exploit this property for mitochondrial gene therapy and chemotherapy applications. The accumulation or localization of ENMs inside mitochondria can be accomplished through surface functionalization via triphenylphosphonium (TTP), which is a mitochondriotropic moiety. The localization of TPP-conjugated ENMs inside the mitochondria has also been confirmed by experimental reports. It has been demonstrated that polystyrene NPs functionalized with both TPP and poly-L-lysine (PLL) localize in mitochondria due to the hydrophobic properties of TPP (Wang et al., 2013). Doxorubicin-loaded, TPP-functionalized polyethylenimine hyperbranched polymer (PEI) NPs likewise triggered rapid and severe cytotoxicity within a few hours of incubation at low concentrations, in contrast to free doxorubicin (Theodossiou et al., 2013). Another study detailed the design of a multifunctional envelope-coated nanocarrier system, which consists of mesoporous silica nanoparticles (MSNs) functionalized with TPP, the antineoplastic drug topotecan (TPT), and a charge-conversional shielding layer. This system was shown to target tumor cells and initiate apoptosis through mitochondrial damage (Luo et al., 2014). Mitochondrion-mediated apoptotic cell death was also achieved in drug-resistant breast cancer cells via NP-directed efficient subcellular localization of doxorubicin into the mitochondria (Zeng et al., 2014). The surface chemistry and size of ENMs are important parameters for mitochondrial targeting and localization: While carboxylated Polystyrene NPs with 20 nm diameters could be located in the mitochondria of different hepatocyte cell lines, the mitochondrial uptake and localizations of 200 nm diameter Polystyrene NPs were limited (Johnston et al., 2010b).

Lysosomes can also be targeted depending on the functionalization moiety of NPs. Glycol chitosan NPs decorated with hydrophobic moieties have been reported to eventually localize into the lysosome (Nam et al., 2009). The intracellular fate of AuNPs has also been reported in different studies: Protein-coated AuNPs are generally entrapped in endosomes but can be directed to several cellular compartments by the addition of targeting moieties such as TAT cell-penetrating peptides or a liposome coating (Krpetic et al., 2011; Nativo et al., 2008). The uptake and localization of ENMs allow these molecules to interfere with intracellular mechanisms, potentially resulting in organelle damage, membrane leakage, or the production of reactive oxygen species (ROS). These effects may be medically beneficial if the ENMs are designed to specifically target cancer cells or other diseased cell types. SPIONs covalently coated with the lysosomal protein marker LAMP1 (LAMP1-SPION), for example, were used to apply mechanical stress to the lysosomes of rat insulinoma tumor cells. The activation of the SPIONs was triggered using a dynamic field generator, decreasing the sizes and numbers of the lysosomes and resulting in a corresponding increase in cellular apoptosis. The remote control of apoptosis was therefore achieved through the activation of magnetic ENMs (Zhang et al., 2014).

The excretion and clearance of ENMs will be discussed in the *in vivo* part of this chapter, but here it should be noted that cells themselves respond to ENMs by activating their exocytosis mechanisms, thus extruding or degrading ENMs at the subcellular level. The exocytosis of ENMs has been reported for a variety of materials. For example, Bartczak et al. characterized the exocytosis trends of AuNPs functionalized with angiogenesis-inhibiting ("inhibitor") or similar but noninhibitory ("mutant") peptides and made the interesting observation that mutant peptides were first exocytosed and subsequently reuptaken and reexocytosed from the cell, while inhibitors were progressively removed with no reuptake (Bartczak et al., 2012).

The intracellular localization of ENMs determines whether they can be exocytosed through the activity of lysosomes or the Golgi apparatus. The exocytosis of phosphate-modified mesoporous nanoparticles (P-MSNs) showed a good correlation with the rates of lysosomal exocytosis in different cell lines, which has significant implications for the development of efficient drug delivery agents (Yanes et al., 2013). The degradation of ENMs is a big concern for the delivery of biological cargos. Sahay et al. reported that siRNA-conjugated cationic lipid NPs are eventually degraded in the lysosome (Sahay et al., 2013), which is a common drawback of biodegradable ENMs utilized for gene therapy, cancer treatment, or drug delivery.

Various methods exist for the evaluation of the intracellular activity of ENMs: (i) Deactivation of cellular uptake mechanisms (phagocytosis, macropinocytosis, clathrin-mediated endocytosis, caveolae-mediated endocytosis, and transcytosis); (ii) assays for cell viability/proliferation (MTT (3-(4,5-dimethylthiazol-2-yl)-2,5-diphenyltetrazolium bromide), alamar blue, XTT, lactade dehydrogenase (LDH), calcein AM/ethidium homodimer-2staining,5-ethynyl-20-deoxyuridine(EdU),and5-bromo-2′-deoxyuridine (BrdU)); (iii) mechanistic assays (detection of ROS by dichlorofluorescein (DCF) and GSH level assays, apoptosis (caspase activity assays), necrosis); (iv) microscopic evaluation of intracellular localization (transmission electron microscopy (TEM) or scanning electron microscopy (SEM), confocal laser scanning microscopy (CLSM), atomic force microscopy, and magnetic resonance imaging); (v) quantification of ENM uptake and accumulation (inductively coupled plasma spectroscopy techniques: ICP-MS, ICP-OES, and flow cytometry); (vi) genotoxicity assays (DNA damage, micronuclei formation, DNA fragmentation).

TABLE 11.1 Inhibitors Used to Study the Uptake Mechanisms of ENMs

Inhibitor	Mechanism of Action	Affected Uptake Mechanism
Nystatin	Disrupts caveolar function	Caveolae-mediated endocytosis (Greulich et al., 2011)
Amyloride hydrochloride	Blocker of epithelial Na^+ channels	Macropinocytosis (Gitrowski et al., 2014; Weissleder et al., 2014)
Cytochalasin A/B/D	Polymerization of actin	Caveolae-mediated endocytosis (Gao et al., 2013)
Chlorpromazine	Reversible translocation of clathrin from cell membrane	Clathrin-mediated endocytosis (Gao et al., 2013)
Genistein	Tyrosine kinase inhibitor	Caveolae-mediated endocytosis
Lovastatin	Cholesterol synthesis	Macropinocytosis and clathrin- and caveolae-mediated endocytosis
Nocodazole	Polymerization of microtubules	Clathrin-mediated endocytosis (Gao et al., 2013)
Sodium azide	Energy	Gao et al. (2013)
Sucrose	Blocking agents of clathrin-coated pit formation	Clathrin-mediated endocytosis (Gao et al., 2013)
Filipin	Inhibitor of caveolae-associated endocytosis	Caveolae-mediated endocytosis (Gao et al., 2013)
BFA	Disruption of Golgi apparatus	Intracellular transportation of ENMs (Gao et al., 2013)
Monensin	Lysosome inhibitor	Intracellular cargo transport and disposition (Gao et al., 2013)
Phenylarsine oxide	Endocytosis	Gao et al. (2013)
Wortmannin	Inhibits phosphoinositide-3 kinases (PI3Ks) through covalent interactions	Macropinocytosis (Sokolova et al., 2013)

11.3.2 Cellular Uptake Mechanisms

Imaging and spectroscopy techniques are widely used in monitoring the cellular uptake and intracellular localization of ENMs. Selective targeting is an important design concern for ENMs, and the ability to engineer the localization of a nanostructure within the cell significantly improves its capacity for selective delivery. Some applications require the retention of ENMs in specific organelles or necessitate their uptake by specific cell types, which can be achieved through the activation of different uptake mechanisms. As such, pathway-specific inhibitors are commonly used to determine the methods of uptake used by ENMs. Some of these inhibitors are listed in Table 11.1.

11.4 ASSAYS FOR CELL VIABILITY/PROLIFERATION

The evaluation of the short- and long-term effects of ENMs is commonly conducted through cell viability and proliferation assays. ENM toxicity is mainly quantified by metabolic activity assays (e.g., MTT, XTT, alamar blue, LDH) and direct cell death assays (calcein AM/ethidium homodimer-2 staining). Cytotoxicity screening of an ENM is often a preliminary step that determines whether the material is suitable for further testing. As such, comparative assays have been developed for this process. One particularly stringent example is provided by Kroll et al., who have developed a matrix-based test system to evaluate the cytotoxic effects of 23 ENMs against a set of sensitive cell lines and suggest that such a system can be used to predict *in vivo* nanostructure behavior, minimizing the number of animals used in toxicity assessments by preselecting against cytotoxic ENMs (Kroll et al., 2011). The matrix-based, high-throughput cytotoxicity screening of ENMs is of great importance for the development standardized cytotoxicity assessments and modeling studies.

Cell proliferation assays are also performed frequently in ENM research, often to investigate the short-term detrimental effects of ENMs or to determine whether the material can trigger cell cycle arrest in cancer cells. Although ENMs may not necessarily cause short-term cytotoxicity in cells, they may nonetheless interfere with DNA synthesis and cell division, which is a desirable property for materials intended for use in anticancer treatments (Kim et al., 2013). EdU and BrdU, synthetic nucleosides that replace the thymidine moiety in DNA during replication, are commonly used to analyze cell proliferation, cell division, DNA synthesis, and cell cycle arrest.

11.4.1 Assays for Oxidative Stress and Apoptosis Mechanisms

The generation of ROS or reactive nitrogen species (RNS) may alter the homeostatic oxidative states of cells and induce cell death through apoptosis, necrosis, or other mechanisms. ENMs may affect the cellular oxidative state in various ways. They typically cause generation of excessive ROS or RNS at higher concentrations, which is considered a major factor in the severity of ENM cytotoxicity. On the other hand, ENMs with reductive properties can also be fabricated through surface modification methods to act as free radical scavengers and decrease oxidative damage in cells (Karakoti et al., 2010). The most widely used method for the quantification of ROS is the DCF assay, which relies on the oxidation of nonfluorescent DCF to fluorescent DCF by various intracellular oxidants. ROS generation can also be monitored by the evaluation of GSH level, free radical content, mitochondrial membrane potential, and stress-related gene/protein expression analyses.

The apoptotic state of cells can be analyzed by annexin V staining, cytochrome c release, caspase-3 activity, and other caspase activity assays. Annexin V staining is the most widely used method for the evaluation of early apoptosis in ENM-treated cells. Normally, phosphatidylserine is located on the cytoplasmic surface of the cell membrane. However, after the initiation of apoptosis, Polystyrene translocates from the inner surface and is found on the cell exterior, which can be detected by fluorescent dye-tagged annexin V.

11.4.1.1 Microscopic Evaluation of Intracellular Localization of ENMs CLSM or electron microscopy (SEM/TEM) can also be used for qualitative, semiquantitative, and quantitative evaluations of the uptake, localization, intracellular fate, accumulation, and exocytosis profiles of ENMs (Bartneck et al., 2010; Havrdova et al., 2014; Krpetic et al., 2011; Lesniak et al., 2013; Ma et al., 2011; Nativo et al., 2008; Rodriguez-Lorenzo et al., 2014; Soenen et al., 2012; Ye et al., 2014).

11.4.2 Evaluation of Uptake and Accumulation of ENMs

The evaluation of nanomaterial uptake, accumulation, and exocytosis might also be performed through inductively coupled plasma spectroscopy techniques, laser scanning microscopy techniques, flow cytometry (in the case of fluorescent nanomaterials), and TEM. Inductively coupled plasma spectrometry (i.e., ICP-AES (Bartczak et al., 2012; Krpetic et al., 2011), ICP-MS

(Wang et al., 2011), and ICP-OES (Ferrati et al., 2010)) are the most widely used methods for quantitative assessment of endocytosis, accumulation, and exocytosis. However, ICP-based techniques cannot distinguish ENMs and their breakdown products (i.e., metal ions) from naturally existing ions and elements within the cells. As such, this quantification method is applicable only to ENMs composed of materials not normally found in biological systems. In addition, the reproducibility of ICP spectroscopy measurements requires the ENMs to have a monodispersed size distribution, well-defined geometry, and well-characterized crystalline lattice.

Fluorescence-based techniques (i.e., fluorescence spectroscopy, flow cytometry, and laser scanning microscopy) can be used for the quantitative and qualitative evaluations of internalization, localization, accumulation, and exocytosis of ENMs (Rodriguez-Lorenzo et al., 2014; Slowing et al., 2011). Although these techniques are very precise, they can only be used with ENMs displaying fluorescent activity, either as an intrinsic property (as in quantum dots) or through surface modification with fluorescent dyes. However, it must be kept in mind that functionalization of an ENM with a fluorescent dye might change its surface properties and biological activities. In addition, weakly conjugated fluorescent dyes may detach from the ENM in biological environments and result in major errors in the quantification of ENM uptake and localization.

11.4.3 Genotoxicity Assays

ENMs are capable of localizing into the nucleus, where they interact with and potentially damage genomic DNA. The genotoxicity of ENMs can be analyzed through techniques such as chromosome aberration test, comet assay, and micronucleus assay. The chromosome aberration test is an indicator of chromosomal alterations induced by ENMs and involves visual evaluation following the isolation and Giemsa staining of metaphasic chromosomes. The comet assay, also known as single-cell gel electrophoresis (SCGE), is the most widely used assay for evaluating DNA damage at the level of individual cells. The underlying mechanism of comet assay is that damaged DNA migrates faster than undamaged DNA in an agarose matrix. Quantitative analysis of DNA damage is performed by staining and imaging the comet-like structures formed following electrophoresis process. However, as the comet assay is low throughput and often faces reproducibility issues, high-throughput screening tools are necessary to reliably assess ENM-mediated DNA damage (Watson et al., 2014).

11.5 ANIMAL MODELS AND LONG-TERM RISK ASSESSMENT

Despite the widespread production and application of ENMs, there is lack of satisfactory information regarding their impact on humans, animals, and the ecosystem. Exposure to ENMs may directly or indirectly cause detrimental effects on the human health. ENMs can be introduced to the human body through inhalation, ingestion, dermal penetration, or injection, either unintentionally or through the course of an ENM-based therapy (Oberdorster et al., 2005). As with any foreign material, ENMs may trigger adverse effects following their administration. As such, the biocompatibility, biodistribution, biodegradation, excretion, and acute and long-term toxicities of ENMs must be carefully assessed before their use in clinical trials. Although the *in vitro* tests previously outlined in this chapter provide means of predicting the behavior of nanomaterials, it is nonetheless possible that unforeseen *in vivo* effects may occur. In addition, the behavior of a nanomaterial can change through the spontaneous and rapid formation of protein coronae in biological media, which has a major impact on the interaction of ENMs with cells and organelles. It has also been reported that the immune system can detect NPs in the same way that it senses microbes and other foreign agents (Feliu and Fadeel, 2010; Mahmoudi et al., 2011). It is difficult to see many of these effects under *in vitro* conditions, which necessitates *in vivo* experiments.

ENMs that are intended for biomedical purposes are almost universally administered directly into the human body. Examples of such ENM designs include SPIONs and dendrimers for MRI studies; gold NPs for photothermal therapy, imaging, and drug and gene delivery; MSNs for MRI studies, drug delivery, and topical treatment; and carbon nanotubes for drug delivery and tumor detection and treatment. These materials are promising for the development of efficient and innovative solutions for the major challenges faced by the biomedical field, especially with regards to the treatment of common disorders such as cancer, obesity, and neurodegenerative diseases. Nevertheless, as with any material intended for human use, their safety must first and foremost be investigated to sufficient detail, through both *in vitro* experiments and animal studies, before they can be administered to human patients.

11.5.1 The Blood–Brain Barrier

Diseases of the central nervous system (CNS) can severely affect the health and comfort of patients through deterioration of either the spinal cord (myelopathy) or brain (encephalopathy). Development of ENMs for monitoring and treatment of CNS-related diseases is now a topic of great interest, as surface functionalization allows nanomaterials to display properties necessary for both diagnosis and therapy (this combined approach is also called *theranostics*). Although the fundamental mechanisms of certain CNS-related disorders are still not fully understood, there has been some progress in developing effective diagnosis-and-treatment methods. However, many of these efforts are hampered due to the protective role of the blood–brain barrier (BBB), which blocks the transport of drugs and other large molecules into the brain. ENMs may facilitate the passage of their cargo through the BBB, which makes them attractive for improving drug targeting, increasing theranostic efficacy and reducing drug toxicity in CNS-related disorders (Krol et al., 2013). ENMs can therefore be used for the early diagnosis, imaging, or therapy of CNS-related disorders or taken as a prophylactic measure against the advancement of potential disease.

The mechanisms ENMs employ to pass through the airway epithelia and BBB into the secondary target organs (STOs: liver, spleen, kidneys, heart, and brain) are not fully understood. Oberdorster and his group performed an inhalation study in rats with iridium (2–4 nm) and carbon (5–10 nm) NPs, as well as their aggregated forms (mean mobility diameters between 20 and 80 nm) and found the translocation of iridium NPs into the STOs to be significantly higher than that of carbon NPs. Furthermore, the translocation and accumulation of 80 nm aggregates of iridium were lower than 20 nm aggregates, and inhaled NPs could also be detected in the skeleton and soft tissues. However, the underlying mechanism for cerebral localization following NP inhalation remains uncertain (Kreyling et al., 2009).

Two types of strategies are used to enhance the transport of drugs into the brain: BBB manipulation and drug manipulation (Krol et al., 2013). The utilization of ENMs is classified as drug manipulation, and its importance has increased during the past few years. Surface functionalization of ENMs can increase their ability to cross the BBB (depicted in Fig. 11.2), although the precise mechanisms involved in surface functionalization-mediated BBB transport are an issue of contention. The stimulation of endocytosis or low-density lipoprotein (LDL)-mediated transcytosis, the weakening of the BBB due to toxicity, and the loosening of tight junctions between endothelial cells are among the mechanisms proposed for this phenomenon (Gidwani and Singh, 2014). Regardless of the mechanism involved, surface modification of ENMs is nonetheless advantageous for drug delivery into the CNS, as nanomaterials provide a high drug-loading capacity, prevent the degradation of the drug in the biological milieu, enhance the retention of the drug in the bloodstream, avoid efflux pump action

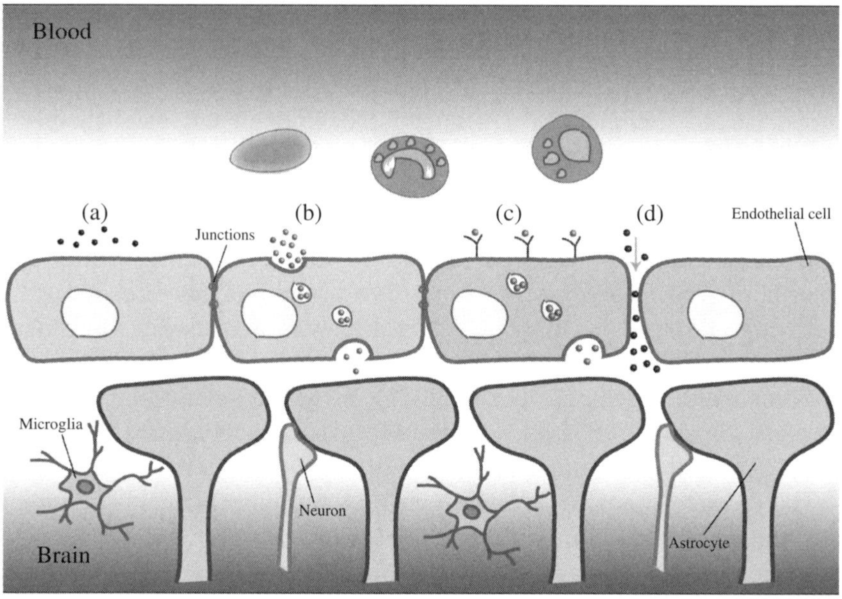

FIGURE 11.2 Schematic representation of the blood–brain barrier (BBB). Microvascular endothelial cells are connected through tight junctions (TJs) and adherens junctions (AJs) that prevent the leakage of foreign material into the brain. Nanoparticles ordinarily cannot pass through the BBB (a) and must be functionalized with targeting moieties that facilitate BBB transport through endocytosis (b), receptor-mediated transcytosis (c), or the loosening of the junctions between endothelial cells (d).

at the brain microvessels, and can be synthesized in the size ranges required for efficient transport across the BBB.

A variety of targeting moieties have been studied for the enhancement of BBB-crossing efficiency. The targeting molecule is often involved in receptor-mediated transport pathways; surface functionalization with materials that bind to the cellular receptors for insulin (Ulbrich et al., 2011), LDL (Bertrand et al., 2010; Wagner et al., 2012), and glutathione (Gaillard et al., 2012) has been reported in the literature. In addition, the BBB manipulation and drug manipulation strategies can be combined to increase the penetration efficiency of ENMs. Surfactants such as dimethylsulfoxide (DMSO), sodium dodecylsulphate (SDS), and polysorbate 80 (PS-80) are commonly used to loosen the adherent and tight junctions between the endothelial cells of the BBB and can be used in conjunction with ENMs to increase targeting efficiency. PS-80 is a commonly employed reagent for the coating of NPs intended to target brain parenchyma (Kreuter, 2013).

ENMs designed for BBB penetration typically do not display drug activity on their own and are instead loaded with a drug cargo that is released at the site of interest. ENMs have been loaded with well-known chemotherapeutics such as doxorubicin (Li et al., 2014; Wohlfart et al., 2011), paclitaxel (Dilnawaz et al., 2012; Zhang et al., 2012), and prodrugs like curcumin (Tsai et al., 2011). It has been shown that doxorubicin-loaded PS-80 functionalized poly(methacrylic acid) (PMAA) NPs accumulated in brain tumors and brain metastases of breast cancer and reduced the growth of brain tumors to a greater extent than free Dox. Similarly, Dox-loaded PS-PMAA NPs could be detected in tumor-bearing brain tissue and induced apoptosis in cancer cells while leaving the rest of the brain unharmed, suggesting cancer cell-specific targeting capacity (Li et al., 2014). Another study with doxorubicin was performed with rats treated intravenously with doxorubicin-loaded poly(*n*-butyl cyanoacrylate) (PBCA) NPs with and without a 1% polysorbate 80 coating, and PS-80 coating was shown to significantly improve the transcytosis of the drug into the brain parenchyma compared to free Dox and noncoated NPs (Wohlfart et al., 2011). Paclitaxel-loaded magnetic NPs were also reported to be transported to the STOs and the brain stem, cerebellum, and cortex more compared to native paclitaxel. In addition, Pac-MNPs displayed effective MRI contrast agent properties and can be used to acquire MRI images of the rat brain and liver (Dilnawaz et al., 2012).

An effective chemotherapy for brain tumors requires a nanoformulated drug that can both penetrate the BBB and selectively target cancer cells. As such, ENMs have also been functionalized with peptides and proteins that target membrane receptors that are overexpressed in cancer cells. It has been shown that transferrin- (Tf) modified, cyclo-[Arg-Gly-Asp-D-Phe-Lys] (c[RGDfK])-paclitaxel conjugate loaded micelles (TRPM) accumulate in the brain to a greater extent than free taxol, paclitaxel-loaded micelles (PM), and transferrin-modified, PM. More importantly, the nanoformulated paclitaxel (TRPM) exhibited the strongest anti-glioma activity, as measured through an *in vivo* survival experiment. The mean survival time of mice treated with TRPM (42.8 days) was significantly longer that those treated with TPM (39.5 days), PM (34.8 days), taxol (33.6 days), and saline (34.5 days) (Zhang et al., 2012). Likewise, Tsai et al. showed that curcumin-loaded PLGA NPs were accumulated in all STOs, except the heart, to a greater extent than free curcumin. In addition, PLGA NPs loaded with curcumin also accumulated in the hippocampus and displayed extended retention times in both the hippocampus (increased by 83%) and the cerebral cortex (increased by 96%) (Tsai et al., 2011).

Although nanomaterials are highly efficient platforms for the treatment of a great variety of disorders, their short- and long-term effects should be investigated in great detail prior to their *in vivo* application. Recent reports demonstrate that the presence of the protein corona on the surface of ENPs can alter the theranostic behaviors of nanomaterials; for instance, the formation of amyloid β fibrils is inhibited in the case of corona-coated NPs, whereas the rate of amyloid β fibrillation is accelerated in the presence of uncoated NPs (Ghavami et al., 2013). As such, greater attention should be given to the therapeutic behaviors of ENMs and how these behaviors may be altered in biological environments.

11.6 CONCLUSIONS AND FUTURE PERSPECTIVES

In this chapter, we presented some of the emerging research regarding the considerable impact that ENMs may have on living systems. The potential toxicity of ENMs is a significant concern throughout their design, modification, and functionalization, and the fabrication of ENMs with well-defined physicochemical properties is critical for efficient targeting with minimal side effects. The uptake and cellular localization of ENMs are affected by their physicochemical properties and chemical and biological identities. While the former two can be controlled during fabrication, the biological identity of an ENM depends on the biological environments it encounters and may change considerably inside living organisms. Interactions at cell–ENM interface are also dependent on physicochemical properties (e.g., size, shape, surface charge, etc.) and the biological identity of nanomaterials (e.g., protein corona formation) and cell/organism types. The entry and localization of nanomaterials within the cells are also important for their function, as well as for risk assessment studies. Methods listed in this chapter are the most widely used techniques and assays of the investigation of ENM localization and toxicity. In general, *in vitro* models and computational simulations are used for short-term toxicity analyses, while *in vivo* animal studies are preferred for long-term toxicity and risk assessments. In addition to experimental research, advanced computational simulations and models can facilitate the investigation of the possible effects of nanomaterials, through the simulation of nanomaterial–cell interactions for the prediction of ENM efficiency and potential toxicity. In addition, the development of reliable and high-throughput *in vitro* models may also allow the *in vitro* evaluation of the long-term impacts of ENMs. The combination of advanced *in vitro*, *in vivo*, and *in silico* methods will no doubt further our understanding of nanomaterial function and may allow the design of safer and more effective ENMs for the treatment of a wide range of disorders.

REFERENCES

Agarwal, R., Singh, V., Jurney, P., Shi, L., Sreenivasan, S.V., and Roy, K. (2013). Mammalian cells preferentially internalize hydrogel nanodiscs over nanorods and use shape-specific uptake mechanisms. Proc Natl Acad Sci U S A *110*, 17247–17252.

Arora, S., Rajwade, J.M., and Paknikar, K.M. (2012). Nanotoxicology and in vitro studies: the need of the hour. Toxicol Appl Pharmacol *258*, 151–165.

Bartczak, D., Nitti, S., Millar, T.M., and Kanaras, A.G. (2012). Exocytosis of peptide functionalized gold nanoparticles in endothelial cells. Nanoscale *4*, 4470–4472.

Bartneck, M., Keul, H.A., Singh, S., Czaja, K., Bornemann, J., Bockstaller, M., Moeller, M., Zwadlo-Klarwasser, G., and Groll, J. (2010). Rapid uptake of gold nanorods by primary human blood phagocytes and immunomodulatory effects of surface chemistry. ACS Nano *4*, 3073–3086.

Barua, S., Yoo, J.W., Kolhar, P., Wakankar, A., Gokarn, Y.R., and Mitragotri, S. (2013). Particle shape enhances specificity of antibody-displaying nanoparticles. Proc Natl Acad Sci U S A *110*, 3270–3275.

Bertrand, Y., Currie, J.C., Demeule, M., Regina, A., Che, C., Abulrob, A., Fatehi, D., Sartelet, H., Gabathuler, R., Castaigne, J.P., et al. (2010). Transport characteristics of a novel peptide platform for CNS therapeutics. J Cell Mol Med *14*, 2827–2839.

Browning, L.M., Lee, K.J., Nallathamby, P.D., and Xu, X.-H.N. (2013). Silver nanoparticles incite size- and dose-dependent developmental phenotypes and nanotoxicity in zebrafish embryos. Chem Res Toxicol *26*, 1503–1513.

Cartiera, M.S., Johnson, K.M., Rajendran, V., Caplan, M.J., and Saltzman, W.M. (2009). The uptake and intracellular fate of PLGA nanoparticles in epithelial cells. Biomaterials *30*, 2790–2798.

Choi, J. and Wang, N.S. (2011). Nanoparticles in Biomedical Applications and Their Safety Concerns. In R. Fazel (ed.) Biomedical Engineering—From Theory to Applications. InTech, Maastricht.

Clark, K.A., White, R.H., and Silbergeld, E.K. (2011). Predictive models for nanotoxicology: current challenges and future opportunities. Regul Toxicol Pharmacol *59*, 361–363.

DeLoid, G., Cohen, J.M., Darrah, T., Derk, R., Rojanasakul, L., Pyrgiotakis, G., Wohlleben, W., and Demokritou, P. (2014). Estimating the effective density of engineered nanomaterials for in vitro dosimetry. Nat Commun *5*, 3514.

Demokritou, P., Gass, S., Pyrgiotakis, G., Cohen, J.M., Goldsmith, W., McKinney, W., Frazer, D., Ma, J., Schwegler-Berry, D., Brain, J., et al. (2013). An in vivo and in vitro toxicological characterisation of realistic nanoscale CeO(2) inhalation exposures. Nanotoxicology *7*, 1338–1350.

Dilnawaz, F., Singh, A., Mewar, S., Sharma, U., Jagannathan, N.R., and Sahoo, S.K. (2012). The transport of non-surfactant based paclitaxel loaded magnetic nanoparticles across the blood brain barrier in a rat model. Biomaterials *33*, 2936–2951.

Donaldson, K. and Poland, C.A. (2013). Nanotoxicity: challenging the myth of nano-specific toxicity. Curr Opin Biotechnol *24*, 724–734.

Dykman, L.A. and Khlebtsov, N.G. (2014). Uptake of engineered gold nanoparticles into mammalian cells. Chem Rev *114*, 1258–1288.

Feliu, N. and Fadeel, B. (2010). Nanotoxicology: no small matter. Nanoscale *2*, 2514–2520.

Ferrati, S., Mack, A., Chiappini, C., Liu, X., Bean, A.J., Ferrari, M., and Serda, R.E. (2010). Intracellular trafficking of silicon particles and logic-embedded vectors. Nanoscale *2*, 1512–1520.

Gaillard, P.J., Appeldoorn, C.C., Rip, J., Dorland, R., van der Pol, S.M., Kooij, G., de Vries, H.E., and Reijerkerk, A. (2012). Enhanced brain delivery of liposomal methylprednisolone improved therapeutic efficacy in a model of neuroinflammation. J Control Release *164*, 364–369.

Gangwal, S., Brown, J.S., Wang, A., Houck, K.A., Dix, D.J., Kavlock, R.J., and Hubal, E.A. (2011). Informing selection of nanomaterial concentrations for ToxCast in vitro testing based on occupational exposure potential. Environ Health Perspect *119*, 1539–1546.

Gao, H., Yang, Z., Zhang, S., Cao, S., Shen, S., Pang, Z., and Jiang, X. (2013). Ligand modified nanoparticles increases cell uptake, alters endocytosis and elevates glioma distribution and internalization. Sci Rep *3*, 2534.

García-Remesal, M., García-Ruiz, A., Pérez-Rey, D., de la Iglesia, D., and Maojo, V. (2013). Using nanoinformatics methods for automatically identifying relevant nanotoxicology entities from the literature. BioMed Res Int *2013*, 9.

Ghavami, M., Rezaei, M., Ejtehadi, R., Lotfi, M., Shokrgozar, M.A., Abd Emamy, B., Raush, J., and Mahmoudi, M. (2013). Physiological temperature has a crucial role in amyloid beta in the absence and presence of hydrophobic and hydrophilic nanoparticles. ACS Chem Neurosci *4*, 375–378.

Ghosh Chaudhuri, R. and Paria, S. (2011). Core/shell nanoparticles: classes, properties, synthesis mechanisms, characterization, and applications. Chem Rev *112*, 2373–2433.

Gidwani, M. and Singh, A.V. (2014). Nanoparticle enabled drug delivery across the blood brain barrier: in vivo and in vitro models, opportunities and challenges. Curr Pharm Biotechnol *14*, 1201–1212.

Gitrowski, C., Al-Jubory, A.R., and Handy, R.D. (2014). Uptake of different crystal structures of TiO(2) nanoparticles by Caco-2 intestinal cells. Toxicol Lett *226*, 264–276.

Granqvist, C.G., Buhrman, R.A., Wyns, J., and Sievers, A.J. (1976). Far-infrared absorption in ultrafine Al particles. Phys Rev Lett *37*, 625–629.

Grassian, V.H., O'Shaughnessy, P.T., Adamcakova-Dodd, A., Pettibone, J.M., and Thorne, P.S. (2007). Inhalation exposure study of titanium dioxide nanoparticles with a primary particle size of 2 to 5 nm. Environ Health Perspect *115*, 397–402.

Greulich, C., Diendorf, J., Simon, T., Eggeler, G., Epple, M., and Koller, M. (2011). Uptake and intracellular distribution of silver nanoparticles in human mesenchymal stem cells. Acta Biomater *7*, 347–354.

Han, Y., Xu, X., Yin, L., Long, L., Liu, R., and Liu, S. (2011). Dynamical systems: an effective way in nanotoxicology study. J Algorithms Comput Technol *5*, 79–94.

Han, X., Corson, N., Wade-Mercer, P., Gelein, R., Jiang, J., Sahu, M., Biswas, P., Finkelstein, J.N., Elder, A., and Oberdorster, G. (2012). Assessing the relevance of in vitro studies in nanotoxicology by examining correlations between in vitro and in vivo data. Toxicology *297*, 1–9.

Havrdova, M., Polakova, K., Skopalik, J., Vujtek, M., Mokdad, A., Homolkova, M., Tucek, J., Nebesarova, J., and Zboril, R. (2014). Field emission scanning electron microscopy (FE-SEM) as an approach for nanoparticle detection inside cells. Micron *67*, 149–154.

Holt, B.D., Short, P.A., Rape, A.D., Wang, Y.L., Islam, M.F., and Dahl, K.N. (2010). Carbon nanotubes reorganize actin structures in cells and ex vivo. ACS Nano *4*, 4872–4878.

Johnston, H.J., Hutchison, G.R., Christensen, F.M., Aschberger, K., and Stone, V. (2010a). The biological mechanisms and physicochemical characteristics responsible for driving fullerene toxicity. Toxicol Sci *114*, 162–182.

Johnston, H.J., Semmler-Behnke, M., Brown, D.M., Kreyling, W., Tran, L., and Stone, V. (2010b). Evaluating the uptake and intracellular fate of polystyrene nanoparticles by primary and hepatocyte cell lines in vitro. Toxicol Appl Pharmacol *242*, 66–78.

Karakoti, A., Singh, S., Dowding, J.M., Seal, S., and Self, W.T. (2010). Redox-active radical scavenging nanomaterials. Chem Soc Rev *39*, 4422–4432.

Kim, J.A., Aberg, C., de Carcer, G., Malumbres, M., Salvati, A., and Dawson, K.A. (2013). Low dose of amino-modified nanoparticles induces cell cycle arrest. ACS Nano *7*, 7483–7494.

Kreuter, J. (2013). Mechanism of polymeric nanoparticle-based drug transport across the blood-brain barrier (BBB). J Microencapsul *30*, 49–54.

Krewski, D., Acosta, D., Jr., Andersen, M., Anderson, H., Bailar, J.C., 3rd, Boekelheide, K., Brent, R., Charnley, G., Cheung, V.G., Green, S., Jr., et al. (2010). Toxicity testing in the 21st century: a vision and a strategy. J Toxicol Environ Health B Crit Rev *13*, 51–138.

Kreyling, W.G., Semmler-Behnke, M., Seitz, J., Scymczak, W., Wenk, A., Mayer, P., Takenaka, S., and Oberdorster, G. (2009). Size dependence of the translocation of inhaled iridium and carbon nanoparticle aggregates from the lung of rats to the blood and secondary target organs. Inhal Toxicol *21* Suppl 1, 55–60.

Krol, S., Macrez, R., Docagne, F., Defer, G., Laurent, S., Rahman, M., Hajipour, M.J., Kehoe, P.G., and Mahmoudi, M. (2013). Therapeutic benefits from nanoparticles: the potential significance of nanoscience in diseases with compromise to the blood brain barrier. Chem Rev *113*, 1877–1903.

Kroll, A., Pillukat, M.H., Hahn, D., and Schnekenburger, J. (2009). Current in vitro methods in nanoparticle risk assessment: limitations and challenges. Eur J Pharm Biopharm *72*, 370–377.

Kroll, A., Dierker, C., Rommel, C., Hahn, D., Wohlleben, W., Schulze-Isfort, C., Gobbert, C., Voetz, M., Hardinghaus, F., and Schnekenburger, J. (2011). Cytotoxicity screening of 23 engineered nanomaterials using a test matrix of ten cell lines and three different assays. Part Fibre Toxicol *8*, 9.

Krpetic, Z., Saleemi, S., Prior, I.A., See, V., Qureshi, R., and Brust, M. (2011). Negotiation of intracellular membrane barriers by TAT-modified gold nanoparticles. ACS Nano *5*, 5195–5201.

Kuhn, D.A., Vanhecke, D., Michen, B., Blank, F., Gehr, P., Petri-Fink, A., and Rothen-Rutishauser, B. (2014). Different endocytotic uptake mechanisms for nanoparticles in epithelial cells and macrophages. Beilstein J Nanotechnol *5*, 1625–1636.

Lesniak, A., Salvati, A., Santos-Martinez, M.J., Radomski, M.W., Dawson, K.A., and Aberg, C. (2013). Nanoparticle adhesion to the cell membrane and its effect on nanoparticle uptake efficiency. J Am Chem Soc *135*, 1438–1444.

Li, F., Sun, J.-y., Wang, J.-y., Du, S.-l., Lu, W.-y., Liu, M., Xie, C., and Shi, J.-y. (2008). Effect of hepatocyte growth factor encapsulated in targeted liposomes on liver cirrhosis. J Control Release *131*, 77–82.

Li, J., Cai, P., Shalviri, A., Henderson, J.T., He, C., Foltz, W.D., Prasad, P., Brodersen, P.M., Chen, Y., DaCosta, R., et al. (2014). A multifunctional polymeric nanotheranostic system delivers doxorubicin and imaging agents across the blood-brain barrier targeting brain metastases of breast cancer. ACS Nano *8*, 9925–9940.

Luo, G.-F., Chen, W.-H., Liu, Y., Lei, Q., Zhuo, R.-X., and Zhang, X.-Z. (2014). Multifunctional enveloped mesoporous silica nanoparticles for subcellular co-delivery of drug and therapeutic peptide. Sci Rep *4*, 6064.

Lynch, I., Salvati, A., and Dawson, K.A. (2009). Protein-nanoparticle interactions: what does the cell see? Nat Nanotechnol *4*, 546–547.

Ma, X., Wu, Y., Jin, S., Tian, Y., Zhang, X., Zhao, Y., Yu, L., and Liang, X.J. (2011). Gold nanoparticles induce autophagosome accumulation through size-dependent nanoparticle uptake and lysosome impairment. ACS Nano *5*, 8629–8639.

Mahmoudi, M., Sahraian, M.A., Shokrgozar, M.A., and Laurent, S. (2011). Superparamagnetic iron oxide nanoparticles: promises for diagnosis and treatment of multiple sclerosis. ACS Chem Neurosci *2*, 118–140.

Matsuo, T., Sugita, T., Kubo, T., Yasunaga, Y., Ochi, M., and Murakami, T. (2003). Injectable magnetic liposomes as a novel carrier of recombinant human BMP-2 for bone formation in a rat bone-defect model. J Biomed Mater Res A *66A*, 747–754.

Nam, H.Y., Kwon, S.M., Chung, H., Lee, S.Y., Kwon, S.H., Jeon, H., Kim, Y., Park, J.H., Kim, J., Her, S., et al. (2009). Cellular uptake mechanism and intracellular fate of hydrophobically modified glycol chitosan nanoparticles. J Control Release *135*, 259–267.

Nativo, P., Prior, I.A., and Brust, M. (2008). Uptake and intracellular fate of surface-modified gold nanoparticles. ACS Nano *2*, 1639–1644.

Nel, A., Xia, T., Mädler, L., and Li, N. (2006). Toxic potential of materials at the nanolevel. Science *311*, 622–627.

Nel, A.E., Madler, L., Velegol, D., Xia, T., Hoek, E.M., Somasundaran, P., Klaessig, F., Castranova, V., and Thompson, M. (2009). Understanding biophysicochemical interactions at the nano-bio interface. Nat Mater *8*, 543–557.

Neuberger, T., Schöpf, B., Hofmann, H., Hofmann, M., and von Rechenberg, B. (2005). Superparamagnetic nanoparticles for biomedical applications: possibilities and limitations of a new drug delivery system. J Magn Magn Mater *293*, 483–496.

Oberdorster, G. (2012). Nanotoxicology: in vitro-in vivo dosimetry. Environ Health Perspect *120*, A13; author reply A13.

Oberdorster, G., Oberdorster, E., and Oberdorster, J. (2005). Nanotoxicology: an emerging discipline evolving from studies of ultrafine particles. Environ Health Perspect *113*, 823–839.

Park, M.V.D.Z. (2012). Nanotoxicology—an in vitro approach. Proefschrift Maastricht (Maastricht University Maastricht).

Park, M.V.D.Z., Neigh, A.M., Vermeulen, J.P., de la Fonteyne, L.J.J., Verharen, H.W., Briede, J.J., van Loveren, H., and de Jong, W.H. (2011). The effect of particle size on the cytotoxicity, inflammation, developmental toxicity and genotoxicity of silver nanoparticles. Biomaterials *32*, 9810–9817.

Powers, K.W., Palazuelos, M., Moudgil, B.M., and Roberts, S.M. (2007). Characterization of the size, shape, and state of dispersion of nanoparticles for toxicological studies. Nanotoxicology *1*, 42–51.

Prahalad, A.K., Soukup, J.M., Inmon, J., Willis, R., Ghio, A.J., Becker, S., and Gallagher, J.E. (1999). Ambient air particles: effects on cellular oxidant radical generation in relation to particulate elemental chemistry. Toxicol Appl Pharmacol *158*, 81–91.

Prasad, G.L. (2009). Biomedical Applications of Nanoparticles. In Safety of Nanoparticles, T.J. Webster, ed. (Springer New York), pp. 89–109.

Rahman, M., Laurent, S., Tawil, N., Yahia, L., and Mahmoudi, M. (2013). Protein-Nanoparticle Interactions (Springer Series in Biophysics 15), Springer-Verlag, Berlin/Heidelberg.

Rodriguez-Lorenzo, L., Fytianos, K., Blank, F., von Garnier, C., Rothen-Rutishauser, B., and Petri-Fink, A. (2014). Fluorescence-encoded gold nanoparticles: library design and modulation of cellular uptake into dendritic cells. Small *10*, 1341–1350.

Sahay, G., Querbes, W., Alabi, C., Eltoukhy, A., Sarkar, S., Zurenko, C., Karagiannis, E., Love, K., Chen, D., Zoncu, R., et al. (2013). Efficiency of siRNA delivery by lipid nanoparticles is limited by endocytic recycling. Nat Biotechnol *31*, 653–658.

dos Santos, T., Varela, J., Lynch, I., Salvati, A., and Dawson, K.A. (2011). Effects of transport inhibitors on the cellular uptake of carboxylated polystyrene nanoparticles in different cell lines. PLoS One *6*, e24438.

Saravanan, P., Gopalan, R., and Chandrasekaran, V. (2008). Synthesis and characterisation of nanomaterials. Def Sci J *58*, 504516.

Sayes, C. and Ivanov, I. (2010). Comparative study of predictive computational models for nanoparticle-induced cytotoxicity. Risk Anal *30*, 1723–1734.

Schaeublin, N.M., Braydich-Stolle, L.K., Maurer, E.I., Park, K., MacCuspie, R.I., Afrooz, A.R.M.N., Vaia, R.A., Saleh, N.B., and Hussain, S.M. (2012). Does shape matter? bioeffects of gold nanomaterials in a human skin cell model. Langmuir *28*, 3248–3258.

Slowing, II, Vivero-Escoto, J.L., Zhao, Y., Kandel, K., Peeraphatdit, C., Trewyn, B.G., and Lin, V.S. (2011). Exocytosis of mesoporous silica nanoparticles from mammalian cells: from asymmetric cell-to-cell transfer to protein harvesting. Small *7*, 1526–1532.

Soenen, S.J., Nuytten, N., De Meyer, S.F., De Smedt, S.C., and De Cuyper, M. (2010). High intracellular iron oxide nanoparticle concentrations affect cellular cytoskeleton and focal adhesion kinase-mediated signaling. Small *6*, 832–842.

Soenen, S.J., Manshian, B., Montenegro, J.M., Amin, F., Meermann, B., Thiron, T., Cornelissen, M., Vanhaecke, F., Doak, S., Parak, W.J., et al. (2012). Cytotoxic effects of gold nanoparticles: a multiparametric study. ACS Nano *6*, 5767–5783.

Sokolova, V., Kozlova, D., Knuschke, T., Buer, J., Westendorf, A.M., and Epple, M. (2013). Mechanism of the uptake of cationic and anionic calcium phosphate nanoparticles by cells. Acta Biomater *9*, 7527–7535.

Suresh, A.K., Pelletier, D.A., Wang, W., Morrell-Falvey, J.L., Gu, B., and Doktycz, M.J. (2012). Cytotoxicity induced by engineered silver nanocrystallites is dependent on surface coatings and cell types. Langmuir *28*, 2727–2735.

Szabo, I., Leanza, L., Gulbins, E., and Zoratti, M. (2012). Physiology of potassium channels in the inner membrane of mitochondria. Pflugers Arch *463*, 231–246.

Theodossiou, T.A., Sideratou, Z., Katsarou, M.E., and Tsiourvas, D. (2013). Mitochondrial delivery of doxorubicin by triphenylphosphonium-functionalized hyperbranched nanocarriers results in rapid and severe cytotoxicity. Pharm Res *30*, 2832–2842.

Tsai, Y.M., Chien, C.F., Lin, L.C., and Tsai, T.H. (2011). Curcumin and its nano-formulation: the kinetics of tissue distribution and blood-brain barrier penetration. Int J Pharm *416*, 331–338.

Ulbrich, K., Knobloch, T., and Kreuter, J. (2011). Targeting the insulin receptor: nanoparticles for drug delivery across the blood-brain barrier (BBB). J Drug Target *19*, 125–132.

Wagner, S., Zensi, A., Wien, S.L., Tschickardt, S.E., Maier, W., Vogel, T., Worek, F., Pietrzik, C.U., Kreuter, J., and von Briesen, H. (2012). Uptake mechanism of ApoE-modified nanoparticles on brain capillary endothelial cells as a blood-brain barrier model. PLoS One *7*, e32568.

Walkey, C.D. and Chan, W.C.W. (2012). Understanding and controlling the interaction of nanomaterials with proteins in a physiological environment. Chem Soc Rev *41*, 2780–2799.

Wang, L., Liu, Y., Li, W., Jiang, X., Ji, Y., Wu, X., Xu, L., Qiu, Y., Zhao, K., Wei, T., et al. (2011). Selective targeting of gold nanorods at the mitochondria of cancer cells: implications for cancer therapy. Nano Lett *11*, 772–780.

Wang, X.-H., Peng, H.-S., Yang, L., You, F.-T., Teng, F., Tang, A.-W., Zhang, F.-J., and Li, X.-H. (2013). Poly-l-lysine assisted synthesis of core-shell nanoparticles and conjugation with triphenylphosphonium to target mitochondria. J Mater Chem B *1*, 5143–5152.

Watson, C., Ge, J., Cohen, J., Pyrgiotakis, G., Engelward, B.P., and Demokritou, P. (2014). High-throughput screening platform for engineered nanoparticle-mediated genotoxicity using CometChip technology. ACS Nano *8*, 2118–2133.

Weissleder, R., Nahrendorf, M., and Pittet, M.J. (2014). Imaging macrophages with nanoparticles. Nat Mater *13*, 125–138.

Williams, D.F. (2008). On the mechanisms of biocompatibility. Biomaterials *29*, 2941–2953.

Wohlfart, S., Khalansky, A.S., Gelperina, S., Begley, D., and Kreuter, J. (2011). Kinetics of transport of doxorubicin bound to nanoparticles across the blood-brain barrier. J Control Release *154*, 103–107.

Wu, X.A., Choi, C.H., Zhang, C., Hao, L., and Mirkin, C.A. (2014). Intracellular fate of spherical nucleic acid nanoparticle conjugates. J Am Chem Soc *136*, 7726–7733.

Xia, T., Kovochich, M., Liong, M., Madler, L., Gilbert, B., Shi, H., Yeh, J.I., Zink, J.I., and Nel, A.E. (2008). Comparison of the mechanism of toxicity of zinc oxide and cerium oxide nanoparticles based on dissolution and oxidative stress properties. ACS Nano *2*, 2121–2134.

Xie, Y., Ye, L., Zhang, X., Cui, W., Lou, J., Nagai, T., and Hou, X. (2005). Transport of nerve growth factor encapsulated into liposomes across the blood, Äìbrain barrier: in vitro and in vivo studies. J Control Release *105*, 106–119.

Yanes, R.E., Tarn, D., Hwang, A.A., Ferris, D.P., Sherman, S.P., Thomas, C.R., Lu, J., Pyle, A.D., Zink, J.I., and Tamanoi, F. (2013). Involvement of lysosomal exocytosis in the excretion of mesoporous silica nanoparticles and enhancement of the drug delivery effect by exocytosis inhibition. Small *9*, 697–704.

Yang, J.A., Lohse, S.E., and Murphy, C.J. (2014). Tuning cellular response to nanoparticles via surface chemistry and aggregation. Small *10*, 1642–1651.

Ye, D., Dawson, K.A., and Lynch, I. (2014). A TEM protocol for quality assurance of in vitro cellular barrier models and its application to the assessment of nanoparticle transport mechanisms across barriers. Analyst.

Zeng, X., Morgenstern, R., and Nystrom, A.M. (2014). Nanoparticle-directed sub-cellular localization of doxorubicin and the sensitization breast cancer cells by circumventing GST-mediated drug resistance. Biomaterials *35*, 1227–1239.

Zhang, P., Hu, L., Yin, Q., Feng, L., and Li, Y. (2012). Transferrin-modified c[RGDfK]-paclitaxel loaded hybrid micelle for sequential blood-brain barrier penetration and glioma targeting therapy. Mol Pharm *9*, 1590–1598.

Zhang, E., Kircher, M.F., Koch, M., Eliasson, L., Goldberg, S.N., and Renstrom, E. (2014). Dynamic magnetic fields remote-control apoptosis via nanoparticle rotation. ACS Nano *8*, 3192–3201.

Zhu, Z.J., Posati, T., Moyano, D.F., Tang, R., Yan, B., Vachet, R.W., and Rotello, V.M. (2012). The interplay of monolayer structure and serum protein interactions on the cellular uptake of gold nanoparticles. Small *8*, 2659–2663.

INDEX

Therapeutic Nanomaterials, First Edition. Edited by Mustafa O. Guler and Ayse B. Tekinay.